BRITISH GEOLOGICAL SURVEY

RESEARCH REPORT RR/11/05

A scenario for systems geology

Suggestions concerning the emerging geoscience knowledge system and the future geological map

T V Loudon

Scientific editor

J L Laxton

Keywords

Knowledge systems; cyberinfrastructure; solid Earth systems model; spatial models; morphometrics; geological framework.

Acknowledgments

The author wishes to thank the British Geological Survey, in particular Holger Kessler, John Laxton and Bob McIntosh, for help, encouragement and advice in the preparation of this book. It reflects the author's views and not necessarily those of the Survey.

Bibliographical reference

LOUDON, T.V., 2011. A scenario for systems geology: Suggestions concerning the emerging geosciences knowledge system and the future geological map. *British Geological Survey Research Report*, RR/11/05. 373pp.

ISBN 978 0 85272 710 2

Keyworth, Nottingham British Geological Survey 2011

BRITISH GEOLOGICAL SURVEY

The full range of our publications is available from BGS shops at Nottingham, Edinburgh, London and Cardiff (Welsh publications only) see contact details below or shop online at www.geologyshop.com

The London Information Office also maintains a reference collection of BGS publications, including maps, for consultation.

We publish an annual catalogue of our maps and other publications; this catalogue is available online or from any of the BGS shops.

The British Geological Survey carries out the geological survey of Great Britain and Northern Ireland (the latter as an agency service for the government of Northern Ireland), and of the surrounding continental shelf, as well as basic research projects. It also undertakes programmes of technical aid in geology in developing countries.

The British Geological Survey is a component body of the Natural Environment Research Council.

British Geological Survey offices

BGS Central Enquiries Desk

Tel 0115 936 3143 Fax 0115 936 3276

email enquiries@bgs.ac.uk

Kingsley Dunham Centre, Keyworth, Nottingham NG12 5GG

Tel 0115 936 3241 Fax 0115 936 3488

email sales@bgs.ac.uk

Murchison House, West Mains Road, Edinburgh EH9 3LA

Tel 0131 667 1000 Fax 0131 668 2683

email scotsales@bgs.ac.uk

Natural History Museum, Cromwell Road, London SW7 5BD

Tel 020 7589 4090 Fax 020 7584 8270

Tel 020 7942 5344/45 email

bgslondon@bgs.ac.uk

Columbus House, Greenmeadow Springs, Tongwynlais, Cardiff CF15 7NE

Tel 029 2052 1962 Fax 029 2052 1963

Maclean Building, Crowmarsh Gifford, Wallingford OX10 8BB

Tel 01491 838800 Fax 01491 692345

Geological Survey of Northern Ireland, Colby House, Stranmillis Court, Belfast BT9 5BF

Tel 028 9038 8462 Fax 028 9038 8461

www.bgs.ac.uk/gsni/

Parent Body

Natural Environment Research Council, Polaris House, North Star Avenue, Swindon SN2 1EU

Tel 01793 411500 Fax 01793 411501

www.nerc.ac.uk

Website www.bgs.ac.uk

Shop online at www.geologyshop.com

Preface

Κυβερνήτης, in ancient Greece, referred to a helmsman and how he steered a craft towards its intended destination. The word reappeared as cybernetics – the study of mechanisms for feedback and control – in organisms, machines, and systems in general. As we build more intricate representations of systems (sets of interacting parts that function as a whole), we need to understand and integrate their feedback, control and other mechanisms, as in financial and fiscal systems, automotive and aerospace systems, hardware and software systems, or biomedical and environment systems. This is not readily achieved. Systems can and do fail. Yet, complicated representations of systems as sets of organised, interconnected components are an essential means of exploring, depicting and living with the complexity of the real world.

Cyber- (or e-) has become a prefix to identify information systems built on the 'cyberinfrastructure' of computing, information and communications technologies, as opposed to the conventional infrastructure based on pen, paper and printing press. E-scientists have set out a vision that is transforming the entire knowledge system. And organisations long embedded in methodical accumulation of knowledge must climb slippery slopes of innovation to redefine their role in the future system.

The e-scientists' vision provides an opportunity to include a more complete representation of geological knowledge in a more comprehensive systems model. This is considered in 'The emerging geoscience knowledge system'. It in turn opens new possibilities, considered in 'The future geological map', for geologists to develop their objectives, legacy of knowledge, and ways of working, and extend the techniques of geoinformatics to build integrated systems of geological processes and their consequences.

Knowing where you want to go helps to decide how to get there. But the future is unknown, and all we can do is assemble ideas and consider where they might lead. We can develop a scenario – a description of a feasible but uncertain outcome. This tentative scenario is an initial response to the challenge of the e-scientists vision and the opportunities of geoinformatics. Its objective is to stimulate discussion and criticism that can lead to its revision and improvement. A revised scenario could in due course underpin strategic planning for system design (mapping a route to the uncertain goal), its implementation (building the facilities to get there), and migration (making the journey and moving the information).

Navigating 'A Scenario for Systems Geology'

Click on **red** to follow links, or turn to page number

Table of contents 1	Introduction 2	Overview 11
A list of the top three levels of section heading, with links to each	What is systems geology, **why does it matter, and how** can a scenario help?	An introductory summary of systems geology, its main ideas and components

The emerging geoscience knowledge system 42 Improving and extending the representation of geological knowledge as a comprehensive systems model	
Stages of concept development (cyberinfrastructure) 45	**Stages of concept development** (geological thinking) 56
The solid Earth systems model (sEsm) 71 A repository of information on the systems of the solid Earth	**The geological cyberenvironment (gce)** 85 End-to-end support for operations on the solid Earth systems model
The geological business model 93 Determines priorities and controls quality	**The geological investigation model** 98 Populates and tests the model
The geological framework model 105 Structures the content, coordinates modules	**The geological infrastructure model** 117 Maintains and provides infrastructure support

The future geological map 125 How geologists view their subject, how cyber-based methods change their approach, and how it may develop in future	
Reasoning, models and reality 127 How geological surveying depends on reasoning and observation	**From map to digital model** 145 How spatial models overcome limitations of the conventional map
Reconfiguration 165 Integrating diverse models as a single system	**An object-oriented approach** 178 An appropriate structure for geological surveying
The geometry of the spatial model 203 Digital representation of forms and shapes	**Transforming space** 224 An evolving stack of geological shapes
Seeking shared concepts 247 Multidisciplinary approaches to spatial knowledge and uncertainty	**Mapping geology into the knowledge system** 282 Bringing systems geology into a comprehensive knowledge system

References 314	Glossary 328	Some related initiatives 341	Index 359

A scenario for systems geology

SUGGESTIONS CONCERNING
THE EMERGING GEOSCIENCE KNOWLEDGE SYSTEM
AND THE FUTURE GEOLOGICAL MAP

Table of contents

A scenario for systems geology 1

A scenario for systems geology 1
Table of contents 1

Introduction and overview 1
Introduction 2
Systems geology 2
This scenario 4
Geological survey documentation 6
This e-book 7
Responding to the changing infrastructure 8
Costs, benefits and risks 9
Overview 11
An overview of systems geology 12
Three views of systems geology 15
Overview of the main components 18
Overview of concept development 19
Overview of the solid Earth systems model 26
Overview of the geological cyberenvironment 28
Overview of four implementation models 31
Overview of the geological business model 32
Overview of the geological investigation model 34
Overview of the geological framework model 35
Overview of the infrastructure model 37
Overview of the future geological map 38

The emerging geoscience knowledge system 42
Stages of concept development (summary) 44
Stages of concept development (cyberinfrastructure) 45
Mechanisms 45
Models and frameworks 47
Objects, ontologies and systems 48
Semantic Web and Grid 50
The service-oriented knowledge utility 52
Workflows, collaborative networks and Linked Data 53
Stages of concept development (geological thinking) 56
Invariance and processes 56

Mechanising the database ... *58*
The surveyor's holistic view ... *60*
Integrating information types ... *62*
Unexpressed knowledge ... *63*
The systems approach to Earth science ... *65*
The next steps ... *68*
The solid Earth systems model (sEsm) ... 71
Objectives of the solid Earth systems model ... *72*
Scope of the sEsm ... *73*
Remodelling the map ... *74*
The sEsm as a predictive machine ... *76*
Geological surveying as reinforcement learning ... *79*
Design requirements for the sEsm ... *83*
The geological cyberenvironment (gce) ... 85
Objectives of the geological cyberenvironment ... *86*
Scope of the gce ... *87*
Structure of the geological cyberenvironment ... *88*
Design requirements for the gce ... *90*
The geological business model ... 93
Objectives of the business model ... *94*
Scope of the business model ... *95*
Design requirements for the business model ... *96*
The geological investigation model ... 98
Objectives of the geological investigation model ... *99*
Scope of the geological investigation model ... *100*
Remodelling geological investigation ... *101*
Phases of investigational activity ... *103*
Design requirements for the investigation model ... *104*
The geological framework model ... 105
Objectives of the geological framework model ... *106*
Scope of the geological framework model ... *107*
Remodelling the systems framework ... *108*
The solid Earth systems metamodel (sEsmm) ... *110*
Linking beyond the sEsm ... *115*
The geological infrastructure model ... 117
Objectives of the infrastructure model ... *118*
Scope of the geological infrastructure model ... *118*
From document orientation to systems orientation ... *119*

The future geological map ... 125
Reasoning, models and reality ... 127
The need to look again ... *127*
The dialectic model ... *129*
Abstracting from reality to model ... *131*
At the interface ... *133*

A framework for the reasoning 135
The stratigraphical framework 139
Stratigraphical units in space and time 141
From map to digital model 145
The imperfect map 145
Ambiguity and map representation 148
Diverse objectives and products 150
Forward and inverse models 154
The imperfect model 156
Complex and emergent systems 159
Reconfiguration 165
Many models, one system 166
Projects and information communities 167
The importance of space and visualisation 169
Reconfiguring the system 172
Representing spatial information and relationships 173
Mark-up and metadata 175
An object-oriented approach 178
The object-oriented perspective 179
Object instances and classes 181
Relationships between objects 183
Reconciliation 186
Microdocuments and the threads of reasoning 190
Object-oriented survey 195
Benefits of an object-oriented system 199
The geometry of the spatial model 203
The need to harmonise the geometry 203
Making a mesh 207
Drawing the line 209
Estimation by interpolation 212
Continuity, fractals, octrees and wavelets 217
A wish list for integrated geometry 221
Transforming space 224
Rationale 224
Geometrical transformations 228
Invariant properties and classification 232
DSIs, FEMs and their geometrical significance 235
Unevenly spaced data 238
Spatial variation and uncertainty 241
The geometry of interpolation 243
Seeking shared concepts 247
Zoom 247
Grain, set and patch 251
Scale-space 255
Multiresolution survey 259

Boundaries: discontinuities and zero-crossings *264*
Shape *266*
Morphometrics *268*
Deformable models *274*
Reconsidering geological mapping *277*
Mapping geology into the knowledge system 282
Representing wider knowledge *283*
The role of the dynamic model *287*
Broadening the framework *291*
The general geoscience spatial model *293*
The multifaceted model *297*
The field survey model *300*
The digital geoscience spatial index *305*
The conceptual model *307*
The system framework *310*
Conclusions on mapping to the knowledge system *313*
References 314
Appendices 327
Glossary *328*
Some related initiatives *341*
Index 359

List of figures:

Figure 1: Decide where to go before planning how to get there 8
Figure 2: The S-shaped curve of methodological development 13
Figure 3: A crossing curve of new technology 13
Figure 4: Four main models in the geological knowledge system (duplicate of Figure 6) 19
Figure 5: Suggested targets for systems geology. 30
Figure 6: Aspects of the geological knowledge system 31
Figure 7: Business model 32
Figure 8: Geological investigation model 34
Figure 9: A proposal for a geological framework 36
Figure 10: Stages of investigation in the geological cyberenvironment (duplicate of Figure 17). 37
Figure 11: Four models in the geoscience knowledge system (duplicate of Figure 6) 43
Figure 12: Five information types and their representations in four different contexts. 62
Figure 13: Four component models in the geoscience knowledge system (duplicate of Figure 6). The solid Earth systems model to which they relate is outlined in red. 71
Figure 14: The solid Earth systems model. 73
Figure 15: Four models in the geoscience knowledge system (duplicate of Figure 6). The geological cyberenvironment is outlined in red. 85
Figure 16: Supporting specialised knowledge. 87
Figure 17: Stages of investigation, repeated frequently in whole or in part during a project 89
Figure 18: The cyberenvironment aims to provide end-to-end support for the user through the cycle of phases of investigational activity. (Duplicates of Figure 22 and Figure 17). 90
Figure 19: Four models in the geoscience knowledge system (duplicate of Figure 6). The business model is outlined in red. 93
Figure 20: Business model. 94
Figure 21: Four models in the geoscience knowledge system (duplicate of Figure 6). The geological investigation model is outlined in red. 98
Figure 22: The geological investigation model. 99
Figure 23: Duplicates of Figure 22 and Figure 17. The cyberenvironment aims to provide end-to-end support for the individual user through the cycle of phases of investigational activity. 103
Figure 24: Four models in the geoscience knowledge system (duplicate of Figure 6). The geological framework model is outlined in red. 105
Figure 25: Extract from metamodel in Figure 9 110
Figure 26: Extract from metamodel in Figure 9 111
Figure 27: Extract from metamodel in Figure 9 112
Figure 28: Extract from metamodel in Figure 9 113
Figure 29: A proposal for a geological framework (duplicate of Figure 9) 114
Figure 30: Supporting specialised knowledge (duplicate of Figure 16) 116
Figure 31: Four models in the geoscience knowledge system (duplicate of Figure 6). The infrastructure model is outlined in red. 117

Figure 32: Comparison of documentation structures 120
Figure 33: Flow of information in a modular information system 122
Figure 34: Stages of geological investigation (duplicate of Figure 17) 192
Figure 35: Interpolating a line or surface implies that any point on it can be estimated. 211
Figure 36: Generation of polynomial curves 214
Figure 37: A complex periodic curve 215
Figure 38: Semi-variance versus separation distance (from Loudon, 2000) 242
Figure 39: The bell curve used in Gaussian filtering 257
Figure 40: Topological elements in scale-space (From fig.10, Stewart et al., 2004) 258

Introduction and overview

<<Table of contents 1
Introduction and overview 1
Introduction 2
Overview 11
>>The emerging geoscience knowledge system 42
>>The future geological map 125
>>References 314
>>Glossary 328
>>Index 359

'A Scenario for Systems Geology' outlines some developments in e-science and geoinformatics that have a bearing on geology as a whole. Most accounts of geoinformatics are concerned with specific applications. But geoscience should move in parallel with the e-science vision of a more coherent and comprehensive representation of knowledge. With this in mind, systems geology takes a more comprehensive view of geology: rebased on a cyberinfrastructure of information, computing and communication technologies and viewed as a system (a set of interacting parts that function as a whole) embedded in the wider knowledge system.

This scenario considers the nature of the emerging geoscience knowledge system and the future geological map. They are seen as evolving towards a multiresolution, multidimensional, multimedia, comprehensive, quantifiable, predictive digital record, contributing to a whole Earth knowledge system that responds flexibly to user needs: a system of interoperable geological models, supported by mutually reinforcing techniques from geoinformatics, in harmony with legacy systems, human thought processes and geological thinking. Developments such as the semantic Grid, the Service-Oriented Knowledge Utility, and the unbounded Web of Data, offer support for geologists' holistic view of the Earth as a system.

The scenario describes a feasible but uncertain outcome. It starts from the view that it is helpful to consider where you want to go, before planning how to get there. It explores the possibilities and benefits of a coherent, coordinated, global system, and looks ahead to a system where old boundaries are irrelevant, and geologists share their knowledge more efficiently and express it more rigorously, precisely and comprehensively. As an explicit scenario, it is exposed to discussion, criticism and improvement; it can be modified as we learn more; it can reveal uncertainties that we can investigate; it can identify benefits and hazards; and it is a basis for collaboration. The Overview 11 is a condensed version of the material in The emerging geoscience knowledge system 42 and The future geological map 125.

Introduction

<<Table of contents 1
<<Introduction and overview 1
Introduction 2
Systems geology 2
This scenario 4
Geological survey documentation 6
This e-book 7
Responding to the changing infrastructure 8
Costs, benefits and risks 9
>>Overview 11

Summary: Systems geology is a view of geology re-based on the developing cyberinfrastructure and regarded as a system (a set of interacting parts that function as a whole) embedded in the wider knowledge system. This e-book sets out a scenario for development towards the e-science vision, and suggests a strategy to respond to the opportunities and reduce the risks.

Systems geology

<<Introduction 2

E-scientists have set out a clear vision of the future of the global knowledge system (see Overview of concept development 19). This scenario considers how systems geology might enable geology and geological surveying to conform to that vision. It considers how we might express geological knowledge in its new setting, and it makes some tentative suggestions on its future form and scope, as a contribution to the necessary debate on its longer-term development. Conventional (pre-digital) documents and maps cannot fully represent geological knowledge (see Representing spatial information and relationships 173). Digitisation makes information more flexible and accessible, and many geologists may feel that their work has made full use of geoinformatics[1] for many years. But that is not the point. Correct answers in one context may elsewhere refer to the wrong question. For example, cartographical generalisation techniques are useful for changing the scale of a geological map, but are not relevant where multiresolution geological knowledge is collected in the field, independently of later visualisation (see Multiresolution survey 259).

[1] Geoinformatics: The application of information science and technology to geography and geoscience.

Developing coherent objectives for systems geology requires an overall review of geological thinking and methodology. Because future developments build on existing knowledge, the scenario must relate future objectives to long-embedded geological ideas. Advances in geoinformatics, e-science and the technologies of information, computing and communication are creating the so-called advanced cyberinfrastructure[2], which can provide geologists with better ways to obtain, process, interpret and communicate their knowledge. It can represent their thinking more fully and extend their understanding, notably by supporting the systems view. It can make the various techniques of geoinformatics available for more systematic geological investigation, tapping into the synergy of mutually reinforcing techniques. Systems geology opens geology to these developments.

The eventual target is seen as a comprehensive system[3] where all the component parts work with one another and with the wider systems in which they are embedded. It aims to provide users with more powerful scientific methods, more comprehensive information resources, and rapid delivery of information to meet user requirements for relevance and presentation. It should help to overcome inappropriate barriers between types of information, and between regions, disciplines, and organisations.

The benefits should include:

- a holistic[4], object-oriented[5], systems representation of geology, forming an important component of Earth systems science
- a systematic framework to collect, organise, exchange and integrate geological information, following shared standards to ensure wide compatibility
- more extensive, accurate and comprehensive representation of geological thinking, across all information types, dimensionality and levels of detail
- comprehensively documented, evaluated, predictive geology, substantiated by records of investigational procedures and observational evidence
- integration of methods to collect, filter[6], analyse and simulate data
- consolidation and integration of information across objectives, disciplines, organisations and geographical regions
- more flexible delivery, depiction and visualisation of information, selectable by users to match their specific needs

The anticipated benefits stem from infrastructure developments, but must be based on a carefully planned strategy reflecting the needs of the geological community. Global geology can be understood in detail only because world-wide stratigraphical and map conventions have been established. Similarly, systems geology can make the relevant knowledge of

[2] Cyberinfrastructure: An integrated assemblage of computing, information and communication facilities, deploying the combined capacity of multiple sites to provide a framework to underpin research and discovery, typically with broad access and end-to-end coordination.

[3] System: A set of interacting parts that function as a whole. The systems approach involves study of linkages or interfaces between the component activities.

[4] Holistic: A view of a system that emphasises its properties and interrelationships acting as a whole, as opposed to the reductionist approach of studying its components in isolation as distinct entities.

[5] Object-oriented: An approach to analysis, design, and classification of the objects of interest and their behaviour.

[6] Filtering: A process that selectively enhances or reduces specified components of the information stream.

geologists available throughout the emerging knowledge system only if appropriate geological systems are designed and implemented. The involvement of informed geologists in many countries, organisations and disciplines will be essential for the successful development of systems geology.

This scenario

<<Introduction 2

A cyber-strategy[7] that describes how geologists and geological organisations might plan to respond to the advancing infrastructure can at present refer only to a scenario[8], because the path of future developments is uncertain. Nevertheless, for forward planning we must 'take a view' of the future, as a basis for interoperability[9] of the mutually dependent components in an overall system. The scenario must be explicit, so that it can be evaluated, tested, discussed, criticised, amended, extended and adjusted, before a strategy is formulated for system implementation.

This account (summarised in the Overview 11) outlines some past and present developments, their scientific and technical basis and possible future directions. It describes two aspects of systems geology – the advancing technology of the geoscience knowledge system (see The emerging geoscience knowledge system 42); and how geologists can, in the mathematical sense, map[10] their findings into it (see The future geological map 125). It describes various methods that may require further investigation and evaluation by geologists. It suggests strategies and standards that geological organisations can develop to position themselves within the mainstream of the advancing infrastructure and exploit its support for systems science. It aims to stimulate discussion that can influence and help to coordinate future development.

The solid Earth systems model[11] and the geological cyberenvironment[12] are seen as two main structures defining a possible knowledge system for systems geology. One describes the science; the other describes the supporting technology. The solid Earth systems model

---

[7] Cyber-strategy: A plan or scenario describing how an organisation or individual intends to respond to the current and future development of the infrastructure.

[8] Scenario: A description of a plausible, though uncertain, outcome.

[9] Interoperability of information is the ability of concepts, terms or models from various sources to work together, by meeting standards that enable sharing and reuse of information.

[10] Mapping: Conventionally, geological mapping leads to a graphical depiction, usually on a flat surface, of spatial relationships and forms of geological features or properties in a selected area of the Earth's surface or subsurface. Mathematically, mapping relates the elements of one set to those of another. A broader definition of geological mapping could be 'relating elements of geological observation or interpretation of the solid Earth to corresponding elements in an appropriate model in the geoscience knowledge system'.

[11] Solid Earth systems model (sEsm): An approach to structuring distributed knowledge of the science of geology to provide an integrated view in the context of sciences of the solid Earth as a whole.

[12] Geological cyberenvironment: The cyberinfrastructure for end-to-end support of geological investigation, for example, in the context of a solid Earth systems model.

(sEsm) 71 might be seen as the systems geology equivalent of the system of geological maps and map explanations provided by surveying organisations worldwide. It is proposed as part of a comprehensive solution based on geological objects[13], their spatial distributions and relationships, their properties and composition, their origin and geological history, the underlying reasoning, and the source and evaluation (provenance[14]) of the information. The model might integrate information conventionally held in geological maps and a wide range of related documents, more recently supplemented by geographical information systems and databases. It aims to align procedures and representations across geographical and disciplinary boundaries (see Seeking shared concepts 247). It calls for shared methods of classification and boundary selection (see Boundaries: discontinuities and zero-crossings 264); links from the observed outcome back to the causative process (see From map to digital model 145); and quantitative sampling and interpolation techniques that conform to those in related fields (see The geometry of the spatial model 203). A global view is desirable for both the system design and the supporting software (see The geological cyberenvironment (gce) 85).

The proposed framework for the systems model is an organised, interconnected set of widely accepted geological concepts (see Overview of the geological framework model 35), linked to indexes, which in turn are linked through identifiers (URI's) to distributed information in the computing 'cloud[15]'. It should provide the user with access to comprehensive, authoritative information evaluated and supplied by trusted organisations, as well as linking to a wide range of other material from diverse sources. Like Survey maps, it could develop into a powerful shared resource for a wide range of applications related to solid Earth sciences. The infrastructure support of information, computing and communications technologies can be provided in the geological cyberenvironment (see Overview of the geological cyberenvironment 28), aiming to provide end-to-end support for geologists, matching their familiar working practices and extending to new requirements in systems geology.

Changes to the geoscience knowledge system bring opportunities to explore new methods in geological surveying, recording shareable information that currently remains unexpressed as knowledge in geologists' minds. The possibilities are considered in detail in The future geological map 125, which describes the evolution of the geological map from a paper document to a means of mapping observation and interpretation of the real world as digital representations in the solid Earth systems model.

---

[13] Objects: Representations of real-world or conceptual things or entities of interest in a particular context.

[14] Provenance (of information): The source, origin or derivation of items of information, which might be formalised in terms of, for example, project, originator, date, place, collection method, archive or database identifier, authorisation.

[15] Cloud computing: Distributed computing, supplying services, such as data and processes, to the desktop or mobile device from the 'cloud' of large, distributed data centres.

Geological survey documentation

<<Introduction 2

Geological survey organisations provide a consistent authoritative view of core elements of geoscience. Systems geology alters the means of documenting their findings. Conventionally, they rely on maps, map explanations or memoirs, scientific papers, and informal field notes, logs, and datasets. These pre-digital documents were well fitted for many purposes, for which they had been specifically designed. But they lack flexibility of presentation and updating, and are constrained by the mechanics of the infrastructure. For example, the traditional underlying structure of much geological survey publication was a set of rectangular map sheets, unrelated to the geology and inappropriate for wider integration with Earth and environmental sciences. Rapid progress has been made in implementing seamless digital methods, and it now seems appropriate to look ahead to a more comprehensive systems structure to collect and analyse survey information and codify[16] the survey results.

Documentation of geoscience survey findings (seen as the postulated outcome of systems in the solid Earth) has important features in common with documenting other complicated systems, such as the design of a large commercial aircraft or a major software system, where conventional manuals are no longer adequate. For example, an electrician investigating a failure in an engine in Bahrain Airport of an aircraft designed, built and customised in Seattle, might require immediate access to full wiring diagrams, data and descriptions of components in the starboard wing, as modified in the specific aircraft, along with information on the wing structure for access purposes, and the means of simulating the overall consequences of failure of specific parts of its electrical system.

The requirements include:

- rapid remote access to the most recent and relevant information
- thorough, authoritative, evaluated documentation of many linked topics combining text, data, maps and images
- a tightly organised structure of documentation in 'minimum revisable units[17]', showing the current version unless another is requested
- a database that indexes comprehensively archived historical records, their provenance and linkages, amended with rigorous version control
- support for analysis, visualisation and interpretation
- background information to explore ideas and collect further evidence, including simulation ('what-if') procedures
- immediate feedback and updating when errors or inconsistencies are found

Similar needs arise in geological surveying. They call for reconsideration of documentation practices in the light of systems documentation developing elsewhere. Systems geology

---

[16] Codify: Create a representation or record of something in a form appropriate to the organised system of which it becomes a part.

[17] Minimum revisable unit: A self-contained subset of information in a documentation system: a component of the system designed to be revised as necessary without endangering the integrity of the system as a whole.

requires a more comprehensive framework, but must still present information in familiar formats.

This e-book

<<Introduction 2

The digital-media equivalent of a conventional printed book is an e-book, possibly enhanced with hyperlinks and multimedia, available to view, complete or in part, on screen or printed on demand. At all levels of detail, an e-book has the potential to interweave hypertext threads of narrative, data, visualisations[18], images, analyses and workflows[19]. This e-book is structured as self-contained topics described at various levels of detail, as shown in the Table of contents 1 (which lists a hierarchy of three levels of revisable units). At the expense of some repetition, this format can generate various presentations in meaningful, readable sequences by means of filtering by topic, level of detail, relevance in a particular context, or by invoking chained information trails or workflows. The aim is to match the reader's specific needs more precisely, and to explore documentation structures of possible relevance in systems geology. To retain flexibility, titles and headings (as well as page numbers) are referred to in cross references. The definitions of terms, as used here, are listed in the Glossary 328 and, where it seemed useful, repeated as footnotes. Similarly bibliographical references appear as footnotes for ease of reference as well as in the cumulated References 314.

Cross-references (Ctrl+click) are shown in red type in the text and diagrams. Readers should be able to select their own information trails (see Blazing a trail 20) by traversing the content following pointers in the figures and text, including the headings in the Table of contents 1. There is also an opportunity to restructure the e-book content as components of a Web site, gaining connectivity and flexibility at the expense of an enduring coherent view, or reorganising fragments to provide background information within a solid Earth systems model, analogous to 'Help' information in a software system.

This e-book sets out one scenario for the development of systems geology. As with an edition of a conventional book, its content is fixed and archived on publication. The aim is to present a coherent set of interconnected concepts that persist in a stable form, where they can be discussed and cited by others and are open to disproof. However, the usefulness of the scenario depends on how far it is accepted by geologists, geological organisations, the wider geological community, the users of their products, and workers in related disciplines from Earth science to e-science. The e-book can therefore be seen as one of many static components embedded in an evolving, collaborative, Web-based structure, where they can be linked to comment, discussion, correction and extension. Like all formal publications, it

---

[18] Visualisation: Transforming quantitative data (including the results of interpolation) into sensory information – images that the eye and brain can interpret and manipulate as a mental picture.

[19] Workflow: The representation of a process or procedure in terms of a sequence of operations in a task or event.

remains a stable element in this fluid environment until it is revised, superseded or forgotten.

Responding to the changing infrastructure

<<Introduction 2

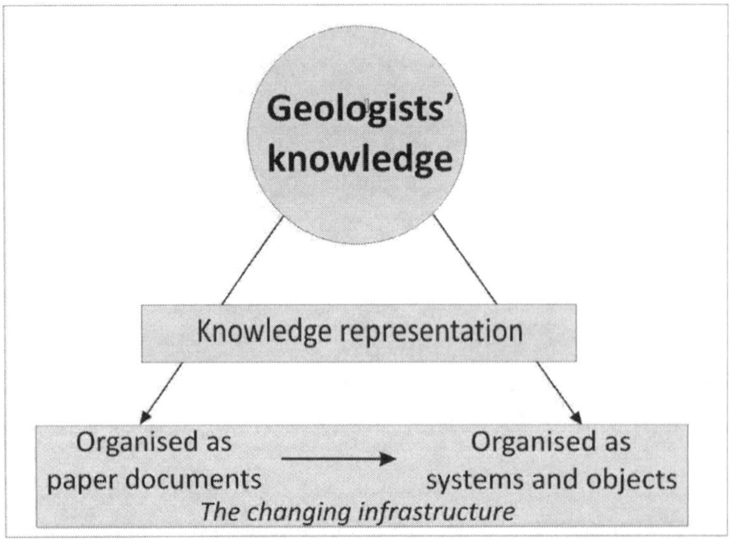

Figure 1: Decide where to go before planning how to get there

Most geological knowledge is held unrecorded in the collective human memory, as background acquired by training, education and experience. The changing infrastructure offers the promise of extending the representation and sharing of this knowledge. Geological information communities differ in their response (*Figure 1*). At one extreme (on the left of the diagram), they might take the view that the bulk of information is, and should remain, in conventional (pre-digital) form, enhanced only when new systems, such as word processing or digital cartography, are widely accepted and can be justified by scientific or cost benefits. At the other extreme (on the right), they might consider that the full benefits can be achieved only by planning to take full advantage of proposed future systems and contribute to their development, even where this requires a direct route from unexpressed geological knowledge to its representation as digital information, omitting any conventional documentation. The view taken here is that both approaches should work together.

On the one hand, there is a need to define the longer-term goals of systems geology, taking full advantage of the benefits of the advanced infrastructure, not constrained by earlier knowledge systems, but in tune with the knowledge and ways of thinking of geologists, supporting the systems approach and quantitative methods made practicable by new technology, embedded in a broader view of overlapping areas of science and its applications. This scenario is an initial attempt to clarify those possibilities. On the other hand, there is a need to plan ahead for accessing conventional information in a more comprehensive, evolving knowledge system, and for its reorganisation, enhancement and migration as and when appropriate (lower arrow). Effective planning is based on an opinion (explicit or implied) of the future goal. It is argued here that an explicit scenario that is open to criticism, evaluation and modification, and which relates geological aspects to their multi-disciplinary, global context, is becoming an essential planning tool.

The emerging technology can be explored in small pilot projects that organise systems geology as systems[20] and objects[21], with content collected specifically to match the longer-term objectives (right-hand arrow). The immediate aims would be to explore new techniques, guide an assessment of the longer term prospects, their value and feasibility, and define priorities. Taken together, they should influence, extend, enhance and perhaps eventually replace existing procedures of recording information (left-hand arrow).

>>See also An overview of systems geology 12

Costs, benefits and risks

<<Introduction 2

Proposals for similar work in other fields, such as systems biology and medicine (see BBSRC (2007)[22], Other relevant fields 355, Overview of the solid Earth systems model 26) suggest that the development costs of a full systems geology knowledge system would exceed the resources of most geological organisations. Also, the need for international and interdisciplinary collaboration among many organisations (to the scientific and financial benefit of all) indicates that support from shared external funding might be appropriate. 'Collaboratories[23]' can improve the efficiency and reduce the costs of wider collaboration. And even limited extension of existing work on development of international standards across organisations and disciplines will encourage the development of commercial and open-source implementations. Early adoption of international standards reduces the difficulties and costs of later back-tracking and migration to future systems, and provides easier access to work done elsewhere.

Within an organisation where information technology and geoinformatics are widely applied, a cyber-strategy[24] might initially aim to extend standards for a structural

---

[20] System: A set of interacting parts that function as a whole. The systems approach involves study of linkages or interfaces between the component activities.

[21] Objects: Representations of real-world or conceptual things or entities of interest in a particular context.

[22] BBSRC, 2007. Systems biology. http://www.bbsrc.ac.uk/publications/corporate/systems-biology.aspx

[23] Collaboratory: A networked system linking scientists for formal and informal communication across locations and organisations to share and discuss their investigations and collaborate in such tasks as system design or research projects.

[24] Cyber-strategy: A plan or scenario describing how an organisation or individual intends to respond to the current and future development of the infrastructure.

framework, a metamodel, ontologies for solid Earth systems, and components and interfaces in geological cyberenvironments. A small initial investment in pilot studies to establish the scope, relevance and feasibility of a geoscience cyber-strategy could clarify the value of a range of new techniques and developments in Earth systems science and systems geology, and help to determine why, how and when geologists might integrate the various aspects within their organisational structure. A clear systems design should clarify the scope of different aspects of systems geology and the interfaces between them. This should help to control costs by reducing overlap and duplication among groups working on current and future applications, and help to ensure that the various components can work together.

The risks of ignoring the implications of major developments in the technology and the science it supports are self-evident. Not least is the risk of undermining existing knowledge resources (including legacy information, and the coherent organisation and expertise for its extension, preservation and delivery) by failing to ensure that they migrate in tune and in time with the advancing infrastructure. A particular risk is misunderstandings arising from the different background knowledge and outlook of those involved in management, in geological investigation and in development of the infrastructure. These aspects are viewed here as separate but closely connected subsystems. Progress relies on complicated systems of unfamiliar analytical methods, notoriously prone to unexpected, catastrophic failure. To avoid this, the sceptical attitude of scientists, insisting on full explanation and understanding, constantly searching for flaws and testing methods and results, is an essential part of the response. Another important priority of the geological community must be to ensure that existing knowledge resources remain fully available and, where appropriate, are enhanced and integrated in step with the changing technology. Costs and risks can both be reduced by an explicit strategic plan that enables components of the strategy to develop and work in unison, and reduces duplicated effort by clarifying long-term objectives.

Overview

<<Table of contents 1
<<Introduction and overview 1
Overview 11
An overview of systems geology 12
Three views of systems geology 15
Overview of the main components 18
Overview of concept development 19
Overview of the solid Earth systems model 26
Overview of the geological cyberenvironment 28
Overview of four implementation models 31
Overview of the geological business model 32
Overview of the geological investigation model 34
Overview of the geological framework model 35
Overview of the infrastructure model 37
Overview of the future geological map 38
>>The emerging geoscience knowledge system 42
>>The future geological map 125
>>References 314
>>Glossary 328
>>Index 359

Summary: The overview summarises some ideas developed later at greater length. This account of systems geology is built around the development of two interacting themes. The first is the representation of what we think we know about geology[25] (the solid Earth systems model or sEsm); the second is the tools to assemble and process it (the geological cyberenvironment or gce). They might simply be called geology and its infrastructure, but that would miss an important change of emphasis. It is suggested here that the sEsm and the gce could work together (aim for interoperability[26] within the system) by means of shared standards in four models describing aspects of the geoscience knowledge system: the business model; geological framework model; geological investigation model; and geological infrastructure model. Together, they could support a future digital, multiresolution, multidimensional, map that could more accurately portray geological knowledge.

---

[25] Geology: The study of the planet Earth, the materials of which it is made, the processes that act on these materials, the products formed, and the history of the planet and its life forms since its origin.
[26] Interoperability of information: the ability of concepts, terms or models from various sources to work together, by meeting standards that enable sharing and reuse of information.

An overview of systems geology

<<Overview 11

What is systems geology?

Systems geology is a view of geology re-based on the developing cyberinfrastructure and regarded as a system (a set of interacting parts that function as a whole) embedded in the wider knowledge system.

What is the advanced cyberinfrastructure?

The **cyberinfrastructure** is the developing network of support services based on computing, information and communications technology. It is an important external influence on future directions of geology. E-scientists are creating an '**advanced cyberinfrastructure**' that:

- handles and supplies information as a commodity
- breaks down artificial barriers between geographical areas or scientific disciplines
- liberates information (through high connectivity and electronic delivery) from rigid packaging into distinct formats, such as maps, datasets, text explanations and advisory services

Why is it a vital concern in future planning for geology?

The advanced cyberinfrastructure supports a **systems approach** to science, which:

- affects all aspects of the **geological knowledge system** – the system that collects, organises, evaluates, assembles and supplies knowledge of the solid Earth
- provides a **holistic** view that integrates and connects wide-ranging aspects of knowledge
- considers an entity, such as the solid Earth, as a single, coherent **system** of related, organised and interacting parts, processes and feedback that function as a whole (with properties that cannot be reduced to those of its components)
- is well-matched to patterns of geological thinking and reasoning

The need to look ahead

The advanced cyberinfrastructure and the systems approach, taken together, have far-reaching consequences throughout geology. Therefore:

- a priority in future planning within an international and interdisciplinary context is to develop an **explicit strategic vision** for the future of geology and geological survey
- the strategic vision reflects a changed paradigm (in the sense of an exemplary pattern or model) for the knowledge system as a whole, with inevitable knock-on effects in how geologists handle their science
- of course, the strategy does **not** concern the underlying scientific paradigm of the theory and concepts of geology – its aim is to strengthen the supporting procedures

How will the procedures change?

The development of ideas typically follows an **S-shaped curve**. Think, for example, of the development of the geological map from 1800 to date. The suitability (or **fitness**

for purpose) of the map improves rapidly at first, then settles to a routine surveying process involving **normal science**, where consistency is all important (see *Figure 2*). Normal science follows an established paradigm and the time-curve of suitability is almost flat. "It is a profoundly erroneous truism that we should cultivate the habit of thinking about what we are doing. The precise opposite is the case. Civilisation advances by extending the number of important operations which we can perform without thinking about them." *A N Whitehead, 1911 An introduction to mathematics.*

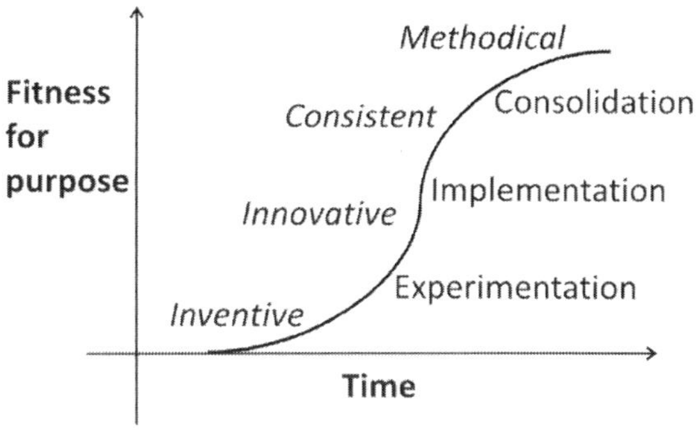

Figure 2: The S-shaped curve of methodological development

Adjusting to new technology

"Normal science is based on a well-established view of a science in which the practitioners share the same exemplars or paradigm – universally recognised scientific achievements that for a time provide model problems and solutions to a community of practitioners. The individual takes the paradigm for granted, and need no longer start from first principles and justify the use of each concept introduced." ... "Discoveries (novelties of fact) and inventions (novelties of theory or instrumentation) provide scientists with the rules of the game... When the profession can no longer evade anomalies that subvert the existing tradition of normal science – *then* begin the extraordinary investigations that lead the profession at last to a new set of commitments, new basis for the practice of science, new gestalt, new paradigm." *T. S. Kuhn, 1962 The Structure of Scientific Revolutions.*

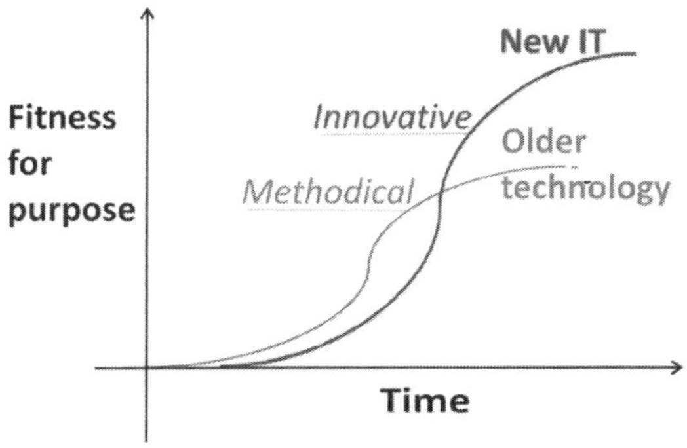

Figure 3: A crossing curve of new technology

What to do?

Faced with adjustment to radically new technology, the 'common-sense' approach is tactical, adding electronic support to existing activities as needs arise. This leads to ineffective compartmentalisation, followed in due course by the organisation's displacement by fitter competitors riding the new curve, resulting in much discomfort, and loss of continuity, existing knowledge and expertise. "... common sense is a bad master. Its sole criterion for judgment is that the new ideas should look like the old one" *A N Whitehead, 1929 Process and Reality.* Organisations long embedded in familiar ways of methodical accumulation of knowledge, must leap across to the slippery slopes of innovation on the steeply rising S-curve of new information technology (*Figure 3*).

Maintaining a coherent overview

The technology leap applies, not just to individual components like geological maps, but to the Earth-sciences knowledge system as a whole. The role of the map in the system is changing: from the means of assembling and communicating the findings of geological surveying (which will both be done in better ways) to being just one visualisation of our spatial understanding of the Earth. Each such change brings its own widespread knock-on effects, and amending system components in separate, isolated compartments cannot succeed.

A better way:

- aim to unleash the potential of the cyberinfrastructure, not merely to replicate existing procedures electronically
- maintain services with **parallel systems** (conventional geology and systems geology, each on its own S-curve) as an essential interim measure
- ensure that the test-bed for systems geology is based on and contributes to appropriate global, multidisciplinary standards
- avoid quirky systems that do not meet wider industry standards and protocols
- build processes to guide and assist migration from old to new
- rebuild a knowledge system within which processes, concepts and ideas can evolve freely as the science develops, coordinated within the structure of the new paradigm
- develop an **explicit strategic vision and framework**; implement it step by step, modifying the strategy as experience is gained

Why explicit?

"All constructive thought, on the various topics of scientific interest, is dominated by a scheme of ideas, unacknowledged but no less influential in guiding the imagination. Philosophy is a sustained effort to make such schemes explicit, and thereby capable of criticism and improvement." *A N Whitehead, 1929 Process and Reality.*

The **geological framework**, our scheme of ideas, must be explicit: open not only for discussion, criticism and improvement, but also for building interfaces to the artefacts of technology.

But why do we need a strategic vision?

- to position geology to gain the full benefits of the cyberinfrastructure
- to inform policy-makers of the relevance of geology when allocating resources within the environmental sciences and e-science
- to develop the rationale for a **strategic framework** within which tactical studies, despite the inherent unpredictability of their outcome, can coherently build on and extend geological knowledge as a whole

What is the function of the strategic geological framework?

- to structure geological knowledge as a component of Earth systems science
- to clarify its intricate network of relationships
- to assist the design of components that can work together
- to identify relevant content for users to select, retrieve and evaluate
- to connect items to appropriate analytical methods
- to connect items to their provenance and context
- to consolidate and facilitate the parallel evolution of the science and the supporting technology
- to form the foundation on which a coherent cyber-based **geological knowledge system** can be built from the unpredictable results of scientific investigations

The emerging **framework** is therefore a key development in geological science within its wider context, and fundamental to the future of geological surveys and other geological organisations.

Three views of systems geology

<<Overview 11

An important benefit of the systems approach to Geological survey documentation 6 is the possibility of generating on demand many documents that meet specific user needs (for example, for area, type of geology, level of detail and form of presentation or visualisation), all derived from the same pool of revisable modules. This e-book was similarly organised in the hope that readers could derive (and perhaps print) a sequence of extracts to meet their specific interests, or could follow their own information trails on screen, by clicking on the cross-references (see Blazing a trail 20 or Mechanisms 45). As an example, three information areas are outlined here, reflecting views from different aspects of expertise and responsibility: The management view 15, The geological investigation view 16 and The infrastructure view 17. Of course, they overlap to some extent and, in a small project, the same individual or team may share all these viewpoints.

The management view

Geological managers may define a business model on which is based the objectives of their geological investigations or projects and the means of achieving them: why the projects are undertaken, and therefore what geological aspects are of particular interest, and how and where the investigation is carried out. This project plan is an important outcome of the business model. Management may also have to address issues of efficiency (getting most benefit at least cost from projects and from the system as a whole), and legal issues, such as mandatory deposition of information, and intellectual property rights and exploitation. The project plan and methods of investigation involve managers as well as surveyors, because the tools, techniques and procedures affect both costs and benefits. The form of presentation and dissemination of results involves managers along with experts on the infrastructure. Evaluation of the results in terms of originality, predictive power, relevance and accuracy is also a management concern, both for in-house studies and, through a collaborative process, for results that are more widely shared.

>> See also Overview of the geological business model 32

Other sections particularly relevant to the management view include:

Introduction 2
Geological survey documentation 6
Responding to the changing infrastructure 8
Costs, benefits and risks 9
An overview of systems geology 12
Overview of concept development 19
Overview of the solid Earth systems model 26
Overview of the geological cyberenvironment 28
Overview of the future geological map 38
The geological business model 93

The geological investigation view

An integrated view of geology could aim to develop a symbiotic[27] relationship (see What about unexpressed thought? 23) between geological thinking (a function of the human mind) and computer-based handling of geological models, methods and procedures (a function of the infrastructure). New approaches must incorporate and build on what already exists and, if it is to develop successfully, geologists must drive the system and contribute to it.

A geological framework based on generally accepted scientific ideas is required to structure the system. It depicts and clarifies the principal relationships among the findings of geology, and is thus the concern of the geological community as a whole. During an investigation, various phases of the project cycle can be identified. Parts or all of this cycle are repeated numerous times during an investigation, and may or may not eventually result in formal contributions being shared by adding to the geological knowledge stored in the infrastructure. Each phase should be supported by the cyberenvironment.

[27] Symbiosis: A close interdependence or association (in the literal sense, of animals or plants of different species) often of mutual benefit.

>> See also Overview of the geological investigation model 34
Overview of the geological framework model 35
The geological investigation model 97

Other sections relevant to the geological investigation view include:

Introduction and overview 1
Overview of the geological cyberenvironment 28
Stages of concept development (geological thinking) 56
The solid Earth systems model (sEsm) 71
The geological cyberenvironment (gce) 85
The geological framework model 105
From document orientation to systems orientation 119
The future geological map 125

The infrastructure view

The infrastructure of tools and mechanisms that capture, store, process and share geological information must (as in The geological investigation view 16) be organised systematically to meet the needs of geological investigation. A standard open-ended framework should reduce unnecessary duplication within a system where disciplines and organisations can connect and work together through on-demand services – supplying access to knowledge and processing power as a utility, like water or electricity. The intended result is a geological cyberenvironment in tune with the working practices and familiar concepts and methodologies of geology, along with end-to-end support for geological investigations and interoperability with other disciplines. The implementation of the geological cyberinfrastructure is primarily a task for e-scientists, but its specification depends on geologists and Earth scientists.

Geological information is mostly recorded as conventional representations, such as maps, datasets, scientific papers, and text-books. Much of this information can be, and has been, digitised for easier access. But it is probably outweighed by unrecorded knowledge held collectively in the minds of geologists, gained through education and experience, partly shared by teaching, discussion and demonstration. The advancing cyberinfrastructure can incorporate parts of this previously unexpressed knowledge. It has the potential to formalise, codify, quantify and integrate aspects of geological thinking in previously unavailable ways, which could clarify geological interpretations and add significantly to our understanding of Earth systems (see The future geological map 125). Its greater connectivity encourages more systematic integration of ideas, weaving together threads of thought, identifying and interpreting what has been observed and predicting what has not. It can bring geological investigation into an established region (artificial intelligence) of e-science by viewing it as a process of reinforcement learning, prediction and generalisation, accessible to investigators and users through the geological cyberenvironment.

>> See also Overview of the infrastructure model 37
The geological infrastructure model 117

Other sections relevant to the infrastructure view include:

Introduction and overview 1
Stages of concept development (cyberinfrastructure) 45
Overview of the solid Earth systems model 26
Overview of the geological cyberenvironment 28
Overview of the future geological map 38
The solid Earth systems model (sEsm) 71
The solid Earth systems metamodel (sEsmm) 110
Mapping geology into the knowledge system 282

Overview of the main components

<<Overview 11

Two themes

The geological knowledge[28] system[29] is considered in terms of two linked themes (see This scenario 4).

- The **solid Earth systems model[30]** refers to an integrated view of distributed knowledge of the science of geology, in the context of sciences of the solid Earth as a whole
- The **geological cyberenvironment** supplies the infrastructure for end-to-end support of geological investigation, in the holistic context of the solid Earth systems model

Four models

Four main models (see *Figure 4*) support the two main themes (see Loudon[31], 2009), calling on different areas of expertise but working together to support the changing representation of geological knowledge.

- The **business model** defines the objectives of geological investigation: primarily the concern of geological management
- The **geological investigation model** describes the procedures by which new and existing information is obtained: the concern of both geological surveyors contributing new information and users accessing existing information
- The **geological framework model** depicts and clarifies the principal relationships among the findings of geology, standardising the organisation of the geoscience knowledge system: the concern of the geological community as a whole

[28] Information, knowledge: As used here, information is a representation of knowledge, which is regarded as what is known (and possibly recorded) about a topic, gained through learning, experience and familiarity.

[29] System: A set of interacting parts that function as a whole. The systems approach involves study of linkages or interfaces between the component activities.

[30] Model: A formalised representation giving a simplified view of aspects of the real (or of an imaginary) world relevant to the purposes in hand.

[31] Loudon, T.V., 2009 Four interacting aspects of a geological survey knowledge system. *Computers & Geosciences*, 35 (4). 700-705. 10.1016/j.cageo.2007.12.009 or http://nora.nerc.ac.uk/7258/

- The **infrastructure model** describes the structure, facilities and mechanisms to store, process and share information. The **cyberinfrastructure** is the emerging infrastructure based on information and communication technology: primarily the concern of e-scientists.

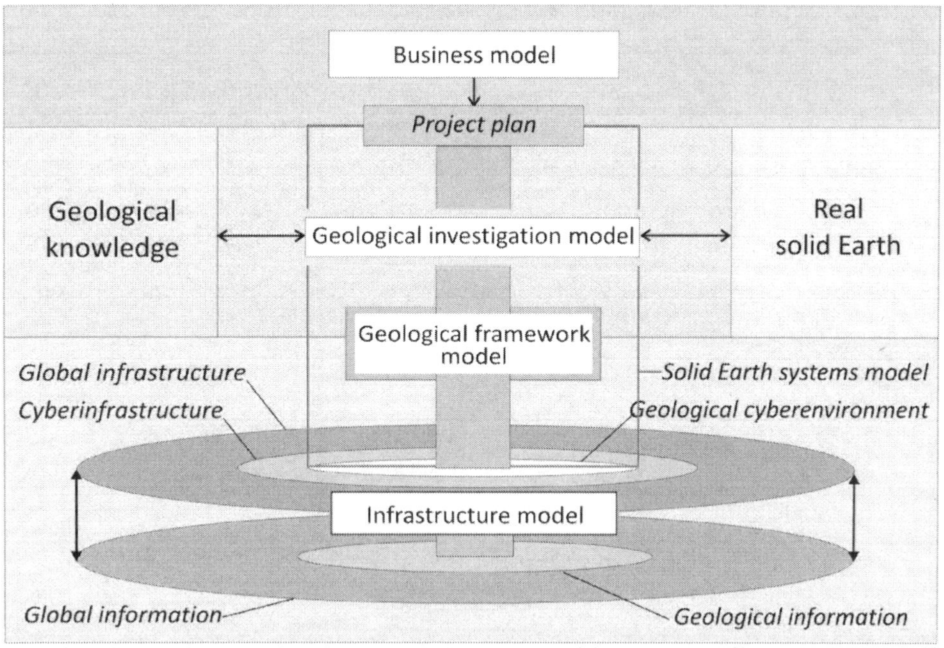

Figure based on Loudon, T.V., 2009. *Computers & Geosciences,* **35**(4), 700-705 http://nora.nerc.ac.uk/7258/

Figure 4: Four main models in the geological knowledge system (duplicate of **Figure 6**)

Overview of concept development

<<Overview 11

'As we may think'

In 1945, Vannevar Bush[32] reviewed the prospects, means and consequences of mechanising aspects of human thought.

Mechanising repetitive thought (Bush, 1945)

"...every time one combines and records facts in accordance with established logical processes, the creative aspect of thinking is concerned only with the selection of the data and the process to be employed

"...the manipulation thereafter is repetitive in nature and hence a fit matter to be relegated to the machine

"...whenever logical processes of thought are employed – that is, whenever thought runs for a time along an accepted groove – there is opportunity for the machine."

---

[32] Bush, V., 1945. As we may think. The Atlantic Monthly (July). http://www.theatlantic.com/doc/print/194507/bush

Mass production methods and interoperable components

Bush pointed out that "...machines with interchangeable parts can be constructed with great economy of effort and, in spite of much complexity, can perform reliably."

There is, Bush argued, good reason to borrow ideas from mass production methods:

- break the system down into small components
- share identical, fully tested, robust components that work together reliably and economically
- avoid reinventing wheels

The cyberenvironment, then, must be built from tested parts that work well with one another (interoperable components) and with the user. This calls for a rigorous system design and clearly defined subsystems, linked through simple interfaces.

Can a machine conform to patterns of human thought?

Bush suggested that our ineptitude in finding relevant information in the scientific record is largely a result of the artificiality of systems of indexing, where information is filed alphabetically or numerically and found (when it is) by tracing it down from subclass to subclass. "The human mind does not work that way. It operates by association of thoughts, in accordance with some intricate web of thoughts carried by the cells of the brain."

Blazing a trail

Bush visualised a machine where the scientist builds (and potentially shares) trails of thought linking many items through webs of information located in archives, with facilities to add comments and links to side trails on related topics. He also envisaged "skip trails" which "stop only on the salient items", providing an overview, with a means of drilling down for more detail when the user needs it.

Helping to forget

Mechanical recording of the trails would enable the user to "reacquire the privilege of forgetting the manifold things he does not have immediately at hand, with some assurance that he can find them again if they prove important"

Recorded trails

The trails foreseen by Bush are now being implemented as hypertext threads and scientific workflows. The trails or threads of information clarify the reasoning and its local context, and record patterns of investigation and analysis. They match one pattern of human thinking. But unconnected, they leave loose ends.

How can we avoid a tangled web?

Too busy to think

Licklider (1960)[33] calculated that 85% of his 'thinking' time was spent on clerical and mechanical activities: searching, calculating, plotting, transforming, and determining the logical or dynamic consequences of a set of assumptions or hypotheses. The tasks he attempted were selected to a great extent by clerical feasibility, not intellectual capability.

Scattered brains

Licklider also pointed out that in many creative endeavours only a few people have the potential to contribute effectively. The most creative may be independent thinkers rather than the best team players. The time scale of their communications may stretch out as each builds their own empire "and devotes more time to the role of emperor than the role of problem solver."

Symbiosis

Licklider (1960) saw interactive communication as a means of addressing these problems of handling and sharing information, through a symbiotic partnership of human brains and computers (and now the internet and World Wide Web). The symbiosis is mediated by a shared model, in the sense of a shared conceptual construct that represents reality for a particular purpose. The model is accessible to both the human user and the machine. It also has the potential to map and to relate the trails envisaged by Bush.

Revealing models

Modelling, Licklider (1968)[34] suggested, is basic and central to communication. "Any communication between people about the same thing is a common revelatory experience about models of that thing. Each model is a conceptual structure of abstractions formulated initially in the mind of one of the persons who would communicate, and if the concepts in the mind of one would-be communicator are very different from those in the mind of another, there is no common model and no communication."

Cooperative modelling

Licklider (1968) noted that the individual model can be observed and manipulated only by its originator. For wider applications "society rightly distrusts the modeling done by a single mind. Society demands consensus, agreement, at least majority... individual models must be compared and brought into... accord... [by] cooperation in the construction, maintenance and use of a model."

[33] Licklider, J.C.R., 1960. "Man-computer symbiosis" http://memex.org/licklider.pdf

[34] Licklider, J.C.R., 1968. "The computer as a communications device" http://memex.org/licklider.pdf

To tie loose ends: to weave a web

His proposed solution was technology to maintain, share and manipulate an explicit external model – an agreed structured framework where many minds can work together, weaving threads of thought into the web of the knowledge system.

The solid Earth systems model is a tentative candidate for such a framework in geology, with the potential to link diverse information at all granularities (levels of detail or resolution).

Sharing a framework

The framework must reflect a distilled consensus of expert views, in the same spirit as the Stratigraphic Guide: "...agreement on stratigraphic principles, terminology, and classificatory procedure is essential to attaining a common language of stratigraphy that will serve geologists worldwide. It will allow their efforts to be concentrated effectively on the many real scientific problems of stratigraphy, rather than being wastefully dissipated in futile argument and fruitless controversy arising because of discrepant basic principles, divergent usage of terms, and other unnecessary impediments to mutual understanding." (Hedberg, 1976[35], page v)

Sharing an environment

- a framework reflects a shared paradigm, providing a solid, stable base for normal science, and is necessarily resistant to change
- a cyberenvironment is more fluid, reflecting the growing and ever-changing methods of scientific investigation and communication of the results
- but, like biology and ecology, the framework and cyberenvironment intertwine and must develop together, with well-defined systems interfaces, and procedures for evaluating and accepting change

One framework, how many webs?

Ways of thinking

New information interacts in various ways with existing knowledge in our minds, spinning many webs of thought. For example, specialised mechanisms in our brains handle different information types, which enable us to think about things in distinct ways. At a higher level, our brains can reconcile and coordinate the resulting streams of thought to build an overall view.

Types of information

Thus, we use different information types when thinking about: underlying general knowledge of the science; ways of doing things; narrative text accounts and descriptions; spatial location, arrangement and form; observations and measurements. Details and references are in http://nora.nerc.ac.uk/2405/ . See also Integrating information types 62.

[35] Hedberg, H.D. (editor), 1976. International stratigraphic guide: a guide to stratigraphic classification, terminology and procedure. Wiley-Interscience, New York.

Assembling types of information

Conventionally, geological information is assembled in inflexible documents, each centred on a particular information type, such as maps and cross-sections (spatial), scientific books and papers (narrative), data files and registers (tabular). In a digital environment, each information type still requires its own tools and representations, at all levels of detail, for tasks such as visualisation (spatial), understanding sequences of events (narrative), statistical analysis and testing agreement between interpretations and underlying observations (algorithmic). The cyberenvironment must provide the flexible tool kit to link and coordinate representations of all the information types referring to the same entity.

What about unexpressed thought?

Licklider (1960) envisaged human beings and machines collaborating in a symbiotic partnership. For direct communication among them, information is recorded externally (outside the human brain – electronically, on paper, or any shareable medium). But geologists often cannot or may not wish to externalise all aspects of their investigations.

Indirect communication

Much of a geologist's knowledge involves winnowing out invariant attributes (aspects that stay the same) from a multitude of situations. Unlike the conclusions, the skills and procedures cannot necessarily be put into words or pictures. But demonstrating the procedures in the field can convey information that enables another geologist, with shared experience and background, to follow the same processes of observation and thought, and maybe reach the same conclusions. Confirmation by experts in this way provides scientific validation of unexpressed knowledge.

Communicating the inexpressible

The cyberenvironment can assist this process by recording the workflow and video demonstrations of the procedures that led to the conclusions, thereby enabling others to repeat them, or an experienced geologist could remotely guide a novice in the field or core store, discussing and clarifying unrecorded aspects.

A multi-dimensional fabric

Investigation of even a single outcrop generates records of many diverse items of information. Within a conventional, self-contained document (such as a map or scientific paper) the significance of the recorded item depends on its context within the document. In a digital environment, the significance of the recorded item depends on its many links in webs within the geological knowledge system as a whole. And the cyberenvironment must handle not just the full range of conventional and digital representations, but unexpressed knowledge as well.

How can we model this complex fabric of thought?

Details and references at http://nora.nerc.ac.uk/2405/

Objects of thought

Coad and Yourdon[36] (1991) suggested that three methods of organisation pervade our thinking about the real world:

- differentiation of experience into particular objects and their attributes
- distinction between whole objects and their component parts, and
- distinction between different classes of objects

They advocated an object-oriented approach to information technology that reflects this.

Varieties of object

Objects are simply things of interest – computer representations of real-world or interpreted entities. They may form a hierarchy, little objects within bigger objects, within yet bigger ones, and so on. They might, for example, be:

- geological (fossils, outcrops, formations, terranes)
- documents (maps, memoirs, logs, indexes)
- document contents (individual symbols, words, paragraphs, chapters)
- system contents (sub-systems, hypertext threads, ontologies, databases, expert sources)

Recording objects

In geoscience, this approach might record:

- objects (things of interest)
- object classes (the categories in some classification scheme, usually hierarchical, to which the objects belong)
- object instances (specific occurrences of an object)
- their attributes (properties, composition, relationships and behaviour)
- processes (which cause things to change)

Ontologies

Ontologies[37] can also be constructed, to provide a controlled machine-readable specification, identifying objects, processes, and their definitions, characteristics and relationships.

Only connect

The object-oriented view breaks down information into individual items, and records their relationships. These items can be identified to ensure that many information types and many threads of thought can refer consistently to the same items. They can be categorised by class and metadata to locate them in the framework, and reveal ontological implications and constraints on their analysis. Unlike conventional

---

[36] Coad, P. and Yourdon, E., 1991. *Object-oriented design.* Yourdon Press, Englewood Cliffs, NJ. 197pp.

[37] Ontology: A formal representation and shared vocabulary describing concepts, entities and relationships in a domain of knowledge, typically providing a more detailed and rigorous machine-readable specification than a thesaurus or taxonomy.

records, but like the human mind, the cyberenvironment can accommodate the huge network of interconnections linking the items.

Linked Data

As opposed to seeing a fixed set of data sources, 'Linked Data' aims to create an unbound global data space (Bizer et al.[38], 2009). It links, not documents, but things in the world described by data on the Web, relating individual entities within structured data, in ways similar to those currently used to query a local database.

How does this relate to geology and the advanced cyberinfrastructure?

The early concepts are now leading to a systems view supported by the on-demand services of a knowledge utility (see Semantic Web and Grid 50 and Next Generation Grids Expert Group[39], 2006). The infrastructure can overcome limitations of conventional techniques by improving and extending the representation of ideas as a comprehensive systems model.

Flexible thinking

Geological objects are intricately interlinked in a complex fabric, given form by the accepted geoscience paradigm, expressed as a shared model of the systems of the solid Earth. The object-oriented approach can represent many aspects of thinking about an individual object, linking it to webs of thought defined by geological classification and properties, information type, granularity, location, provenance, and place in the workflows of investigation or reasoning. It can support better information, more rigorous analysis, flexible and efficient access. But can you believe it?

Systemic risk

The increase in connectivity of information in the knowledge economy brings risks that threads of reasoning, linkages, analogies, correlations, blanket applications of statistical predictions, and risk assessments, fully understood only by their developers, become so entangled that failure in one part has catastrophic results for the integrity of the knowledge system as a whole. A successful system depends on a sceptical attitude, with thorough and open evaluation and testing of the scientific findings.

The role of Geological Survey Organisations

A Survey uses detailed local knowledge to assess all available sources and test them in the field to provide a coherent authoritative view of the geology. World-wide, Surveys collaborate to set benchmarks for local and regional geology against which other sources can be judged. They are well positioned to make a major contribution

[38] Bizer, C., Heath, T., Berners-Lee, T., 2009. Linked data – the story so far. International Journal on Semantic Web and Information Systems (IJSWIS), **5**(3), 1-22. DOI: 10.4018/ijswis.2009070101. http://tomheath.com/papers/bizer-heath-berners-lee-ijswis-linked-data.pdf

[39] Next Generation Grids Expert Group, 2006. Future for European Grids: Grids and Service-Oriented Knowledge Utilities: Vision and research directions 2010 and beyond. Report 3 for the European Commission. ftp://ftp.cordis.europa.eu/pub/ist/docs/grids/ngg3-report_en.pdf

to future models of the systems of the solid Earth and to make evaluated, authoritative knowledge and information readily, transparently, and widely available.

Conclusions

The historical development of the infrastructure influences the strategy for enhancing the knowledge system, and helps to understand the concepts and rationale of current work. Despite the rapid evolution of the geological infrastructure, the bulk of recorded geological information remains in conventional forms, with individual topics and business models at different stages of development. Understanding recorded information depends on geological knowledge held unrecorded in the collective human memory, as background acquired by training, education and experience. Implementation is following hot on the heels of the developing concepts. Only geologists can carry forward their own legacy of knowledge in step with the wider systemic transformation.

>>More at Stages of concept development (cyberinfrastructure) 45
and Stages of concept development (geological thinking) 56

See also Overview of the infrastructure model 37

Overview of the solid Earth systems model

<<Overview 11

What is the solid Earth systems model?

The solid Earth systems model (sEsm) is an approach to structuring geoscience information, such as the authoritative view of regional geology maintained by geological survey organisations, and linking it to the facilities of the cyberinfrastructure. It is intended to provide a comprehensive model of the systems of the solid Earth that integrates relevant knowledge in a coherent, shared, testable, predictive system. Its contents refer to:

- the three-dimensional disposition and configuration of the present-day geological objects of the solid Earth (where things are and how they are arranged)
- their observed and interpreted properties, composition and relationships, at all scales
- interpreting the present-day objects as the outcome of geological processes interacting with historical configurations of objects, resulting in events and historical changes throughout geological time

Why are Earth system processes important?

The geological knowledge system must include a comprehensive understanding of Earth system processes – the forces for change which operate now and shaped the past evolution of successive configurations of the solid Earth as disentangled by geoscientists in their record of Earth history. Their outcome depends on the input, not on when and where they took place (they are invariant[40] under specifiable time and space transformations). Some of the features they leave in the rocks (shapes, spatial relationships) inherit the invariance, thus becoming the key to linking past and present and to deciphering geological history. Geological processes involve (and link to) all branches of Earth science, and are therefore a key component of a holistic solid-Earth systems model.

Why model solid Earth systems?

Solid-Earth sciences

The importance of system interactions among the lithosphere, hydrosphere, atmosphere and biosphere was emphasised by the US National Research Council (1993)[41] in their influential report *Solid-Earth Sciences and Society*. A report by the UK Natural Environment Research Council likewise calls for a holistic view of our planet. "The behaviour [of each component part of the Earth] is critically dependent on other parts of the system... we need to understand the behaviour of the entire Earth, from the core to the upper atmosphere... [and] what we know of past changes and the long-term driving forces that caused them" (NERC, 2007).[42]

Parallels in medicine and biology

The 'Virtual Physiological Human' project views the human body as a single, complex system as "a way to share observations, to derive predictive hypotheses from them, and to integrate them into a constantly improving understanding of human physiology/pathology, by regarding it as a single system." (STEP Consortium, 2007[43], Clapworthy et al., 2008[44]).

The European Science Foundation sees Systems Biology as a 'Grand Challenge' "...recognising that biological systems are far too complex to be solved by classic biological approaches... [Systems biology] gives a central role to predictive mathematical models that integrate all relevant data on the topic of investigation and exploits such models to decide which experiments are most effective."

---

[40] Invariant: An object with the property of invariance, that is, it does not change under a specific set of transformations or sequence of operations.

[41] US National Research Council, 1993. *Solid-Earth Sciences and Society*. National Academies Press. 368pp. ISBN-10: 0309047390 ISBN-13: 978-0309047395

[42] NERC, 2007. Next Generation Science for Planet Earth, NERC Strategy 2007-2012. http://www.nerc.ac.uk/publications/strategicplan/documents/strategy07.pdf

[43] STEP Consortium, 2007. Seeding the EuroPhysiome: A Roadmap to the Virtual Physiological Human. http://www.europhysiome.org/roadmap

[44] Clapworthy, G,. Viceconti, V., Coveney, P.V., Kohl, P., 2008. Editorial. Phil. Trans. R. Soc. A 366, 2975-2978. doi: 10.1098/rsta.2008.0103 http://rsta.royalsocietypublishing.org/content/366/1878/2975.full.pdf+html

European Science Foundation, 2007[45], Anteneodo and Da Luz, 2010[46], BBSRC, 2007[47]).

The context

The examples suggest that the systems model must be developed within a broad context:

- compatible with the advanced cyberinfrastructure as it continues to evolve
- recognising geological aspects as an integral part of Earth systems science
- working in parallel with a wide range of initiatives in other fields of science
- ensuring that global collaboration is achieved

The knowledge system inevitably refers to many sources of information, built on incompatible models. It should encourage compatibility where appropriate, but must also provide metadata to enable users to identify conflicts and to select information appropriate for the application. An integrated knowledge system based on a model of the systems of the solid Earth should be matched by an infrastructure that includes specific provision for geological applications – a geological cyberenvironment.

>> More at The solid Earth systems model (sEsm) 71
and Overview of the geological framework model 35

Overview of the geological cyberenvironment

<<Overview 11

What is a cyberenvironment?

The US National Center for Supercomputing Applications (2008)[48] describes cyberenvironments as "a means of enabling research communities to exploit the resources available on the internet... providing an integrated set of hardware, software tools, and services needed to marshal information resources and analyze, visualize, and model phenomena of interest."

How does it relate to the geoscience knowledge system?

An advanced cyberinfrastructure for geological applications requires:

- a shared framework (probably based on a solid Earth systems metamodel) to structure relevant information as a component of a coherent knowledge system

---

[45] European Science Foundation, 2007. Systems Biology: A grand challenge for Europe. http://www.esf.org/publications/medical-sciences.html

[46] Anteneodo, C. , Da Lux, M.G.E. 2010. Complex dynamics of life at different scales: from genomic to global environmental issues. Phil. Trans. R. Soc. A vol. 368, no.1933, 5561-5568 doi: 10.1098/rsta.2010.0286 http://rsta.royalsocietypublishing.org/content/368/1933/5561.abstract

[47] BBSRC, 2007. Systems biology (UK Biotechnical and Biological Sciences Research Council). http://www.bbsrc.ac.uk/publications/corporate/systems-biology.aspx

[48] US National Center for Supercomputing Applications, 2008. NCSA 2010: The future of NCSA. http://www.ncsa.uiuc.edu/AboutUs/NCSA_2010.pdf

- a geological cyberenvironment that supports the working practices of geological investigation as they build on that framework

What does it do?

- it eases the task of making sense of a great diversity of relevant information
- it provides a unified user environment and integrated view of the infrastructure relevant to a particular field of enquiry (in this case, geology)
- it assembles relevant aspects of the infrastructure to provide end-to-end support in geological investigations

How does it do it?

- it is based on the technologies of information, computing and communications
- it emphasises integration, support and automation of work processes rather than standardisation of software components
- it is accessed through a user interface that matches the users' working practices, and the familiar concepts and methods of geology
- through an 'agent' (software assisting the user's access), it presents the knowledge base and associated tools to users as if centred on their own current interests

Implementation examples

An example of a cyberenvironment is the Water and Environmental Research Systems Network (WATERS)[49], described by Liu et al., (2007)[50]. An example in crystallography is Fennick et al., (2008)[51].

Aims of the cyberenvironment

The cyberenvironment will potentially support access to the complex fabric of information about geology. It aims to bring together tools, services and resources as a unified user environment, tailored to allow researchers and educators to interact with the infrastructure using the familiar concepts and approaches of their specific scientific discipline, while automating the use of the resources.

Gathering the tools

A task for the geological community is to specify requirements for the set of tools that will provide end-to-end support for work processes, throughout all the stages of geological investigation. They must enable users to work on diverse information, represented in several information types, stored as objects in human minds and in conventional and digital records, or woven together as webs in the complex fabric of interpretation of a shared systems model.

---

[49] WATERS network, 2009. http://www.watersnet.org/

[50] Liu, Y., Myers, J., Minsker, B., Futrelle, J., 2007. Leveraging Web 2.0 technologies in a Cyberenvironment for Observatory-centric Environmental Research. http://www.semanticgrid.org/OGF/ogf19/Liu.pdf

[51] Fennick, J. R., Keith, J. B., Leonard, R. H., Truong T. N., Lewis J. P., 2008. A cyberenvironment for crystallography and materials science and an integrated user interface to the Crystallography Open Database and Predicted Crystallography Open Database *J. Appl. Cryst.* (2008). 41, 471-475 doi:10.1107/S0021889808000381

Projects

Geological investigations take place as projects (manageable activities with objectives, resources and structure), possibly following a project design methodology specified by a business model. The project may operate at any level from a brief task for an individual to a long-term collaboration among many organisations, and at any level as a sub-project within a hierarchy of projects.

End-to-end support

Each project goes through a sequence of stages, repeatedly followed, in full or in part, in the course of an investigation. The cyberenvironment aims to provide end-to-end support, and the characteristics, requirements, interactions and products of each stage must knit together within the knowledge system.

A means to an end

A geological cyberenvironment should aim to support:

- a systems view of geology in its wider context
- a framework of models and ontologies that overcome unnecessary impediments to mutual understanding and sharing of information
- a connected comprehensive fabric of information, in symbiotic partnership with human thinking
- automation of routine thought processes, including mathematical analysis, interpolation and visualisation
- robust methods to test, evaluate and regulate the knowledge system and its contents

The target of the 'changing paradigm' might be seen as a digital solid Earth systems model that maintains comprehensive information readily accessible to suppliers and users. The geological cyberenvironment is the means to that end (see Figure 5).

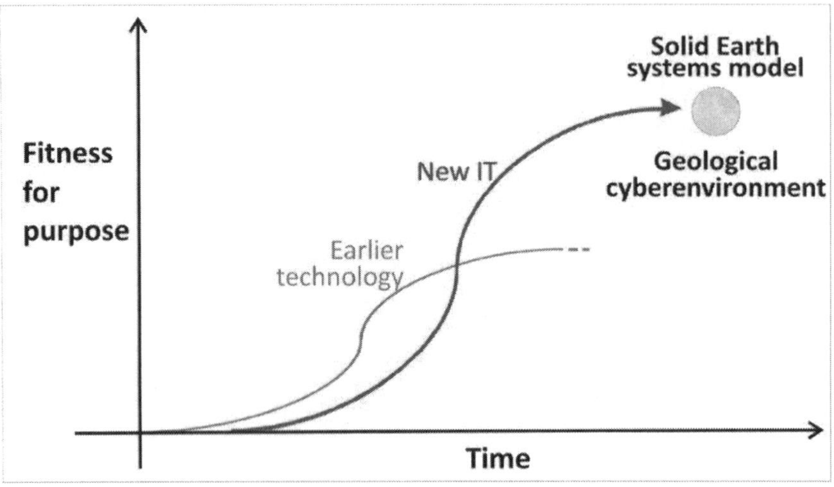

Figure 5: Suggested targets for systems geology.

>> More at The geological cyberenvironment (gce) 85

Overview of four implementation models

<<Overview 11

Four models in the geoscience knowledge system

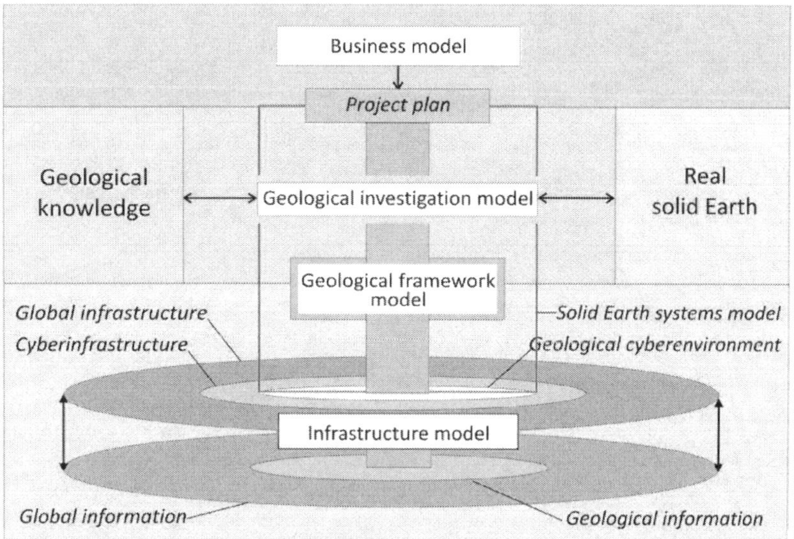

Figure based on Loudon, T.V., 2009. *Computers & Geosciences,* **35**(4), 700-705 http://nora.nerc.ac.uk/7258/

Figure 6: Aspects of the geological knowledge system

Four interacting models (Figure 6) form the main structures brought together in the solid Earth systems model and geological cyberenvironment. The geological business model defines the objectives of a geological investigation and assesses the results: the concern of geological management. The geological investigation model describes the procedures by which information is gathered: primarily the concern of the investigators. The geological framework model depicts and clarifies the principal relationships among the findings of geology, linking all four models through agreed standards: the concern of the geological community as a whole. The geological infrastructure model describes the structure of facilities and mechanisms to store, process and share geological information – the cyberinfrastructure is the emerging infrastructure based on information, computing and communication technology: primarily the concern of e-scientists.
Details and references at Loudon (2009)[52].

Linking the models

The four models, each calling on a different area of expertise, all face radical change to benefit from the cyberinfrastructure. They are considered separately, but must be

[52] Loudon, T.V., 2009 Four interacting aspects of a geological survey knowledge system. *Computers & Geosciences*, 35 (4). 700-705. 10.1016/j.cageo.2007.12.009 or http://nora.nerc.ac.uk/7258/

designed and interfaced to operate in close conjunction with one another, and in a small project, might all be the responsibility of the same individual.

Overview of the geological business model

<<Overview 11

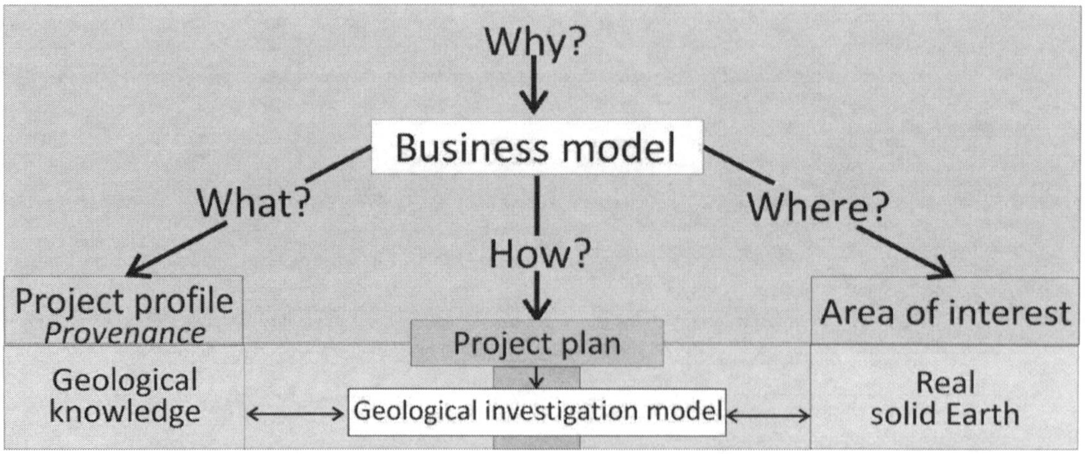

Figure 7: Business model

What is the function of the business model, and how will it change?

The business model (Figure 7) defines why and how a geological investigation is undertaken, who carries it out, what it aims to achieve (the objectives) and where, thus clarifying for end-users the provenance of the results (their source, derivation and reliability). A project plan is defined in collaboration with the investigators, and assigns responsibility for maintaining quality and evaluating results.

The project profile and plan

Each business model and project has its own implicit profile that specifies the relative importance to the project of the various geological aspects and properties. Project profiles might be explicitly defined as regions of the solid Earth systems metamodel. An explicit project plan could guide the investigational process and, as part of the provenance, could help to evaluate the relevance of the results to other applications.

Sharing information

A business objective of many geological investigations (particularly in the academic field) is to share results as part of the body of communicated knowledge. Methods of scientific publication, and links to search engines for information supply, are changing rapidly, with user-specified selection of content and mode of presentation. Business developments within the geological community must relate to these mainstream trends, liaising and outsourcing as appropriate with the information industry.

Range of geological business models

Geological business models describe diverse activities (each with its own profiles and methods of geological investigation) including:

- exploration for and exploitation of natural resources (for example: water, hydrocarbons, minerals, sand and gravel, waste storage)
- civil engineering
- hazard assessment
- research, education
- comprehensive geological survey

Assemble and integrate

Information from many diverse applications and business models must be assembled and integrated to support the holistic approach to the cyberinfrastructure and to Earth systems. For example, survey field work can fill scale-space gaps in hydrocarbons subsurface data. The geological framework and infrastructure can provide a milieu for exchanging and sharing information and its provenance, but business models define how widely it can be shared.

Geological investigation as predictive reinforcement learning

The process of predictive reinforcement learning[53] and generalisation has been studied in the artificial intelligence community (see Geological surveying as reinforcement learning 79), with results that suggest the approach is relevant to describing and sharing geological information and its provenance, and tracking contributions to the predictive scheme from new ideas and observations, thus helping to evaluate them.

Evaluation of project results

The business model implies a project profile or explicit project plan that relates to items in the framework model (such as types of objects, properties, processes or relationships) and their relative importance to the project. The criteria for evaluating the outcome include its relevance, accuracy and predictive power.

How does this apply to the case of the Geological Survey business model?

The general aim of a geological survey organisation is to develop, record and communicate an authoritative, coherent, evaluated account of the geology of a defined region. The cyberinfrastructure might not change the general aims defined in the Survey business model. But it changes radically the methods of meeting its objectives, evaluating and supplying results, and meeting the expectations of users. With their emphasis on generality, it is fitting that Geological Surveys should take a lead in these business developments.

>>More at The geological business model 93, The solid Earth systems model (sEsm) 71

---

[53] Predictive reinforcement learning: A means of characterising a learning problem in terms of an agent seeking to achieve a goal by interacting with an uncertain environment.

Overview of the geological investigation model

<<Overview 11

What is the function of the geological investigation model?

The geological investigation model defines how an investigation is carried out (guided by a project plan shared with the business model). It refers to the tools, techniques and procedures by which information referring to the real world is collected, assembled and communicated. It links through the geological framework to the user and to the advanced cyberinfrastructure.

How will it change?

The cyberinfrastructure makes new surveying methods feasible, and alters the representation and communication of the results, thus changing the conventional view of the geology and the perspectives of surveyors. Various stages are tentatively proposed in the geological investigation model (*Figure 8*) that could be matched with support from the geological cyberenvironment (see *Figure 17*). The aim is to provoke geologists to consider their workflow, and agree on a methodical sequence of operations for end to end software support. Of course, the investigator moves repeatedly and unpredictably from one aspect to another, but might navigate the appropriate support more readily where it is organised in a familiar sequence.

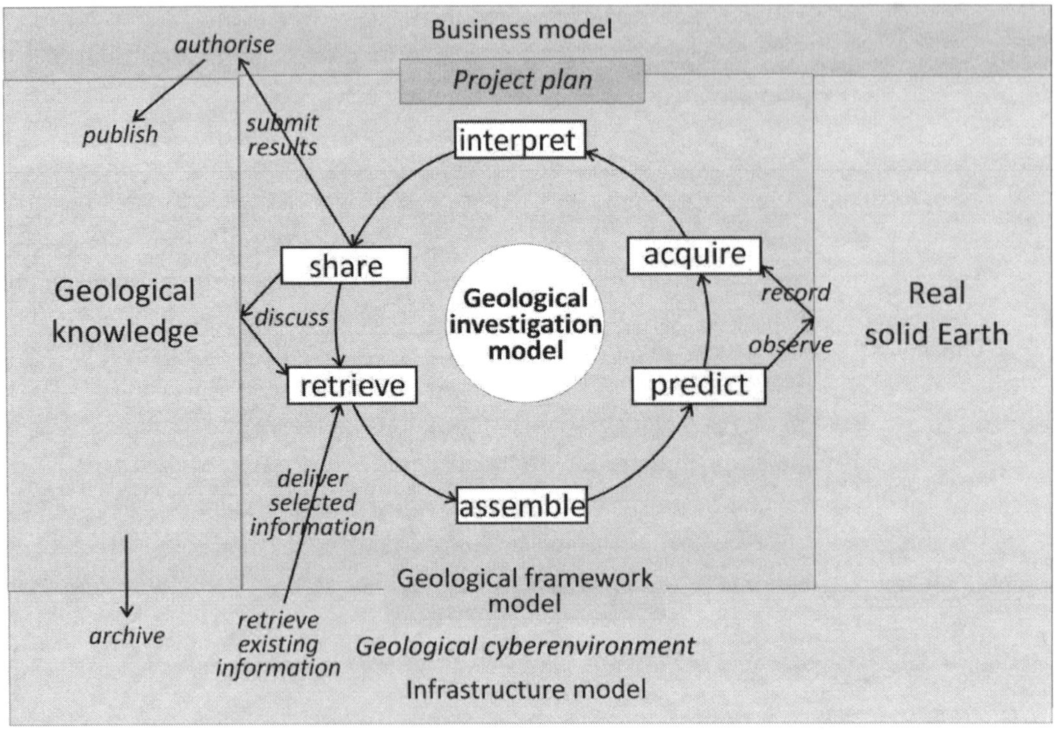

Figure 8: Geological investigation model

What are the major consequences for surveying?

- the cyberinfrastructure, linked by the geological framework model to other models for geological investigation, will support new methods

- the methods offer a more rigorous scientific approach that can integrate information from many sources to provide a more coherent and comprehensive view of Earth systems science
- the constraints of pen and paper for recording sketches, maps, notes and reports are overcome (see, for example , Some related initiatives 341 – Internal BGS - SIGMA 343).
- information can be handled as digital records, referring to many small but highly interconnected items, in a multi-resolution, 3D, geological-time context
- all information types, such as text narrative, images, spatial representations, tabular data, algorithmic processes, statistical relationships and human judgment can be recorded and integrated for individual objects at any scale.

But many methods have been developed in other fields, and so far used in geology largely in exploratory academic studies.

>> More at The geological investigation model 98 and The future geological map 125

Overview of the geological framework model

<<Overview 11

Details and references at http://nora.nerc.ac.uk/1084/

What are the functions of the geological framework model?

- the framework depicts and clarifies the principal relationships among the findings of geology, providing a multidimensional map to locate and connect ideas, concepts, workflows of investigation, and threads of reasoning
- it links the various models of the geoscience knowledge system
- it is a structure for organising and assembling dispersed information relevant to geology, and the basis for a model of systems of the solid Earth (see Overview of the solid Earth systems model 26)

What is the solid Earth systems model about?

- the three-dimensional disposition and configuration of the present-day geological objects of the solid Earth (where things are and how they are arranged)
- their observed and interpreted properties, composition and relationships, at all scales
- geological processes and the outcome of their interactions with configurations of objects
- events and historical changes throughout geological time

A tentative outline for a metamodel[54] of the solid Earth systems model is included in *Figure 9*. It describes the structure and organisation of The solid Earth systems model (sEsm) 71, and is therefore an important part of The geological framework model 105, where it is described in more detail.

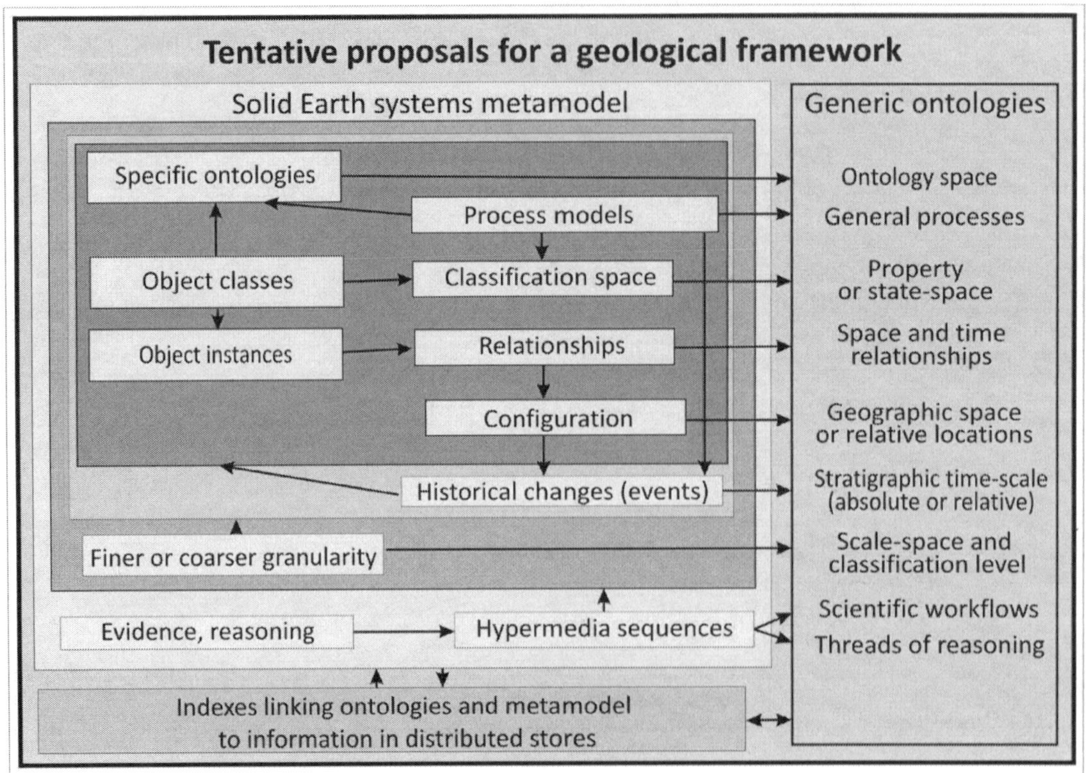

Figure 9: A proposal for a geological framework

See also: Overview of the infrastructure model 37
The geological framework model 105
The solid Earth systems model (sEsm) 71
The solid Earth systems metamodel (sEsmm) 110

[54] Metamodel: A metamodel is a description of the organisation and function of a model, to assist the user or computer to find, manage, control and understand its contents.

Overview of the infrastructure model

<<Overview 11

What is the function of the infrastructure model, and how will it change?

The infrastructure model describes the structure, facilities and mechanisms to collect, store, process and share information. The cyberinfrastructure is the emerging infrastructure based on information, communication and computing technology: primarily the concern of e-scientists.

Service orientation

The internet and World Wide Web are evolving to the Semantic Web and Grid 50 and The service-oriented knowledge utility 52 (SOKU). Provision of on-demand services is taking the place of selling products: "emerging Service-Oriented Architectures will enable the provision of computing, data, information and knowledge capabilities as utility-like services in the future, including services which intersect with the physical world through a wide range of computing devices." Next Generation Grids Expert Group[55] (2006).

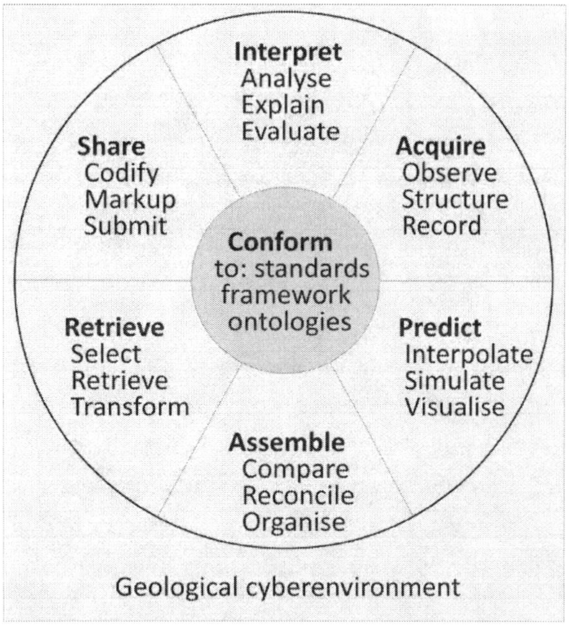

Figure 10: Stages of investigation in the geological cyberenvironment (duplicate of Figure *17).*

Diverse sources for information

"Future business and scientific applications will be built as a complex network of services offered by different providers, on heterogeneous resources, constrained by administrative problems when crossing the borders of different organisations." Next Generation Grids Expert Group (2006).

[55] Next Generation Grids Expert Group, 2006. Future for European Grids: Grids and Service-Oriented Knowledge Utilities: Vision and research directions 2010 and beyond. Report 3 for the European Commission. ftp://ftp.cordis.europa.eu/pub/ist/docs/grids/ngg3-report_en.pdf

Interoperability

Global standards, shared structures and quality assessment of both the science and the technology are required to link information and services from many sources and ensure that they can work together (interoperability). The subsystems and interfaces must be carefully designed as a shared structure for development and use of the cyberenvironment software, and as a means of assembling the relevant processes within the cyberinfrastructure and making them available within the familiar sequence of activities in geological investigation (*Figure 10*).

Elements of a cyber-strategy

Strategic cyber-planning in geology calls for:

- a shared framework – such as the solid Earth systems metamodel
- the development and integration of business, investigation, framework and infrastructure models to create a coherent knowledge system
- a geological cyberenvironment to match users' working practices and define a modular structure for software development

Overview of the future geological map

<<Overview 11

Reasoning, models and reality 127

The advancing infrastructure extends the meaning of 'geological mapping' beyond the concepts of paper maps and geographical information systems. It calls for reconsidering underlying geological objectives, methods and models (taken for granted in a more settled system) in order to map[56] (in a mathematical sense) geological knowledge and observations into a multidimensional, multiresolution, multimedia knowledge base[57]. Geoscientists must control its development, and it is to them that 'The future geological map' 125 (and the summary of its contents here) is primarily addressed. Models are simplified views of reality. They can separate observations, deductions, assumptions and interpretations more effectively than a map. The geological spatial model can condense information from numerous observations as a generic interpretation in harmony with scientific understanding of the historical geology. The interpretation is confronted with reality by locating it in space through surveying procedures based on human knowledge. Short notes in an indexed object store can record the reasoning, linkable as explanations for user-defined areas and topics. The global framework of space and time-sequenced

[56]Mapping: Conventionally, geological mapping leads to a graphical depiction, usually on a flat surface, of spatial relationships and forms of geological features or properties in a selected area of the Earth's surface or subsurface. In the mathematical definition, mapping relates the elements of one set to those of another. A broader definition of geological mapping could be 'relating elements of geological observation or interpretation of the solid Earth to corresponding elements in appropriate models in the geoscience knowledge system'.

[57] Knowledge base: A dynamic repository for information and methods for accessing and processing it. It is generally machine-readable and online, and may include the means to access expert knowledge.

stratigraphical units connects field correlations with historical geology. Top-down surveying extends existing knowledge with directed observations and interpretations, potentially gaining from digital access to related material and rapid updating.

From map to digital model 145

Looking ahead, spatial models can be more rigorous and flexible than the map, avoiding its information overload and meeting a wider range of objectives. Standard maps for wide-ranging spatial comparisons will be joined by spatial models offering visualisations and thematic presentations with user selection of area, content, scale and form of presentation. As an inverse model, the overall geological model must call on diverse approaches to resolve multiple hypotheses. Explanations relate to a general model of Earth history that is complex and fragmentary, involving self-organising processes that restrict deterministic models. Nevertheless, a broad understanding is achieved through hierarchical classification of objects and processes.

Reconfiguration 165

Geological survey organisations provide geological interpretations and located spatial information on which other projects can build – a view of the anatomy of the Earth and the core of reasoning for many topics in geoscience – and therefore have a particular responsibility for its reconfiguration. The concepts and rationale behind surveying must be reconsidered to benefit from the advancing infrastructure in providing a widely integrated, authoritatively evaluated, computer-based representation of the results of geoscience survey. Unlike the traditional map, the representation need not be constrained by the form of visualisation, content, scale, sheet boundaries or number of dimensions. It can provide hypertext links to many sources, with closer integration of text, images, metadata, and computer databases, programs and a wide range of applications and models. Mark-up and metadata are the means of linking diverse sources and information environments.

An object-oriented approach 178

Geological surveying increasingly depends on computer support, where the object-oriented approach fits well with geoscience reasoning. Real-world geoscience entities can be located by surveying, represented as object instances and classified within object-class hierarchies. Relationships between object instances or classes (such as: is a part of, is a kind of, follows, is linked to) can be explicitly recorded to structure the objects as hierarchies and sequences, and join up spatial features and chains of events. The object-based reasoning process may reveal inconsistencies between strands of thought, which must be reconciled. The expected behaviour of object classes, records of the reasoning process, and procedures for visualisation can be incorporated within the model as microdocuments. The full benefits of object-orientation call for a unified approach, leading to more flexible, comprehensive, robust and informative systems.

The geometry of the spatial model 203

A unified system of spatial models requires compatible computer representations of the geometry. Current approaches include interpolation from scattered points to fit surfaces to nodes on a square grid, using a weighted moving average or by fitting mathematical functions to the data. Variation within a volume can be represented by voxels. Fractals offer a model of processes distributed in space at different resolutions. Wavelet analysis breaks down an overall pattern into superimposed local patterns at different scales. An integrated view calls for a coherent overall system with a shared mathematical framework for all formal procedures from initial observation to final presentation.

Transforming space 224

The quest for a comprehensive view of spatial characteristics in geology leads to composite geometrical transformations, built from components like translation, rotation, scaling and projection. The invariance of properties of objects under specific transformations can throw light on their significance and behaviour during abstraction and feedback. Spatially invariant properties such as slope and curvature have a bearing on interpolation, the consequences of unevenly spaced data, and uncertainty envelopes. Considering interpolation as a geometrical operation, linked to algebra for computation, could tie it more closely to geological interpretation, linking the strengths of computer methods and human insights. A systems approach to the knowledge base helps to clarify the complexities of their interactions and the links to dynamic stratigraphy.

Seeking shared concepts 247

Various integrative concepts have emerged within the systems approach of diverse disciplines, such as ecology, landscape diversity, biomedical science and cognitive science. They have been sporadically explored in geoscience, but are not yet part of mainstream geological thinking. There is, however, an obvious case for benefitting from these developments in a solid Earth systems model, both in sharing work done elsewhere, and in widening the interoperability of geological products. They include criteria for defining boundaries of spatial objects (such as stratigraphical units); classifying the extent of processes and object classes in scale-space; their application in a wide range of interpolation models; statistical shape analysis (morphometrics), including changes as a configuration of objects evolves through time; and the integration of spatial knowledge (including uncertainty) in inhomogeneous deformable models – "the confluence of geometry, physics and approximation theory".

Mapping geology into the knowledge system 282

The dynamic spatial model (how things got there) contains much of the geoscience reasoning underpinning the static spatial model (where things are) as shown on a map. Both models are essential components of the geoscience knowledge system. Top-down geological survey would benefit from access to such a system during field and office work, with the ability to capture and adjust the evolving interpretation. The solid Earth systems model (sEsm), dealing conceptually with the history and three-dimensional distribution of geoscience objects, provides a framework that links individual fragments, filtered and projected to fewer dimensions, which are all that we can observe, record in the field, and carry forward as facets of an interpretation. By formally identifying the objects and models, the system could record a reasoning process in the field, which cannot be disentangled from a completed map. Users of the sparse and varied spatial information require access to the sEsm and the Web generally, supported by a spatial index and object store, with browser, database, GIS, and visualisation facilities. A comprehensive, flexible and extensible framework of interlinked sub-models representing fragments of the general model is needed to structure and evaluate information, and to guide searching and browsing. The same framework of systems, metadata, models, objects, attributes and relationships should guide procedures of survey, interpretation, and use of the resulting spatial model, following familiar procedures in the geological cyberenvironment.

The emerging geoscience knowledge system

<<Table of contents 1
<<Introduction and overview 1
The emerging geoscience knowledge system 42
Stages of concept development (summary) 44
The solid Earth systems model (sEsm) 71
The geological cyberenvironment (gce) 85
The geological business model 93
The geological investigation model 98
The geological framework model 105
The geological infrastructure model 117
>>The future geological map 125
>>References 314
>>Glossary 328
>>Index 359

Summary: Despite rapid advances in the knowledge infrastructure, the bulk of recorded geological information remains in conventional forms, with individual topics and business models at different stages of development. Indeed, most geological knowledge may remain unrecorded in the collective human memory, as background acquired by training, education and experience. The cyberinfrastructure can overcome limitations of conventional techniques, by improving and extending the representation of ideas in a comprehensive systems model. It can help to integrate developments in cyber-aided data collection, analysis, interpretation and publication. But the systems model must also tie in to earlier representations of geological knowledge. It should have the potential to extract valuable information from past records to add to the model. It should also maintain links to assist users of the systems model to refer to relevant conventional documents, which they can interpret in the light of their own background knowledge. The emerging geoscience knowledge system must interoperate [58]with existing knowledge systems, to ensure that advances in systems geology contribute to advances in the science as a whole.

It is suggested here that various concepts arising from the cyberinfrastructure (see Stages of concept development (cyberinfrastructure) 45) and from geological thinking (see Stages of concept development (geological thinking) 56 and The future geological map 125) can be brought together in one overall structure. This could comprise a solid Earth systems model, which attempts to relate the various aspects of the geology in a widely acceptable way (see

[58] Interoperability: Interoperability of information is the ability of concepts, terms or models from various sources to work together, by meeting standards that enable sharing and reuse of information.

The solid Earth systems model (sEsm) 71), and a geological cyberenvironment that should provide suitable cyber-support designed to match the pattern of geologists' thinking (see The geological cyberenvironment (gce) 85). The structure is discussed in terms of four main models, reflecting different areas of expertise. The business model defines the objectives of geological investigations: primarily the concern of geological management (see The geological business model 93). The geological investigation model describes the procedures by which information is gathered: the concern of geological surveyors (see The geological investigation model 98). The geological framework model depicts and clarifies the principal relationships among the findings of geology, standardising the organisation of the geoscience knowledge system: the concern of the geological community as a whole (see The geological framework model 105). The infrastructure model describes the structure, facilities and mechanisms to store, process and share information. The cyberinfrastructure is the emerging infrastructure based on information and communication technology: primarily the concern of e-scientists (see The geological infrastructure model 117). These aspects of the geological knowledge system are summarised in the Overview 11, and described in detail as listed above. Their relationships are shown in *Figure 11*.

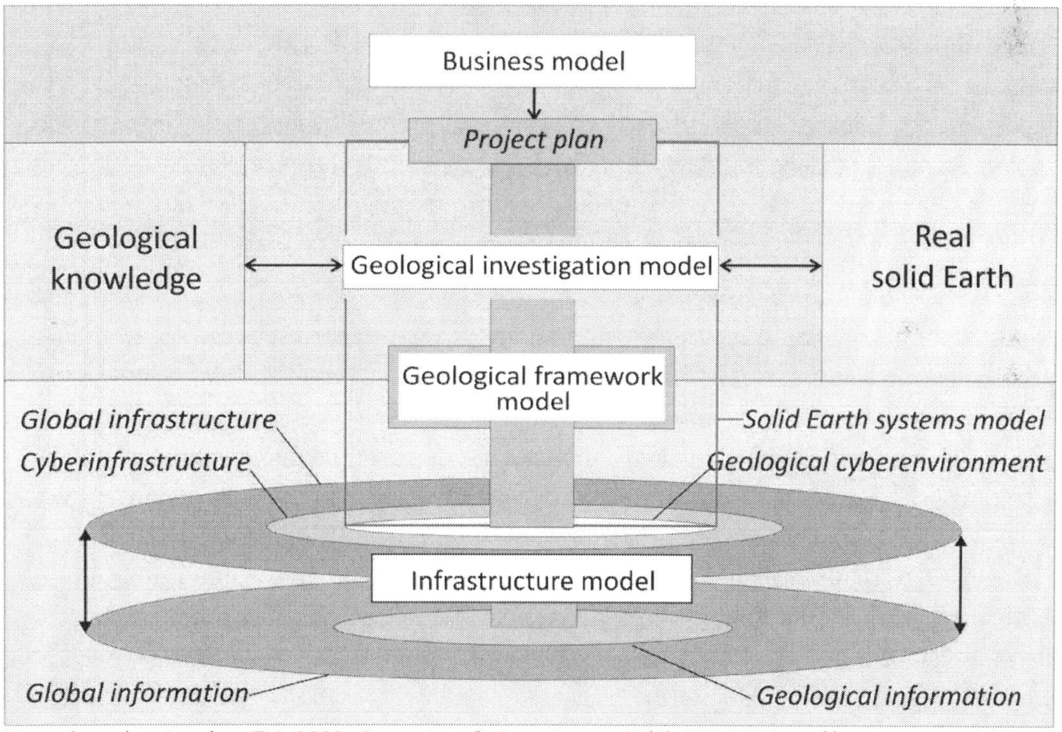

Figure based on Loudon, T.V., 2009. *Computers & Geosciences*, **35**(4), 700-705 http://nora.nerc.ac.uk/7258/

Figure 11: Four models in the geoscience knowledge system (duplicate of Figure 6)

Stages of concept development (summary)

<<Table of contents 1
<<The emerging geoscience knowledge system 42
Stages of concept development (summary) 44
Stages of concept development (cyberinfrastructure) 45
Stages of concept development (geological thinking) 56
>>The solid Earth systems model (sEsm) 71

For a slightly condensed version of this section, see Overview of concept development 19.

The historical development of the infrastructure helps to explain the concepts and rationale of current work and how it influences the strategy for enhancing the knowledge system. The section on Stages of concept development (cyberinfrastructure) 45 traces the development of concepts concerning: mechanising repetitive thought; the importance of a shared model in mediating interactive communication; the organisation of ideas in an object-oriented approach; the sharing of generally accepted concepts through ontologies and systems; the development of business models to service user requirements by integrating information from various sources; and the development of shared workflows and collaborative problem-solving environments. This leads to the more comprehensive aim of Linked Data – a Semantic Web where information about anything can be recorded and shared.

The geological implications are considered in the Stages of concept development (geological thinking) 56. The advanced cyberinfrastructure holds out prospects, not only of creating shareable representations of more of the geologists' internal understanding of the solid Earth, but also of including aspects which previously could not be studied effectively because of insufficient computing and communications power. During geological investigations, such as field survey, geologists must weigh many inputs: a framework of pre-existing definitions, concepts and procedures; observation and measurement; comparison, analogy and correlation; expectations of rock properties and processes; interpretations of Earth history; and reconciliation with topographical evidence and geoscience models from other studies. Geologists take a top-down view, at each stage looking carefully at everything they know so far, testing and improving their current interpretations, imagining how the situation might be in its entirety, and deciding how best to complete the picture by filling space from the available fragments of evidence. Infrastructure advances are leading to more complete and exact representations of geologists' thinking, and the interactive exploration of existing knowledge and the consequences of new observations during both desk study and field survey. The Semantic Web developments indicate how this can be positioned within the broad context of Earth systems and global knowledge.

Stages of concept development (cyberinfrastructure)

<<The emerging geoscience knowledge system 42
<<Stages of concept development (summary) 44
Stages of concept development (cyberinfrastructure) 45
Mechanisms 45
Models and frameworks 47
Objects, ontologies and systems 48
Semantic Web and Grid 50
The service-oriented knowledge utility 52
Workflows, collaborative networks and Linked Data 53
>>Stages of concept development (geological thinking) 56

Mechanisms

<<Stages of concept development (cyberinfrastructure) 45

Vannevar Bush[59] (1945) gave a clear and eloquent account of potential interactions between scientists and the mechanics of their supporting technology. "Much needs to occur...between the collection of data and observations, the extraction of parallel material from the existing record, and the final insertion of new material into the general body of the common record. For mature thought there is no mechanical substitute. But creative thought and essentially repetitive thought are very different things. For the latter there are, and may be, powerful mechanical aids." (Bush, 1945, section 3). He pointed out that the repetitive processes of thought are not confined to matters of arithmetic and statistics. In fact, every time one combines and records facts in accordance with established logical processes, the creative aspect of thinking is concerned only with the selection of the data and the process to be employed and the manipulation thereafter is repetitive in nature and hence a fit matter to be relegated to the machine.

A mathematician, he said, is primarily an individual who is skilled in the use of symbolic logic on a high plane, and especially he is a man of intuitive judgment in the choice of the manipulative processes he employs. "All else he should be able to turn over to his mechanism, just as he confidently turns over the propelling of his car to the intricate mechanism under the hood... Whenever logical processes of thought are employed – that is, whenever thought runs for a time along an accepted groove – there is opportunity for the machine." (his Sections 4, 5).

Bush (1945) pointed out that machines with interchangeable parts can be constructed with great economy of effort and, in spite of much complexity, can perform reliably. There are good reasons, he suggested, for information technology to borrow ideas from mass

[59] Bush, V., 1945. As we may think. The Atlantic Monthly (July). http://www.theatlantic.com/doc/print/194507/bush

production methods, breaking the system down into small interoperable components, and as far as possible sharing identical, fully tested, robust components for greater reliability and less effort in reinventing wheels. Users who were familiar with the data and the system could use it effectively. But without knowledge of background aspects, such as the design of the data collection procedures and the intricate dependencies of recorded items, the results could be misleading. To some extent, background aspects can be included in the data, but there is another approach.

Bush suggested that our ineptitude in finding relevant information in the scientific record is largely a result of the artificiality of systems of indexing, where information is filed alphabetically or numerically and found (when it is) by tracing it down from subclass to subclass. The human mind, he argued, does not work that way. It operates by association of ideas, in accordance with some intricate web of thoughts carried by the cells of the brain. The trails of thought that are not frequently followed are prone to fade, items are not truly permanent, and memory is transitory. "Yet the speed of action, the intricacy of trails, the detail of mental pictures, is awe-inspiring beyond all else in nature... Selection by association, rather than indexing, may yet be mechanized. One cannot hope thus to equal the speed and flexibility with which the mind follows an associative trail but it should be possible to beat the mind decisively in regard to the permanence and clarity of the items resurrected from storage" (his section 6). Bush visualised the scientist marking a trail, accessible at summary and detailed levels, through many items drawn from the record, together with added comments and links to side trails of related topics. He foresaw a new profession of trail blazers, those who find delight in the task of establishing useful trails through the enormous mass of the common record. "Thus science may implement the ways in which man produces, stores, and consults the record of the race" (his section 8).

The trails that Bush envisaged have become the threads of hypertext and the scientific workflows of the internet (see Workflows, collaborative networks and Linked Data 53). They make it possible to place information in its context and to record the design of patterns of analysis that accord with the background situation and knowledge. He also (his section 8) envisaged "skip trails" which stop only on the salient items, providing the generalization mechanism of searching an overview, and drilling down into it for more detail on selected items. Overall, the user can "reacquire the privilege of forgetting the manifold things he does not have immediately at hand, with some assurance that he can find them again if they prove important."

Models and frameworks

<<Stages of concept development (cyberinfrastructure) 45

The role of digital methods in scientific communication was considered by Licklider[60] (1960). He calculated that 85% of his 'thinking' time was spent on clerical and mechanical activities: searching, calculating, plotting, transforming, and determining the logical or dynamic consequences of a set of assumptions or hypotheses. In consequence, the tasks he attempted were selected to a great extent by clerical feasibility, not intellectual capability. He also pointed out that, in many creative endeavours, only a few people have the potential to contribute effectively. The most creative may be independent thinkers rather than the best team players. The time scale of their communications may stretch out as each builds their own empire "and devotes more time to the role of emperor than the role of problem solver".

Licklider (1960) saw interactive communication as a means of addressing these problems of handling and sharing information, through a symbiotic [61]partnership of human brains and computers (to which are now added the internet and World Wide Web). The symbiosis is mediated by a shared model, accessible to both the human user and the machine. Licklider[62] (1968) suggested that modelling is basic and central to communication. "Any communication between people about the same thing is a common revelatory experience about models of that thing. Each model is a conceptual structure of abstractions formulated initially in the mind of one of the persons who would communicate, and if the concepts in the mind of one would-be communicator are very different from those in the mind of another, there is no common model and no communication."

Licklider (1968) noted that the individual model can be observed and manipulated only by its originator. For wider applications "society rightly distrusts the modeling done by a single mind. Society demands consensus, agreement, at least majority... individual models must be compared and brought into... accord... [by] cooperation in the construction, maintenance and use of a model." His proposed solution was technology to maintain, share and manipulate an explicit external model – an agreed structured framework where many minds can work together, weaving threads of thought into the web of the knowledge system. A geological application is proposed in The geological framework model 105. Similarly, Kent[63] (1978) explored (in a database context) the reconciliation of concepts among individuals each of whom has their own unique view of reality. He points out that one individual can hold different views at different times, and can consider more than one view at one time (see Reconciliation 186). This flexibility enables different views to be reconciled, if only to meet the purposes in hand: again, the context (the agreed structured framework) is all important in matching information to interpretation.

---

[60] Licklider, J.C.R., 1960. "Man-computer symbiosis" http://memex.org/licklider.pdf

[61] Symbiosis: A close interdependence or association (in the literal sense, of animals or plants of different species) often of mutual benefit.

[62] Licklider, J.C.R., 1968. "The computer as a communications device" http://memex.org/licklider.pdf

[63] Kent, W., 1978. Data and reality. North-Holland Publishing Company, Amsterdam, 211pp.

Objects, ontologies and systems

<<Stages of concept development (cyberinfrastructure) 45

The nature of the framework in which scientists share information was considered by Coad and Yourdon (1991)[64]. They point out that three methods of organising information pervade our thinking about the real world: differentiation of experience into particular objects and their attributes; distinction between whole objects and their component parts; and distinction between different classes of objects. They advocate An object-oriented approach 178 that reflects this.

Objects are things of interest – computer representations of real-world or conceptual entities. They might, for example, be

- geological (fossils, outcrops, formations, terranes)
- documents (maps, memoirs, logs, indexes)
- document contents (individual map symbols, words, paragraphs, chapters)
- system contents (sub-systems, hypertext threads, ontologies, databases, expert sources)

In geoscience, this approach might record:

- object classes (the categories in some classification scheme, usually hierarchical, to which the objects belong)
- object instances (specific occurrences of an object)
- object attributes (properties, composition, relationships and behaviour)
- processes (which cause things to change)

The object-oriented approach to organising information can be combined with the use of ontologies[65] to clarify its meaning. Gruber[66] (1995) popularised the use of the term 'ontology' in computer science. Defined by him as 'the specification of a conceptualisation', they can be seen as a bridge linking human knowledge and its computer representation. Ontologies provide a controlled vocabulary identifying objects, processes, and their characteristics and relationships. Characteristics can be inherited (within an ontology) from one or more previously defined higher-level concepts, along with specified differences, if any.

The object-oriented view breaks down information into individual identifiable items, and records their relationships. Several information types and many threads of thought can therefore refer consistently to the same items. They can be categorised by object class and metadata to locate them in the framework, to reveal ontological implications and to identify constraints on their analysis. Unlike conventional records, but like the human mind, the

---

[64] Coad, P. and Yourdon, E., 1991. Object-oriented design. Yourdon Press, Englewood Cliffs, NJ, 197 pp.

[65] Ontology: A formal representation and shared vocabulary describing concepts, entities and relationships in a domain of knowledge, typically providing a more detailed and rigorous machine-readable specification than a thesaurus or taxonomy.

[66] Gruber, T. R., 1995. "Toward Principles for the Design of Ontologies Used for Knowledge Sharing". In: *International Journal Human-Computer Studies*, 43(5-6):907-928.

cyberenvironment can accommodate the huge network of interconnections linking the items.

Geological objects are intricately interlinked in a complex fabric, given form by the geoscience paradigm, which might be expressed as a shared model of the systems of the solid Earth. The object-oriented approach can represent many aspects of thinking about an individual object, linking it to webs of thought defined by geological classification and properties, information type, granularity, location, provenance, and place in the workflows of investigation or reasoning. It can support better information, more rigorous analysis, and more flexible and efficient access.

This object-oriented approach to design and analysis owes much to the systems[67] view. A system can be defined as a collection or set of interrelated and interacting objects or entities, including their relationships and behaviour, which can usefully be studied as a whole (see The systems approach to Earth science 65). Early writers in this field, such as Ashby[68] (1964), Beer[69] (1967) and Laszlo[70] (1972) stress the wide applicability of this view. They note the importance in systems of the gestalt principle – that the organised whole is more than the sum of its parts.

For descriptive purposes, the system can be broken down into subsystems. We can think of each of these as a system of smaller extent, selected to include objects and processes that naturally belong together. By incorporating the complexity of behaviour within the subsystems as far as possible, we can simplify the interfaces between them. An interface is the shared boundary between systems or parts of a system, or the means of interaction across the boundary that makes joint operation possible. An interface device, for example, provides compatibility by enabling one item of equipment to communicate with another. Well-chosen interfaces can make the overall system easier to understand, implement and maintain.

System or subsystem boundaries (interfaces) are somewhat arbitrary, and so we must define the scope of the system, that is: its extent, what it consists of and how it works. We need to identify the components of the geoscience information system and their interactions, the participants and their roles, activities and driving forces. The report from the US National Research Council[71] (1993) has influenced universities and research organisations in viewing geological subsystems in the broader context of the past, present, and future behaviour of the whole Earth system. "From the environments where life evolves on the surface to the interaction between the crust and its fluid envelopes (atmosphere and hydrosphere), this interest extends through the mantle and the outer core to the inner core. A major challenge is to use this understanding to maintain an environment in which the biosphere and humankind will continue to flourish."

---

[67] System: A set of interacting parts that function as a whole. The systems approach involves study of linkages or interfaces between the component activities.

[68] Ashby,W.R., 1964. Introduction to cybernetics. London, Methuen.

[69] Beer, S., 1967. Cybernetics and Management, 2nd edn. English Universities Press, London, 240pp.

[70] Laszlo, E., 1972. The Systems View of the World. Braziller, New York, 131pp.

[71] National Research Council, 1993.Solid-Earth sciences and society. National Academy Press, Washington, DC.

An appropriate response to this situation is an object-oriented approach to geology seen in its broader systems context. An early example was the ambitious POSC Epicentre Model. They saw their data model as similar in many ways to a conventional dictionary, listing names and definitions of more than 750 real-world technical and business objects concerned with petroleum exploration (including geology and geophysics) and production, together with their characteristics and interrelationships. Although of undoubted interest, the initial centralised approach may not have been acceptable to the wide diversity of potential users (see Other geological initiatives 343). The Semantic Web and Grid 50 attempt to meet the need for specific responses to particular requirements.

Semantic Web and Grid

<<Stages of concept development (cyberinfrastructure) 45

Berners-Lee et al. (2001)[72], described the Semantic Web as an extension woven into the structure of the existing Web, in which information is given well-defined meaning, improving the ability of computers and people to work in cooperation. The computers must have access to structured collections of information and sets of inference rules to conduct automated reasoning. These have been studied by artificial intelligence researchers, typically looking at centralised knowledge-representation systems which require a shared definition of common concepts. But unlike centralised systems, not only must the Semantic Web provide a language that expresses both data and rules for reasoning about the data, but also it must allow rules to be exported onto the Web from any existing knowledge-representation system.

Two important technologies for developing the Semantic Web are the eXtensible Markup Language (XML) and the Resource Description Framework (RDF). To quote the above article: "XML lets everyone create their own tags: hidden labels that annotate Web pages or sections of text on a page. Scripts, or programs, can make use of these tags in sophisticated ways, but the script writer has to know what the page writer uses each tag for. In short, XML allows users to add arbitrary structure to their documents but says nothing about what the structures mean. Meaning is expressed by RDF, which encodes it in sets of triples, each triple being rather like the subject, verb and object of an elementary sentence. These triples can be written using XML tags. In RDF, a document makes assertions that particular things (people, Web pages or whatever) have properties (such as 'is a sister of,' 'is the author of') with certain values (another person, another Web page). This structure turns out to be a natural way to describe the vast majority of the data processed by machines. Subject and object are each identified by a Universal Resource Identifier (URI), just as used in a link on a Web page (URLs, Uniform Resource Locators, are the most common type of URI). The verbs are also identified by URIs, which enables anyone to define a new concept, a new verb, just

---

[72] Berners-Lee, T., Hendler, J. and Lassila, O., 2001 (May). The Semantic Web. Scientific American, 284 (3). http://www.scientificamerican.com/article.cfm?id=the-semantic-web

by defining a URI for it somewhere on the Web... The triples of RDF form webs of information about related things. Because RDF uses URIs to encode this information in a document, the URIs ensure that concepts are not just words in a document but are tied to a unique definition that everyone can find on the Web."

The third basic component of the Semantic Web is a set of ontologies. A typical ontology for the Web has a taxonomy and a set of inference rules (maybe linked through RDF). The meaning of terms or XML codes used on a Web page can be defined by pointers from the page to an ontology (see Objects, ontologies and systems 48). "Classes, subclasses and relations among entities are a very powerful tool for Web use. We can express a large number of relations among entities by assigning properties to classes and allowing subclasses to inherit such properties...Inference rules in ontologies supply further power. An ontology may express the rule 'If a city code is associated with a state code, and an address uses that city code, then that address has the associated state code.'" Such information enables the computer to manipulate the terms much more effectively in ways that are useful and meaningful to the human user, and can be extended to more advanced cyberinfrastructures. An example is the Semantic Web for Earth and Environmental Terminology (SWEET), described by Raskin (2006)[73], see also Earth Sciences 349 (SWEET 354).

The Grid is seen as a more powerful infrastructure than the Web – an emerging infrastructure that will fundamentally change the way we think about and use computing. An analogy is with the power grid, which provides pervasive access to electricity with dramatic impact on human capabilities and society. "Many believe that by allowing all components of our information technology infrastructure – computational capabilities, databases, sensors and people – to be shared flexibly as true collaborative tools, the Grid will have a similar transforming effect, allowing new classes of applications to emerge" (Foster and Kesselman, 2004[74]).

The concepts of the Semantic Web are being carried forward into this more powerful infrastructure, as described in the Semantic Grid Community Portal[75]. "The Semantic Grid is an extension of the current Grid in which information and services are given well-defined meaning through machine-processable descriptions which maximize the potential for sharing and reuse. We believe that this approach is essential to achieve the full richness of the Grid vision, with a high degree of easy-to-use and seamless automation enabling flexible collaborations and computations on a global scale." The geological cyberenvironment (gce) 85 focuses on provision of appropriate software support.

The '2020 Science Group' chaired by Emmott (2006)[76] concluded that "A significant change in scientists' ability to analyse data to obtain a better understanding of natural phenomena will be enabled by (i) new ways to manage massive amounts of data from observations and

---

[73] Raskin, R., 2006. Development of ontologies for earth systems science, *in* Sinha, A.K., ed., Geoinformatics: Data to knowledge. *Geological Society of America Special Paper* **397**, 195-199. doi: 10.1130/2006.2397(14)

[74] Foster, I, Kesselman, C., 2004. The Grid: Blueprint for a new computing infrastructure, 2nd ed., Amsterdam, Elsevier. 748 pp.

[75] De Roure, D., 2006. Semantic Grid Community Portal. http://www.semanticgrid.org/

[76] Emmott, S., et al., 2006. Towards 2020 Science. Microsoft Corporation. http://research.microsoft.com/en-us/um/cambridge/projects/towards2020science/downloads.htm

scientific simulations, (ii) integration of powerful analysis tools directly into the database, (iii) improved forms of scientist-computer-data interaction that support visualisation and interactivity, (iv) active data, notification, and workflows to enhance the multistage data analysis among scientists distributed around the globe, and (v) transformation of scientific communication and publishing."

The history of the Grid and its relationship with the World Wide Web are discussed by De Roure et al.[77], (2003a). They discuss the service oriented model and knowledge in De Roure et al.[78], (2003b), stressing the importance of its service orientation (see The service-oriented knowledge utility 52).

The service-oriented knowledge utility

<<Stages of concept development (cyberinfrastructure) 45

A vision for a service-oriented knowledge utility (SOKU) was set out by the Next Generation Grids Expert Group (2006)[79] for the European Commission. Building on existing industry practices, trends and emerging technologies, it is seen as giving rules and methods for combining them into an ecosystem that promotes collaboration and self-organisation.

"The need for developing the SOKU vision stems from the necessity of effectively bringing knowledge and processing capabilities to everybody, thus underpinning the emergence of a competitive knowledge-based economy. It captures three key notions:
• *service oriented* — the architecture comprises services which may be instantiated and assembled dynamically, hence the structure, behaviour and location of software is changing at run-time;
• *knowledge* — SOKU services are knowledge-assisted ('semantic') to facilitate automation and advanced functionality, the knowledge aspect reinforced by the emphasis on delivering high-level services to the user;
• *utility* — a utility is a directly and immediately useable service with established functionality, performance and dependability, illustrating the emphasis on user needs and issues such as trust. The primary difference between the SOKU vision and earlier approaches is a switch from a prescribed layered view to a multi-dimensional mesh of concepts, applying the same mechanisms along each dimension across the traditional layers."

---

[77] De Roure, D., Jennings, N. R. and Shadbolt, N. R., 2003. The evolution of the Grid, *In* Berman, F., Hey, A.J.G, and Fox, G., (editors), 2003. Grid Computing: Making the Global Infrastructure a Reality, pp 65-100. John Wiley & Sons, 1080 pp. ISBN: 0470853190. http://www.semanticgrid.org/documents/evolution/evolution.pdf

[78] De Roure, D., Jennings, N. R. and Shadbolt, N. R., 2003. The Semantic Grid: A Future e-Science Infrastructure. *In* Berman, F., Hey, A.J.G, and Fox, G., (editors), 2003. Grid Computing: Making the Global Infrastructure a Reality, pp 437-470. John Wiley & Sons, 1080 pp. ISBN: 0470853190. http://www.semanticgrid.org/documents/semgrid-journal/semgrid-journal.pdf

[79] Next Generation Grids Expert Group, 2006. Future for European Grids: Grids and Service-Oriented Knowledge Utilities: Vision and research directions 2010 and beyond. Report 3 for the European Commission. ftp://ftp.cordis.europa.eu/pub/ist/docs/grids/ngg3-report_en.pdf

Advances in semantic capabilities were seen as the key to realising the vision of an intelligent connected world with pervasive computing systems providing personalised access to content, applications and services.

Workflows, collaborative networks and Linked Data

<<Stages of concept development (cyberinfrastructure) 45

Ludascher et al. (2006) [80] state that: "A problem for the scientist is how to easily make use of the increasing number of databases, analytical tools, and computational services that are available. Besides making these items generally accessible to scientists, leveraging these resources requires techniques for data integration and system interoperability. Traditionally, research by the database community in this area has focused on problems of heterogeneous systems, data models, and schemas."

They compare this situation to the requirements of, for example, an igneous petrologist interested in the distribution of a certain rock type within a specific region, the three-dimensional geometry of the plutons where it occurs, and their relation to the host rock structures. The scientist can gather valuable information towards answering the scientific question from databases and analytical tools. Geologic maps of the region, geophysical databases with gravity contours, foliation maps, and geochemical databases can all provide pieces of information that need to be brought together in an appropriate form.

Scientific workflows can set out the sequence of operations for comparing observed with predicted data. They can include a wide range of components for querying databases, for data transformations and data mining, for execution of simulation codes on high performance computers, and so on. "Ideally, a scientist should be able to (1) plug-in almost any scientific data resource and computational service into a scientific workflow, (2) inspect and visualize data on-the-fly as it is computed, (3) make parameter changes when necessary and re-run only the affected 'downstream' components, and (4) capture sufficient metadata in the final products. For each run of a scientific workflow, when considered as a computational experiment, the metadata produced should be comprehensive enough to help explain the results of the run and make the results reproducible by the scientist and others. Thus, a scientific workflow system becomes a scientific problem-solving environment, tuned to an increasingly distributed and service-oriented Grid infrastructure."

De Roure et al [81] (2009) note that scientific workflows increasingly support scientists in advancing research through in silico experimentation, while the workflow systems

---

[80] Ludascher, B., Lin, K., Bowers, S., Jaeger-Frank, E., Brodaric, B, Baru, C., 2006 Managing Scientific Data: From Data Integration to Scientific Workflows. In Sinha, A.K. (ed) Geoinformatics: Data to Knowledge, Geological Society of America Special Paper 397, pp 109-129. http://users.sdsc.edu/~ludaesch/Paper/gsa-sms.pdf

[81] De Roure, D., Goble, C. and Stevens, R., 2009. The Design and Realisation of the myExperiment Virtual Research Environment for Social Sharing of Workflows. Future Generation Computer Systems v 25 (5), pp.561-567 http://eprints.ecs.soton.ac.uk/15709/

themselves are the subject of ongoing research and development, as workflow systems address more sophisticated requirements and as workflows are created through collaborative design processes involving many scientists across disciplines. They focus on the dimension of collaboration and sharing.

"Understanding the whole lifecycle of workflow design, prototyping, production, management, publication and discovery is fundamental to developing systems that support the scientist's work and not just the workflow's execution. Supporting that lifecycle can be the factor that means a workflow approach is adopted or not. Workflow design is challenging and labour-intensive, and reusing a body of prior designs through registries or catalogues is highly desirable. Reuse is a particular challenge when scientists are outside a predefined Virtual Organisation or enterprise. These are individuals or small groups, decoupled from each other and acting independently, who are seeking workflows that cover processes outside their expertise from a common pool of components. This latter point arises when workflows are shared across discipline boundaries and when inexperienced scientists need to leverage the expertise of others."

They discuss the motivation, design approach and realisation of the 'myExperiment Virtual Research Environment' for collaboration and sharing of experiments, which aims to provide a 'workflow bazaar' for any workflow management system. While individual workflow systems may provide workflow repository mechanisms, myExperiment is distinctive in that it facilitates the sharing of workflows and these may come from multiple systems. They consider the use of workflows for science, the power of workflows as shared entities, and the requirements for sharing. They envisage: a 'gossip shop' to share and discuss workflows and their related scientific objects, regardless of the workflow system; a bazaar for sharing, reusing and repurposing workflows; a gateway to other established environments, for example depositing into data repositories and journals; and a platform to launch workflows, whatever their system. They emphasise social networking around the workflows, providing gateways to other environments and forming the foundation of a personal or laboratory workbench. "We were drawing inspiration from the Web 2.0 approach, from systems such as Facebook, MySpace and Amazon (see the corresponding .com sites), so we considered our principles in the light of the Web 2.0 design patterns."

A different style of social networks is seen in the massively multiplayer online game[82] in which numerous players compete and collaborate in a video game. The software that makes their interaction possible; the ability to visualise numerous objects interacting as a complex, self-organising emergent[83] system; and the portability of the user devices, suggest a potential relevance to collaborative simulation and visualisation of systems of geological objects and processes, which remains largely unexplored.

Berners-Lee (2007)[84] described how the internet and the World Wide Web simplified the user's view. "The realization was, 'It isn't the cables, it is the computers which are

[82] Wikipedia 'Massively multiplayer online game', 2010. http://en.wikipedia.org/wiki/Massively_multiplayer_online_game
[83] Emergence: Complex patterns, properties and systems resulting from relatively simple interactions.
[84] Berners-Lee, T., 2007. The Giant Global Graph (blog post) http://dig.csail.mit.edu/breadcrumbs/node/215

interesting'. The Net was designed to allow the computers to be seen without having to see the cables... The WWW increases the power we have as users again. The realization was 'It isn't the computers, but the documents which are interesting'... The Net links computers, the Web links documents... Now, people are making another mental move. There is realization now, 'It's not the documents, it is the things they are about which are important'."

The focus on real-world entities (or non-information resources) and the computer objects that represent them is carried forward in the concepts of 'Linked Data' (Bizer et al., 2009[85]). They refer to a set of best practices for publishing and connecting structured data, building directly on the general architecture of the Web. The aim is the creation of an unbounded global data space – the 'Web of Data'. The usual Web format (html) connects documents, but does not relate individual entities mentioned within several documents. Linked Data uses RDF (see Semantic Web and Grid 50) to make typed statements that link arbitrary things in the world, described by data on the Web. "There are Linked Data search engines that crawl the Web of Data by following links between data sources and provide expressive query capabilities over aggregated data, similar to how a local database is queried today. The Web of Data also opens up new possibilities for domain-specific applications. Unlike Web 2.0 mashups which work against a fixed set of data sources, Linked Data applications operate on top of an unbound, global data space. This enables them to deliver more complete answers as new data sources appear on the Web." (Bizer et al., 2009). Vocabularies differ among information communities, and the Web of Data accepts this. Nevertheless, it is good practice to use the well-known RDF vocabularies where practicable. Where new terms are necessary, they should be defined and self-describing.

The Web of Linked Data is open and self-describing, and applications can discover new resources at run time. However, in general, as networks move beyond the control of a responsible, knowledgeable and trusted group, evaluation becomes a major issue (see The geological business model 93). Unless users fully understand the intricate and convoluted systems that can arise from combining information extracted from its original context in many sources, they could be seriously misled by the results. The aim of cyberenvironments (see Overview of the geological cyberenvironment 28) is to provide end-to-end support of geological investigations. They therefore depend on adequate background knowledge and appropriate use of carefully evaluated data and metadata (see Stages of concept development (geological thinking) 56).

---

[85] Bizer, C., Heath, T., Berners-Lee, T., 2009. Linked data – the story so far. International Journal on Semantic Web and Information Systems (IJSWIS), **5**(3), 1-22. DOI: 10.4018/ijswis.2009070101. http://tomheath.com/papers/bizer-heath-berners-lee-ijswis-linked-data.pdf

Stages of concept development (geological thinking)

<<Table of contents 1
<<The emerging geoscience knowledge system 42
<<Stages of concept development (summary) 44
<<Stages of concept development (cyberinfrastructure) 45
Stages of concept development (geological thinking) 56
Invariance and processes 56
Mechanising the database 58
The surveyor's holistic view 60
Integrating information types 62
Unexpressed knowledge 63
The systems approach to Earth science 65
The next steps 68
>>The solid Earth systems model (sEsm) 71

Invariance and processes

<<Stages of concept development (geological thinking) 56

Early geological maps, notably the work of William Smith[86] (1769-1839) in the early nineteenth century, are seen as the forerunner and inspiration for geological survey organizations. "William Smith's conviction that geological mapping is of vital importance at many levels and in many areas of the nation's society, science, and industry is as true today as it was two hundred years ago, when he conceived his original geological map. The methods involved in map production have developed and... the method of map delivery has altered radically, but the fundamental importance of providing accurate geological map data to today's industries is as vital now as it was in Smith's time." (BGS, 2010d[87])

The remarkable skills of the human brain in abstracting invariant[88] (unchanging) properties from innumerable views enable us to recognise, for example, some hundreds of friends, colleagues and acquaintances in many different circumstances. These human skills, which (until recently) no computer could equal, extend to natural features and would have enabled Smith to classify and correlate rock strata in various conditions, times and places on the basis of their invariant aspects despite their internal variability. Assemblages of fossils were identified as invariant features defining a set order in which strata occur. William Smith's

[86] Winchester, S., 2001. The Map that changed the World. Penguin Books, London.

[87] BGS, 2010d. Geoscience archives, William Smith. http://www.bgs.ac.uk/discoveringGeology/geologyOfBritain/archives/williamsmith/home.html

[88] Invariant: Something that has invariance, that is, it does not change under (when subjected to) a specific set of transformations or sequence of operations.

predictive model of the layered strata of the solid Earth – objects defined by their properties and sequence – was simple, informative, and readily depicted on a map. But the geologist's model goes beyond an instinctive response to what is seen, and provides a deeper understanding.

James Hutton (1726-97) linked a set of geological observations to a scenario of erosion and deposition of strata. He imagined the circumstances in which the rocks formed, recognising geological processes as a basis for reasoning – for informing his observations by relating them to a background theory – thus improving his abilities to predict what he had not yet seen. Presumably influenced by David Hume (Daiches et al., 1996)[89], he appears to have viewed time as a process parameter, not (like many of his contemporaries) as a scale prescribed by Holy Writ. Hume held that "Like causes, in like circumstances will always produce like effects... [And] the course of nature will [through time] continue uniformly the same."

Hutton saw the cycle of geological processes as proceeding now, in the same manner and at the same relative rates, as in the geological past. The outcome of processes depends on the input, not on when and where they took place, because the processes are invariant under specifiable time and space transformations. Likewise, some properties, such as overall mass and energy, are neither created nor destroyed. Some of the features that geological processes leave in the rocks (shapes, spatial relationships) inherit the invariance, becoming the key to linking the outcome to the process and to deciphering geological history. For example, spatial relationships shown in the order of a sequence of sedimentary beds, or the implied sequence of crystal growth seen on a microscope slide, reflect the time relationships of past geological events. Such time relationships may ultimately be correlated to processes, such as radioactive decay, that provide an appropriate standard measure to represent time. Invariants in geological processes and their outcomes are a key to geological understanding, and therefore of central interest in quantitative methods (see The imperfect model 156).

Simpson[90] (1963, page 24) emphasised what he saw as an important distinction in historical sciences such as geology. "The unchanging properties of matter and energy and the likewise unchanging processes and principles arising therefrom are *immanent*[91] in the material universe. They are nonhistorical, even though they occur and act in the course of history. The actual state of the universe or of any part of it at a given time, its configuration, is not immanent and is constantly changing . . . History may be defined as configurational change through time, i.e. a sequence of real, individual but interrelated events. These distinctions between the immanent and the configurational and between the nonhistorical and the historical are essential to clear analysis and comprehension of history and of science."

Unlike investigations of immanent laws, the configurational records of geology offer no level playing field where facts are agreed by all and randomly sampled data can be analysed in any context (see Abstracting from reality to model 131). On this view, there is no sharp

---

[89] Daiches, D., Jones, P., Jones, J. (editors), 1996. The Scottish Enlightenment, 1730-1790: A hotbed of genius. The Saltire Society, Edinburgh.

[90] Simpson, G.G., 1963. *in* Albritton, C.C., ed., 1963, The fabric of geology: Stanford, Freeman, Cooper & Company, p. 24-48

[91] Immanent: Indwelling, inherent, pervading.

distinction between interpretation and data, or between assumptions, explanations and observations (Frodeman, 1995[92]). Instead, the broad model is an existing story of interwoven ideas or memes[93] (Blackmore and Dawkins[94], 2000) that geologists aim to test, extend and clarify. It guides what they look for, what they see, what they record and how they interpret it. Each item of information reflects interpretation and observation (or model and reality) in different proportions. But even identifying an outcrop as Permian, or estimating the depth to the top of the Jurassic from downhole logs, involves interpretative skills and drawing analogies with earlier opinions, so that observational records may not be totally reproducible.

The geological map cannot and does not represent the reasoning process – it merely represents its consequences. But the modern geological surveyor's choice and form of objects shown on the map involves reasoning and interpretation – bringing wider scientific knowledge to bear. For example, a siliceous grit directly overlying the granite from which it had been eroded would not be mapped as part of the granite, for despite their contiguity and similar properties, the proximal causes of their formation are quite different processes far removed in time and environment. The geological objects shown by colours and symbols on the map could thus be regarded as the outcome of the scenarios of geological sub-systems that are components of the surveyors' interpretation.

The geological map is a means of visualising the spatial distribution of geological units on the ground, along with the orientation measurements, stratigraphical key, vertical sections and cross-sections depicted in the map margins. These help users to build a mental picture of the three-dimensional configuration of the rocks, enabling them to visualise the geological forms and spatial relationships in a broader context than the details of observation. The geological processes that created the rock units and determined their geographical distribution have their own spatial distributions but these are not readily depicted on a single map. Instead they are described in text and diagrams in an accompanying map explanation or in scientific papers.

Mechanising the database

<<Stages of concept development (geological thinking) 56

In the 1960's, inspired by the visions of Bush (see Stages of concept development (cyberinfrastructure) 45) and many others, pioneering geologists were transcribing datasets (mostly quantitative) to punched cards or tape and writing computer programs for their

---

[92] Frodeman, R., 1995. Geological reasoning: geology as an interpretive and historical science. *GSA Bulletin*, **107**(8), 960-968.

[93] Meme: An idea or concept passed from mind to mind by imitation or explanation, evolving through variation, selection and heredity

[94] Blackmore, S., Dawkins, R., 2000. *The meme machine.* Oxford University Press, Oxford. 264 pages.

analysis (Loudon, 1996)[95]. The geological data were largely field records, digitised long after collection. Geophysical data or the results of laboratory analysis were more likely to be logged mechanically, possibly in computer-readable form. Packages of programs subsequently became available, which enabled users to apply sequences of reusable analytical procedures to their datasets. By the 1970's, techniques for describing the data by means of metadata (data about data) had entered geology. They enabled reuse of the datasets in various applications, and extensive data banks were established in many geological organisations. Database techniques made it possible to combine data from various sources. In 1970, Codd[96] introduced the concept of relational databases, which organised data as linked tables, 'normalised' to avoid repetition (redundancy) of information. Avoiding redundancy ensures that changes to an entry in one record are carried through to all those linked to it. Individual items of data become reusable in different contexts. Users can select and retrieve data items to meet their requirements by searching on 'equal to', 'less than' or 'greater than' criteria for each of a combination of variables[97].

For example, the digitised lines and symbols of a geological map could be regarded as a database of items that were assigned to predetermined categories (stratigraphical boundaries, orientations, annotations, etc.). Appropriate items could then be selected as interchangeable parts reorganised in a wide range of thematic maps all generated from a single database. For example, Edwards et al. (1987)[98] describe selection of items from a database to provide a range of thematic maps for the Southampton area of the UK, including solid geology, drift geology, drift thickness, rockhead contours, sand and gravel resources and end-use analysis, clay resources, worked ground, aquifer distribution, engineering geology, slope stability, landfill and waste disposal, boreholes, and Sites of Special Scientific Interest. The advent of Geographical Information Systems introduced powerful methods to organise and investigate a range of data in a spatial context, while visualisation[99] techniques provided greater flexibility in presentation (see, for example, Bonham-Carter[100], 1994). Commercial systems later provided a robust unifying framework for viewing and retrieving selected spatial and other quantitative properties. The tabular database format encourages the recording of consistent information, definitions and sampling schemes, appropriate for comparing like with like. But ease of retrieval does not overcome the problems that arise when data with diverse sampling schemes are brought together and inappropriately analysed, statistically or geographically. Furthermore, the tabular structure may not be appropriate for handling interpretations, as opposed to structured observations.

---

[95] Loudon, T.V., Commentary on a British Geological Survey computing archive 1965-85 : British Geological Survey, Information & Data Resources Series, Technical Report WO/96/3 - [i],19p (1996). ISBN: X780907719

[96] Codd, E.F., 1970. "A Relational Model of Data for Large Shared Data Banks". *Communications of the ACM* **13** (6): 377–387. doi:10.1145/362384.362685.

[97] Variable: A quantity that can assume any of a set of values

[98] Edwards, R.A. et al., 1987. "Applied geological mapping, Southampton area. British Geological Survey Research Report ICSO/87/2-4. ISBN X780797824

[99] Visualisation: Transforming quantitative data (including the results of interpolation) into sensory information – images that the eye and brain can interpret and visualise.

[100] Bonham-Carter, G.F., 1994. Geographic Information Systems for Geoscientists: modelling with GIS. Elsevier, Oxford. 398 pp.

The surveyor's holistic view

<<Stages of concept development (geological thinking) 56

Bradley[101] (1963, page 16) referring to the geological map, wrote: "because a geologist can see only parts of the features he studies and must forever deal with partial information (he constructs geologic maps primarily to bring large features down to a comprehensible scale at which he can integrate the parts and visualize the whole), it is most essential that he be able to visualize, in three dimensions and with perspective, processes that may have gone on that will help to reconstruct the events of the past ... A geologist who has no imagination is as ineffective as a duck without webs between his toes."

In field or subsurface mapping, the geologist must weigh evidence from many sources, as Harrison[102] (1963, page 227) points out in his outline of the steps in preparing a geological map. His first step is to assemble topographical maps, air photographs, previous studies and available geophysical, geochemical and other data for the area. The next step might be to plan critical traverses and select sections for detailed study. The geologist would examine the rocks and deposits and classify them by physical characteristics and interpretation of their origin. The classification must be made in the field before it can be shown on the map. The geologist is also concerned with structural data – relative significance of unconformities, relationships of cleavage to folds and faults, relative movement on faults, age of structural deformation, age of igneous intrusions, and all the complexities of rock relationships. These data, together with those available from geophysics, geochemistry, borehole records, and other sources help to extend the depiction of lithological and structural units beneath the overburden. Thin sections may refine the classification or distinguish metamorphic facies, leading to conclusions about the geological history. Statistical analyses of hundreds of field observations help to determine the shape of folds, the position of faults, directions of palaeocurrents, and the significance of fossil assemblages. Harrison describes how, finally, geologists must select the symbols and scheme of presentation that best portray their selection of data and interpretations, enabling users to look into the map and visualise the distribution and relationships of the rocks beneath the surface of the Earth.

The mapping process is based on: a framework of pre-existing concepts, definitions, classifications, and procedures; observation and measurement; comparison, analogy and correlation; expectations of rock properties and processes; interpretations of Earth history; and reconciliation with topographical evidence and geoscience models from other studies. As the mapping proceeds, it builds on a *top-down* view or *gestalt* (an analysis working down from the structure of the whole to its relations with its constituent parts and their characteristics). At each stage, geoscientists look carefully at everything they know so far, and imagine how the situation might be in its entirety. Analysis of items in isolation cannot

---

[101] Bradley, W.H., 1963. Geologic laws, *in* Albritton, C.C., ed., 1963, The fabric of geology: Stanford, Freeman, Cooper & Company, p. 12-23.

[102] Harrison, J.M., 1963, Nature and significance of geological maps, *in* Albritton, C.C., ed., 1963, The fabric of geology: Stanford, Freeman, Cooper & Company, p. 225-232.

provide that understanding, for their significance depends on their place, role and function in the whole.

Based on knowledge of geological processes, the rocks are interpreted as the outcome of spatial configurations (evolving through geological time) of objects, their composition and properties, relationships, processes, states and events. The interpretation is tested and extended with additional observations and other evidence, and adjusted to conform to what has been learned. Techniques such as balanced cross-sections may reconcile the geometry with the behaviour expected from the rock properties. The orientation and form of surfaces throw light on the overall structure. Spatial relationships, such as the sub-parallelism of adjacent units and the topographical expressions of the lithologies, may help to build the interpretation, again using analogies with instances that have been seen elsewhere. The details of significant properties and their geographical distributions may follow an acceptable statistical design and be recorded systematically in data tables, arranged to allow easy comparison of values within a set of observations, assisted by statistical analysis and diagrams of their spatial variation. With all the available knowledge in mind, the interpretation guides decisions on how best to complete the picture from the fragmentary evidence, and fill the space on the map.

Discrete, clearly bounded stratigraphical units are objects that are convenient for description and reasoning, with lines marking their boundaries on the map. They inevitably conceal uncertainties of identification, position and correlation, and the difficulties of reducing the obscured complexities of transition and repetition seen in the field to a sharply defined representation. The surveyors, however, are aware of these problems, and while mapping, they actively seek out evidence to test their current interpretations. This constant feedback during field survey means that each observation and record is influenced by the observer's current knowledge of every aspect, and continually modified as more is learned. For example, an initial identification of the stratigraphical level of an outcrop might be changed as the developing map clarifies the geological structure. The conclusions tie the objects to geographical space and to a sequence in geological time. Modification ceases when a study is regarded as complete and the map is published. In time, a new study and a revised edition of the map may supersede it.

Geological thinking is inadequately represented in pre-digital and early digital patterns of communication. The advancing cyberinfrastructure supports a systems framework that can bring more power, rigour and flexibility into the representation and communication of geologists' understanding, including quantitative models and statistically valid sampling. Mapping procedures can be recorded as informal algorithms, in the broad sense of procedures, or sets of well-defined instructions, for accomplishing some task, such as how a geological boundary line is interpolated between outcrops to reflect expectations, based on, say, a particular interpretation and its likely response to the local landform. The actual process of making observations, interpreting them, and recording results can be recorded as a workflow (see Workflows, collaborative networks and Linked Data 53). Workflows could assist users to understand how the surveyor arrived at the recorded conclusions (see Unexpressed knowledge 63) and are capable of Integrating information types 62.

Integrating information types

<<Stages of concept development (geological thinking) 56

New information interacts in various ways with existing knowledge in our minds, spinning many webs of thought. For example, specialised mechanisms in our brains handle different information types, enabling us to think about things in distinct ways. At a higher level, our brains can reconcile and coordinate the resulting streams of thought to build an overall

	1 **Mental processes** *(in the mind)*	2 **Shareable information** *(unrecorded)*	3 **Recorded information** *(conventional)*	4 **Recorded information** *(digital)*
A	Semantic memory	Paradigm	Textbooks, indexes	Framework, ontologies, metadata
B	Procedural memory	Methods, procedures	Procedure manuals	Embedded in digital support
C	Episodic memory	Narrative	Papers, reports, map explanations	Hypertext sequences, workflows
D	Spatial memory	Spatial information	Maps, sections, diagrams	GIS, spatial models, visualisations
E	Short-term memory	Data, field observations	Data files, field notes, ephemera	Databases

Adapted from Figure 2, Loudon and Laxton (2007) http://nora.nerc.ac.uk/1084/

Figure 12: Five information types and their representations in four different contexts.

view. During field survey, for instance, a geologist would use distinct thought mechanisms to: (a) weigh up preconceptions based on widely held beliefs and background knowledge; (b) follow familiar working procedures; (c) explain the origin and history of the rocks; (d) visualise their local and regional spatial distribution; (e) make observations to amend and extend the ideas. The information is represented, manipulated, analysed, recorded, communicated and shared in ways that match these thought mechanisms (*Figure 12*).

More generally (as shown in the rows of *Figure 12*) we use different information types when thinking about:

A. underlying general knowledge of the science
B. ways of doing things
C. narrative text accounts and descriptions
D. spatial location, arrangement and form
E. observations and measurements

The information types affect the way (shown in the columns of *Figure 12*) that we:
1. think and remember (and forget)
2. rework information to better understand and explain it, and open it to discussion
3. record and manipulate it conventionally
4. record and manipulate it digitally in appropriate computer systems

For details and references, see Loudon (2000)[103].

Conventionally, geological information is assembled in separate, inflexible documents, each centred on a particular information type, such as scientific books and papers (narrative), maps and cross-sections (spatial), data files and registers (tabular). In a digital environment, each information type still requires its own representations and processing procedures, at all levels of detail, as shown in column 4 of *Figure 12*. However, digital methods do bring the ability to represent information, based not on documents but on geological entities; not confined on paper to stable, immutable presentations, but referring to geological objects, processes and systems, amalgamated across information types. Each type can be represented in its own file format (such as .html, .jpeg, .gif) and each can be communicated, stored, and manipulated with appropriate computational tools. Users have the inherent ability to integrate in their minds results shown as text, images, maps, tabular data, video, and so on. Various information types in a digital environment can be selected, edited, analysed, combined, visualised, displayed and printed in ways chosen by the end-users. They presumably have the clearest perception of their own requirements for topic, level of detail, and mode of presentation, and can therefore benefit from the flexibility of digital methods.

Unexpressed knowledge

<<Stages of concept development (geological thinking) 56

Licklider (1960)[104] envisaged human beings and machines collaborating in a symbiotic partnership (see Models and frameworks 47). For direct communication among them, information is recorded externally (outside the human brain – on paper, electronically, or in any shareable medium). But geologists often cannot or may not wish to externalise all aspects of their investigations. For example, until recently a computer struggled to calculate that an electronic image included a likeness of a human being. But as mentioned in considering Invariance and processes 56, most of us know, and can recognise at a glance, many hundred acquaintances, regardless of where they are, how they are positioned, lit, dressed, and even at various stages of their lives. We might recall who they are and predict how they are likely to behave, with no hope of externalising the complex procedure of recognition.

---

[103] Loudon, T.V., 2000. Human requirements that shape the evolving geoscience information system. Computers & Geosciences, 26 (3A) April 2000, pp. A87-A97 http://nora.nerc.ac.uk/2405/

[104] Licklider, J.C.R., 1960. Man-computer symbiosis. *IRE Transactions on Human Factors in Electronics,* vol HFE-1, p. 4–11 (March 1960). http://memex.org/licklider.pdf

These remarkable skills can be carried across to recognition of rock types, fossils, stratigraphical formations and other geological phenomena. A large investment in computerised security is now resulting in the ability to identify individual human beings, or verify their identity, by comparisons involving a combination of many biometric characteristics, including fingerprints, facial features and movement patterns (Jain et al[105]., 2004. For up-to-date references search the Web for 'biometric identification' or 'biometric recognition'). Similar approaches have potential long-term value, for example, for fossil identification or stratigraphical correlation. Meanwhile, and when working with legacy information, human skills are unrivalled for recognising invariant characteristics.

Much of a geologist's knowledge is of this kind: winnowing out invariant attributes (aspects that stay the same) from a multitude of situations and transformations; building a narrative account of the underlying explanatory interpretations; visualising spatial relationships of the present and past geology; integrating and recording what seems significant in the context of the investigation; and allowing the short-term memories of the trivial and irrelevant to gradually fade away. Unlike the conclusions, the skills and procedures cannot generally be put into words or pictures. But with electronic support, the geologist can demonstrate procedures in the field to convey information that enables another geologist to follow the same processes of observation and thought, and maybe reach the same conclusions. Confirmation or refutation by experts in this way provides scientific validation of unexpressed knowledge. The cyberenvironment can assist this process by recording an illustrated workflow of the investigation. It can record the geologist describing and demonstrating the procedures that led to the conclusions, thereby enabling others to follow and repeat the critical observations. By similar means, an experienced geologist could remotely guide a novice in the field or core store, discussing and clarifying implicit knowledge that can be shared but cannot be recorded.

Investigation of even a single outcrop generates records of many diverse items of information. Within a conventional, self-contained document (such as a map or scientific paper) the significance of each recorded item depends on its context within the document and its cited sources. By contrast, in a digital environment, the significance of a recorded item (object) at any level of detail can be independent of a map sheet or other document. The context of its observation and recording is instead defined by a model or subsystem, extended by connections to many threads of related information within the geological knowledge system as a whole (see Objects, ontologies and systems 48 and The multifaceted model 297). The cyberenvironment must not only handle the full range of information types and their conventional and digital representations: it must also link to recorded demonstrations of unexpressed knowledge about the objects and their interpretation; reach out to communicate with experts through collaborative networks (see Workflows, collaborative networks and Linked Data 53); accommodate the holistic view (see The surveyor's holistic view 60) by means of The systems approach to Earth science 65; and provide the computational power for quantitative detection and exploitation of invariance.

---

[105] Jain, A.K., Ross, A., Prabhakar, S., 2004. An introduction to biometric recognition. *IEEE Transactions on circuits and systems for video technology. Special issue on Image- and Video-Based Biometrics.* **14**, 1, 4-20. http://www.csee.wvu.edu/~ross/pubs/RossBioIntro_CSVT2004.pdf

The systems approach to Earth science

<<Stages of concept development (geological thinking) 56

Systems theory has a long history (Ashby, 1956)[106]. It regards a set of scientific phenomena as a single, coherent system[107]. Shared concepts bring a fuller understanding of the interactions between parts, and lead to unification of individual systems as subsystems of a larger whole. The approach overcomes limitations of the reductionist approach, in which a problem is analysed in terms of individual component parts following the deterministic laws of the continuous, linear systems of physical science (see Complex and emergent systems 159). Instead, an entity, such as the Earth, can be considered as a single, coherent system of related, organised and interacting objects, processes, feedback mechanisms and subsystems. All are seen as functioning as a whole, with properties that cannot be reduced to those of parts studied in isolation (see Objects, ontologies and systems 48). This approach can represent and record patterns of geological thinking and reasoning more fully and accurately than conventional methods. A recent surge of interest stems from cyberinfrastructure support for the representation and integration of systems covering wide-ranging aspects of knowledge.

In 1993, the US National Research Council set out recommendations for 'Solid-Earth Sciences and Society'[108] developed through wide consultation by 150 earth scientists over a five-year period. Their influential conclusions took the view that study of the whole-Earth system provides an essential research framework for addressing global issues (many of which are vital to human well-being), interweaving many branches of pure and applied Earth sciences. It points the way to a more rigorous, comprehensive and quantitative understanding of the complex interacting systems of the atmosphere, hydrosphere, biosphere and lithosphere.

The study of Earth systems science became a driving concept for some key international scientific programmes (see Some related initiatives 341), and major universities began the laborious process of revising their curricula in the light of the proposals. To quote from Cornell University (2006)[109]: "For humanity to live on Earth in a sustainable manner and to act to minimize the impacts of natural hazards, our societies need to understand the earth system. To achieve that understanding, we must observe the natural experiments recorded in Earth history, describe the processes by which it now operates, and test emerging knowledge by use of models of Earth's operation that link theory to observation"... However, "the earth system may be the most complicated system that humanity will ever seek to understand and predict. Its physical, chemical and biological processes interact in

---

[106] Ashby, W.R., 1964. Introduction to cybernetics. London, Methuen.

[107] System: A set of interacting parts that function as a whole. The systems approach involves study of linkages or interfaces between the component activities.

[108] National Research Council (U.S.), 1993. Solid-Earth Sciences and Society. National Academy of Sciences, Washington, DC, 346pp.

[109] Cornell University, 2006. Department of Earth and Atmospheric Sciences: Strategic Plan. http://www.geo.cornell.edu/lms82/StrategicPlan_Dec.pdf

myriad non-linear ways over intervals of space and time spanning nanometers to tens of thousands of kilometers and seconds to hundreds of millions of years".

One of many examples of the economic and scientific importance of linking the three-dimensional geometry of geological objects to the processes that created them is provided by Cosgrove (1998)[110]. He contrasts the original role of structural geology, which focused on the geometry and spatial organisation of structures, to the current concern of petroleum geologists to relate these aspects to the dynamics of the processes that formed them and the associated interplay of stress and fluid migration.

Graduates trained in the systems approach are now potential staff in geological organisations and customers for geological information, and are familiar with technology well suited to supporting a holistic, interdisciplinary systems approach that integrates and connects wide-ranging aspects of knowledge. Geological survey and other organizations are altering their perspective for similar reasons. The Natural Environment Research Council (2007)[111], for example, stresses the need for 'a holistic view of our planet': "The behaviour [of each component part of the Earth] is critically dependent on other parts of the system... we need to understand the current behaviour of the entire Earth, from the core to the upper atmosphere. This includes quantifying the fundamental forces and feedbacks that drive the Earth system... Our understanding of the Earth system is also based on what we know of past changes and the long-term driving forces that caused them."

Major investments in the development of a systems approach are being proposed in other areas of science. For example, the European Science Foundation (2007)[112] sees Systems Biology as a 'Grand Challenge'. "Systems Biology evolved by recognizing that biological systems are far too complex to be solved by classic biological approaches. Systems Biology tightly integrates expertise from physicists, mathematicians, engineers with biological knowledge. It gives a central role to predictive mathematical models that integrate all relevant data on the topic of investigation and exploits such models to decide which experiments are most effective. In this way, an effective and goal-oriented iterative cycle of model-driven experimentation and experiment-driven modeling is initiated" (page 5). See also Noble (2010)[113].

In Medical Science, development of the Virtual Physiological Human (STEP, 2007)[114] is proposed as "a methodological and technological framework that, once established, will enable collaborative investigation of the human body as a single complex system... It is a way to share observations, to derive predictive hypotheses from them, and to integrate them

---

[110] Cosgrove, J.W., 1998. The role of structural geology in reservoir characterisation. *In* Coward, M.P., Daltaban, T.S., Johnson, H., (eds) *Structural geology in reservoir characterization*. Geological Society, London, Special Publications, **127**, pp. 1-13.

[111] Natural Environment Research Council, 2007. Next Generation Science for Planet Earth: NERC Strategy 2007-2012. http://www.nerc.ac.uk/publications/strategicplan/documents/strategy07.pdf

[112] European Science Foundation, 2007. Systems biology: a Grand Challenge for Europe. http://www.esf.org/publications/medical-sciences.html 54 p.

[113] Noble, D., 2010. Biophysics and systems biology, *Phil. Trans. Roy. Soc. A,* **368**, 1125-1139.

[114] STEP Consortium. Seeding the EuroPhysiome: A Roadmap to the Virtual Physiological Human. 5 July 2007. http://www.europhysiome.org/roadmap

into a constantly improving understanding of human physiology/pathology, by regarding it as a single system" (see also Clapworthy et al., 2008[115]). For similar reasons, there is a case for establishing The solid Earth systems model (sEsm) 71 as a shared, integrative, predictive system within a consistent framework[116]. It could support a more comprehensive understanding of Earth science processes, as they operate now and as they shaped the past evolution of successive configurations of the solid Earth, disentangled by geologists in their record of the Earth's history. The geological cyberenvironment[117] could both lead to and respond to its development.

However, there are risks that must be borne in mind. *The Economist* (19 May 2008) discussed how close Wall Street had come to systemic collapse, and how financial systems should change as a result. Increasingly complex financial instruments had been developed, seen as "contributing to a far more flexible, efficient, and hence resilient financial system". A laissez-faire attitude allowed services to innovate and spread almost unchecked. "This has created a complex, interdependent system prone to conflicts of interest." Fraud has been rampant, fed by the knowledge that, if disaster struck, someone else would end up bearing the losses. Similarly, the increase in connectivity of information in the knowledge economy brings risks that threads of reasoning, linkages, analogies, correlations, blanket applications of statistical predictions, and risk assessments, fully understood only by their developers, become so entangled that failure in one part has catastrophic results for the integrity of the knowledge system as a whole.

Our perceptions of reality change with time, and no two individuals think quite the same. Users must know where information came from (its provenance) to assess its source and evaluate its relevance in their own context. The information system must cope with overlapping information from different sources, and with many, possibly contradictory, versions of the same ideas. Social networking helps many individuals to innovate and contribute to a participatory cyberenvironment, either restricted to an expert group or open to all. Wider collaboration and contributions to the knowledge base clearly add to its value. Ideas evolve and the favoured survive, through variety, inheritance and selection; but unselected survival of substandard components in the knowledge base could destroy its value.

---

[115] Clapworthy, G,. Viceconti, V., Coveney, P.V., Kohl, P., 2008. Editorial. Phil. Trans. R. Soc. A 366, 2975-2978. doi: 10.1098/rsta.2008.0103 http://rsta.royalsocietypublishing.org/content/366/1878/2975.full.pdf+html

[116] Solid Earth systems model (sEsm): An approach to structuring distributed knowledge of the science of geology to provide an integrated view in the context of sciences of the solid Earth as a whole. A model of the systems of the solid Earth, organised within a framework or metamodel that depicts and clarifies the principal relationships among the findings of geology, providing a multidimensional map to locate and connect ideas, concepts, workflows of investigation and threads of reasoning. The content of the model is distributed information referring to: the three-dimensional disposition and configuration of the present-day observable objects of the solid Earth (where things are and how they are arranged); their observed and interpreted properties, composition and relationships, at all scales; geological processes and the outcomes of their interactions with configurations of objects; events and historical changes throughout geological time.

[117] Cyberenvironment: Aspects of the cyberinfrastructure assembled to meet requirements relevant to a particular field of enquiry, aiming to maintain global compatibility while providing access through interfaces that match users' working practices.

The Economist reported that solutions to the ills of the financial system were being sought in better regulation and greater transparency, ensuring that all relevant parties can understand the implications to them of the actions of others. This has resonance in the knowledge system with its ever-growing interconnections, and the intricacies of provenance trails, information evaluation and quality assessments. For most users a plethora of conflicting views of the geology is unsatisfactory. Without detailed knowledge of the contributors and the background they cannot adequately assess the situation. But how can they obtain convenient access to a better alternative? The mainstream cyberinfrastructure for geology must provide well-ordered systems of robust, tested, shared, interoperable components. Scientific evaluation procedures, currently imposed by editorial judgment, peer review, and corporate quality assessment must find counterparts in the new infrastructure[118].

Mechanisms for evaluation and collaboration must change and grow in the geological cyberinfrastructure to maintain the fragile balance of flexibility, coherence, and integrity in scientific knowledge. The need for interoperable, fully tested, robust components extends beyond the infrastructure components of The geological cyberenvironment (gce) 85, and these should therefore conform also to standards in more general knowledge systems. Users of information must be able to understand its context, its limitations and its implications. A Geological Survey uses detailed local knowledge to assess all available sources and test them in the field to provide a coherent authoritative view of the geology. World-wide, Surveys collaborate to set benchmarks for local and regional geology against which other sources can be judged. They are well placed to make a major contribution to future models of the systems of the solid Earth, to set them in their wider context, and to make evaluated, authoritative knowledge and information readily, transparently, and widely available.

The next steps

<<Stages of concept development (geological thinking) 56

The technical benefits of the advanced infrastructure will combine with the scientific benefits of the systems approach to provide a more comprehensive and accessible understanding of the solid Earth and its interactions in the whole Earth system. This is set to change the focus of geological survey organisations – away from publishing maps and supporting documents, towards making a well-defined contribution to a whole-Earth knowledge system that responds flexibly in supplying information to meet user needs (see The future geological map 125). Because the changes in the knowledge system affect many interdependent elements, explicit design is required to create a coherent environment where diverse components fit together. A formal record of the underlying scheme of ideas is needed: initially to open them to discussion, criticism and improvement; when appropriate to guide the design of the infrastructure and the construction of interfaces within it.

[118] Infrastructure: The basic facilities, services and installations needed for a system to function.

Tentative suggestions for design of an information framework, based on a solid Earth systems metamodel, were offered in Loudon and Laxton (2007)[119]. There is a need from a geological viewpoint (see Overview of the geological investigation model 34) to modify and extend that framework to relate it to a geological cyberenvironment that provides the means for handling all relevant information (digital and conventional) throughout a geological investigation. The aim is to create a structure where shared models encourage cross-fertilisation of ideas and broader understanding of solid-Earth systems and their consequences (see The solid Earth systems model (sEsm) 71).

Geoscientists, who know their own business best, must nevertheless decide on their intended objectives before planning how to achieve them. A starting point in defining objectives is to consider what is feasible, what is desirable, what is to be retained from the conventional geoscience knowledge system, and what can be improved.

Geological interpretation and ideas must come from human investigators. The cyberinfrastructure potentially offers a better medium for geologists to express ideas; build interpretations; organise, analyse, summarise and share observational data; explore the consequences of hypotheses; reconcile information from diverse sources with expectations and background knowledge; display and archive the results and share them more widely. Widely shared solid Earth systems models and geological cyberenvironments become realistic objectives, raising questions of what will be lost and gained in their development. The detailed shortcomings of existing methods are discussed in sections on the individual models, but some general points should be mentioned here.

Conventionally, documents go through a complex process of reviewing and amendment before publication in order to determine: that they meet certain standards; provide a coherent and internally consistent view of their chosen topic; are appropriate for the intended class of reader; respect existing conventions; acknowledge sources; are inoffensive; and are a significant contribution to knowledge. If they are deemed appropriate for publication they may then be distributed and permanently archived as a significant contribution to the science. The process is laborious, expensive and time-consuming. The documents may cut across areas of scientific significance. For example, geological map sheets and their explanations may be bounded by geographical coordinates with no geological relevance. Work on new topics may lack a suitable outlet. For example, computer applications in geology were under way for many years before appropriate journals were set up to publish the results. The geological record is subdivided into types of document, such as textbooks, papers in scientific journals, maps, logs and datasets, and tacit knowledge in the minds of scientists. In many respects the pattern of geological thinking (see Reasoning, models and reality 127) is fractured and distorted in order to correspond to a system of information archiving and sharing that was designed to meet the constraints of an earlier infrastructure.

Developments that take the existing scientific literature as their model are clearly essential to preserve the legacy of earlier work. They may be suited to recording the outcome of

---

[119] Loudon, T.V. and Laxton, J.L., 2007. Steps towards Grid-based Geological Survey: suggestions for a systems framework of models, ontologies and workflows. Geosphere, 3 (5), 319-336. http://nora.nerc.ac.uk/1084/

individual projects, and can be enhanced by hyperlinks, for example, to link to subsequent work, detailed descriptions, instrumental data, or even alternative interpretations. They offer the benefits of a familiar context, but extend the mismatch with the underlying pattern of thinking. The advancing cyberinfrastructure offers potential benefits in enabling us to tackle the underlying requirements in new ways; introducing insights from other fields; and encouraging geologists to reconsider their thought processes, communications and working practices. It enlarges the potential extent of the ability to share concepts and information across specialist interests.

Any geoscience investigation or project is undertaken within a specific business setting. This might be oil exploration, land use, water extraction, geological research, education, or some other activity. Each project has its own objectives, priorities, operational definitions, sampling schemes, and the like. These are tailored to the business needs, and determine the procedures and products. This context must be explained to give the results meaning beyond the project. Rather than every project starting from scratch and proceeding independently, however, it can be more efficient to build on a shared foundation. Geological survey and similar organisations offer this through widely applicable views of the basic geoscience of an area. Their core activity is piecing together a picture of the geometrical configuration and disposition of sequences of strata or other rocks, their constituent materials, characteristics, and properties, and relating that picture to ideas of their history and origin, conventionally recorded as reports, maps and cross-sections. This suggests that geological survey organisations have an important part to play in geological aspects of its implementation.

The wider representation of geological thinking and integration of geological knowledge from many sources is considered in

The future geological map 125
The geological investigation model 98
The geological framework model 105

and mechanisms for its wider evaluation and sharing in

The geological business model 93

The solid Earth systems model (sEsm)

<<The emerging geoscience knowledge system 42
<<Stages of concept development (summary) 44
<<Stages of concept development (cyberinfrastructure) 45
The solid Earth systems model (sEsm) 71
Objectives of the solid Earth systems model 72
Scope of the sEsm 73
Remodelling the map 74
The sEsm as a predictive machine 76
Geological surveying as reinforcement learning 79
Design requirements for the sEsm 83
>>The geological cyberenvironment (gce) 85

The solid Earth systems model (sEsm) is an approach to collecting, organising, integrating and sharing geoscience information, linked to requirements specified in the business model and to the facilities of the infrastructure (*Figure 13*). It aims to provide a comprehensive structure for representing information on the systems of the solid Earth, in which relevant knowledge can be integrated as a shared, coherent, predictive system, where like can be compared with like and quantitative relationships assessed (see The solid Earth systems metamodel (sEsmm) 110). The focus here is on the general geoscience model and authoritative view of regional geology maintained by geological survey organisations.

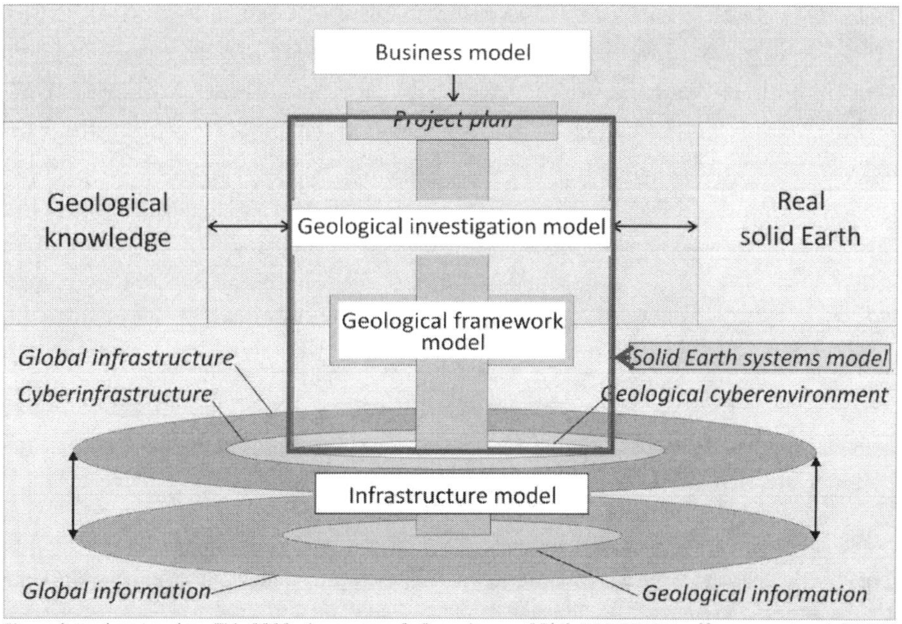

Figure based on Loudon, T.V., 2009. *Computers & Geosciences,* **35**(4), 700-705 http://nora.nerc.ac.uk/7258/

Figure 13: Four component models in the geoscience knowledge system (duplicate of Figure *6). The solid Earth systems model to which they relate is outlined in red.*

The electronic support for populating, maintaining and accessing the model is considered in The geological cyberenvironment (gce) 85. Details of the solid Earth systems model[120] are described from different viewpoints in four models. The geological business model 93 considers the objectives of geological investigations, their planning, and evaluation of the results. The geological investigation model 98 describes procedures by which the information is (and might be) gathered. The geological framework model 105 proposes a systems structure for organising the information. The geological infrastructure model 117 looks at facilities and mechanisms for storing, processing and sharing information. The process of moving from conventional methods to a systems model is described in outline in Remodelling the map 74 and in more detail in Mapping geology into the knowledge system 282.

Objectives of the solid Earth systems model

<<The solid Earth systems model (sEsm) 71

The objective of The solid Earth systems model (sEsm) 71 is to link and integrate relevant knowledge of geological concepts and the results of geoscience investigations as a coherent, shared, predictive system (*Figure 14*). Rather than the geologist mapping observations on the ground onto various two-dimensional sheets of paper, the investigator can map observations and interpretations into a multidimensional digital structure (The geological framework model 105), from which many representations and visualisations[121] can be generated. The proposed contents refer to: the three-dimensional disposition and configuration of the present-day geological objects of the solid Earth (where things are and how they are arranged); their observed and interpreted properties, composition and relationships, at all scales; and their interpretation as the outcome of events and historical changes throughout geological time, as geological processes interact with pre-existing configurations of objects. The model relates to both conventional representations and information collected using cyber-based techniques.

---

[120] Solid Earth systems model (sEsm): An approach to structuring distributed knowledge of the science of geology to provide an integrated view in the context of sciences of the solid Earth as a whole. A model of the systems of the solid Earth, organised within a framework or metamodel that depicts and clarifies the principal relationships among the findings of geology, providing a multidimensional map to locate and connect ideas, concepts, workflows of investigation and threads of reasoning. The content of the model is distributed information referring to: the three-dimensional disposition and configuration of the present-day observable objects of the solid Earth (where things are and how they are arranged); their observed and interpreted properties, composition and relationships, at all scales; geological processes and the outcomes of their interactions with configurations of objects; events and historical changes throughout geological time.

[121] Visualisation: Transforming quantitative data (including the results of interpolation) into sensory information – images that the eye and brain can interpret and visualise.

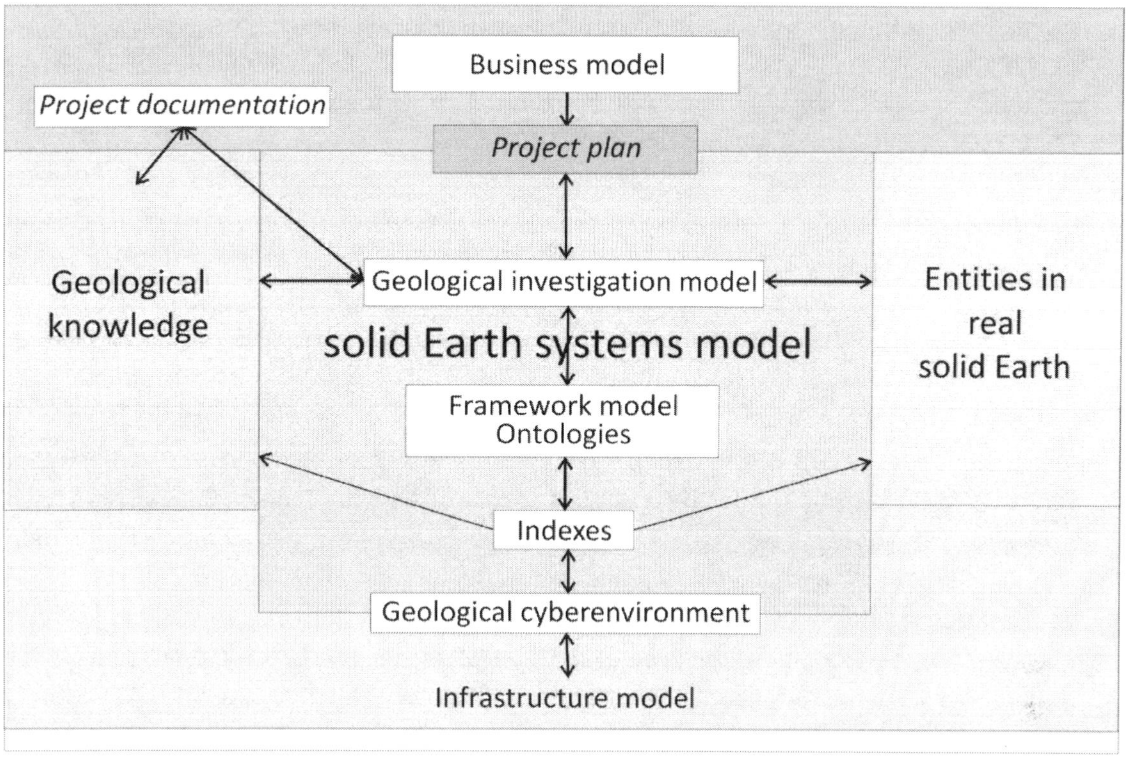

Figure 14: The solid Earth systems model.

Scope of the sEsm

<<The solid Earth systems model (sEsm) 71

The solid Earth systems model aims primarily to support the results of past and future geological surveying. The surveyor can place geological information about material, properties and processes into their geographical and historical context, regarding them as a single system[122] of interacting components (see Reasoning, models and reality 127). The initial emphasis of the sEsm is on near-surface geoscience observations and their explanation in terms of geological systems, many of which relate to processes in the deep Earth. Many explanations also involve processes of considerable complexity in the atmosphere and hydrosphere. It is convenient to consider these as distinct but linked systems, with well-defined interfaces[123] where they meet the solid Earth at the land surface or beneath the sea. System boundaries are inevitably somewhat arbitrary. However, solid Earth systems have many features in common, which may justify regarding them as a distinct system, interfaced with other systems describing related aspects of science or

[122] System: A set of interacting parts that function as a whole. The systems approach involves study of linkages or interfaces between the component activities.

[123] Interface: The shared boundary between systems or parts of a system, or the means of interaction across the boundary that makes joint operation possible.

applications. The sEsm, like the science it represents, must adopt an open-ended holistic[124] view that reaches out beyond the interfaces. Its design shares the aims of Linked Data, as one model in a hierarchy of systems within the 'unbounded space of the Web of Data' (see Workflows, collaborative networks and Linked Data 53).

Remodelling the map

<<The solid Earth systems model (sEsm) 71

Systematic topographical surveying of the UK began in the eighteenth century. Using the results as base maps, the British Geological Survey (since 1835) has undertaken geological surveying of the distribution and relationships of rock bodies throughout Britain. There are now some 150 geological survey organisations worldwide. Widespread adoption of standard map representations and conventions, and internationally agreed classifications of stratigraphical units, minerals, rock types, fossils, and other geological entities, enable their maps and written accounts of the geology to be widely understood. As the OneGeology project (see Some related initiatives 341) illustrates, Geological Surveys are able to provide consistent and authoritative information on what is known of the regional geology, within its global context, based on the conventional infrastructure of published maps and associated documents.

A geological survey map typically superimposes the surveyed extent and boundaries of geological units on a topographical map, together with information at points (such as measurements of orientation of bedding planes) and lines (such as the intersection of faults with the land surface). The maps and reports created by geological survey organisations generally aim for geographical continuity and completeness in establishing the spatial patterns of near-surface rock units. The map may include cross-sections to illustrate the three-dimensional interpretation. Subsurface geological and geophysical maps, providing limited coverage of deeper geology, are maintained internally by major oil companies and regulators. Some geological survey organisations have collaborated with them to include subsurface geology in their systematic surveys, for example, the Geological Atlas of the Western Canada Sedimentary Basin (Mossop and Shetsen, 1994[125]). Subsurface maps typically depict the three-dimensional form of geological surfaces by means of contours and cross-sections.

Geological surveying thus involves field (and possibly subsurface) observation and measurement, severely limited by the accessibility and availability of exposed geology. Both surface and subsurface geological maps include a large element of interpretation to interpolate the form of geological boundaries between scattered outcrops, or surfaces

---

[124] Holistic: A view of a system that emphasises its properties and interrelationships acting as a whole, as opposed to the reductionist approach of studying its components in isolation as distinct entities.
[125] Mossop, G.D., Shetsen, I (comp.), 1994. Geological atlas of the Western Canada Sedimentary Basin. *Canadian Society of Petroleum Geologists and Alberta Research Council*, Special Report 4L http://www.ags.gov.ab.ca/publications/wcsb_atlas/atlas.html

between wells or boreholes. In their minds, geologists recreate past sequences of geological configurations of interacting objects, processes and events by studying their outcome as glimpsed today, taking into account present-day analogues, experimental investigations, theoretical knowledge and logical reasoning. Guided by interpretation and observation, the geologist fills the chosen area or space, representing the geology by lines, coloured areas on the map, symbols and annotations. Although spatial geological objects can be described in terms of their objective properties, the classification of some object classes, such as those used in unravelling geological history, may depend more fundamentally on hypothetical scenarios of their origin (see At the interface133). The surveyor thus makes testable predictions of the distribution of the present-day geology, using background knowledge of spatial attributes and patterns. Geologists are studying a highly complicated system, but the conventional representation of survey results, namely, a series of standardised maps covering wide geographical areas, supported by diagrams and text accounts, is an inadequate representation of their knowledge of that system (see The surveyor's holistic view 60, The imperfect map 145).

Stages of cyber-awakening are described by Berners-Lee (2007)[126]: from 'It isn't the cables, it is the computers which are interesting' to 'It isn't the computers, but the documents which are interesting' and then to 'It's not the documents, it is the things they are about which are important' (see Workflows, collaborative networks and Linked Data 53). On that basis, an obvious starting point for systems geology is the conventional geological survey map, with its well-defined two-way linkage between points and areas on the map and the 'things they are about' on the ground. Also, survey maps offer standardised content and global coverage. But because systems geology must overcome the limitations of the conventional geological map to record a more comprehensive view of geology, it requires an overall framework for modelling and representing the systems of the solid Earth. It must be consistent with the geological paradigm, and be readily understood and accepted, enabling geologists to organise their observations and ideas (see The geological framework model 105).

Digitisation of geological maps, along with systems for database management and geographical information, make it possible to organise the mapped information as objects, such as features or areas that have been categorised geologically. This flexible structure is supported by internationally accepted standards and has led to improved methods for linking objects and maps from different survey organisations (see Other geological initiatives 343– CGI, GeoSciML, GEON, OneGeology and IGME5000). Digital spatial models (see Other geological initiatives 343 – DGSM) enable geologists to record their three-dimensional interpretation more comprehensively, with flexible visualisation of the results. Building on these initiatives, a more radical approach is now feasible.

The cyberinfrastructure can support powerful observational and analytical methods and record more of the geologists' thinking and concepts, mapping them into a wider systems view of the solid Earth (see The solid Earth systems model (sEsm) 71). It can provide mechanisms for a more complete and rigorous representation, bringing it into a formal systems structure (see The geological framework model 105). It provides a widespread,

---

[126] Berners-Lee, T., 2007. The Giant Global Graph (blog post) http://dig.csail.mit.edu/breadcrumbs/node/215

highly connected environment that can not only link to existing information but also add to its value by providing quantitative analysis, flexible visualisation of spatial information, powerful indexing and search tools, and rapid delivery of results.

The objectives of the developing geological cyberinfrastructure extend beyond overcoming limitations in map representation. Systematic geological surveying can provide the starting point for a solid Earth systems model (sEsm), as discussed at length from a geologist's viewpoint in 'The future geological map' 125. Geology is a historical science, and invariant aspects of processes and their outcomes mean that geologists can use reconstructions of past events to improve their understanding and prediction of present-day geology, and vice versa (see Invariance and processes 56). The geological surveyor develops an interpretation based on an understanding of the history of geological processes and events that created the observable geology. Like a physician's directed search of a patient's symptoms, field geologists aim to narrow down their interpretation by looking for syndromes of diagnostic characteristics to categorise, assess, confirm or refute their current view of the observed geology. In so doing, they call on a wide range of knowledge from theory, experiment and related examples.

The sEsm aims to encapsulate the surveyors' understanding of the systems of the solid Earth in an appropriate structure (see The geological framework model 105). Fragments of the model currently exist in the minds of geologists. They are represented, recorded and communicated through maps, books, papers, datasets, training, discussion and demonstration. The advancing cyberinfrastructure has the potential to support more comprehensive representations of the model, joining up the individual fragments. It could develop and maintain a model of the multi-dimensional, multi-resolution, complex system of deeply interconnected geological objects and processes, and their attributes, relationships, historical development and present-day expression, all seen as a consistent, coherent part of whole-Earth systems (see Mapping geology into the knowledge system 282). The sEsm can thus create a more comprehensive record, filling space with an interpretation supported by observation and based on wide-ranging knowledge (see The sEsm as a predictive machine 76).

The sEsm as a predictive machine

<<The solid Earth systems model (sEsm) 71

The core objectives of the solid Earth systems model (sEsm) are to:

- record what has been observed of the geology
- predict what has not
- quantify properties, processes and uncertainties
- record reasoning, evaluation and justification
- communicate results to meet a wide variety of specific requirements

This aspect of the sEsm might be thought of as a predictive machine, with its mechanisms (and 'symbiotic' links to users) contained in The geological cyberenvironment (gce) 85.

Predictive power is important because observations of limited fragmentary evidence (see Remodelling the map 74, Remodelling geological investigation 101) must be expanded to give the comprehensive explanatory view required by the science and its applications support. Prediction is possible because of correlation among geological properties and entities. This implies that they do not behave independently, and therefore knowledge about one aspect can contribute to knowledge of related aspects (see Stages of concept development (geological thinking) 56). The objectives of quantifying, recording and communicating the results along with the underlying evidence and thinking should provide users with more comprehensive, precise and relevant information.

The cyberinfrastructure brings powerful connectivity and Reconciliation 186 mechanisms for integrating model fragments, including those previously restricted by difficulties in handling large data volumes, such as data from remote sensing, simulation[127], visualisation, and detailed, wide-ranging statistical analysis. Data-intensive remote-sensing methods, including satellite imagery, seismic, 3d seismic and downhole logging, generate patterns at a wide range of scales. They refer to properties (such as acoustic impedance, light reflectance or electrical conductivity) that are not primary variables in geological interpretation. But after filtering out artefacts generated by the instrumentation and environment, they reflect underlying patterns created by interacting geological processes operating on specific configurations of pre-existing objects. These secondary variables therefore have wide-ranging predictive value in a wider statistical context.

Data-intensive simulation-based science is ubiquitous in conventional geological survey. On the basis of background knowledge, field geologists may conceive and develop explanatory hypotheses and scenarios, and imagine their possible outcomes (see Abstracting from reality to model 131). These simulated perceptions guide further observations; testing and modifying the interpretation as the survey progresses. With advanced electronic field support, simulations, based on the geologist's current interpretation, could to some extent be formalised, codified, and visualised at appropriate resolutions. Their characteristics could be compared with those of the outcrop and amended while surveying, as part of the 'symbiosis[128]' between surveyor and system (see Models and frameworks 47). As a multi-resolution tool for prediction, testing and verification, simulations would create large quantities of data.

Understanding the complicated interactions of the system, however, poses greater problems than handling the amount of data. The geological framework model 105 includes many separate properties (variables) that can be observed, possibly measured, and recorded on numerous separate dimensions within the database. Their interactions are the basis for understanding the geology. Statistical techniques, such as principal component analysis and factor analysis, can identify the correlation between variables and capture much of the total variation in a much smaller number of synthetic variables. But the results are misleading where like is not being compared with like. Understanding them depends on disentangling

---

[127] Simulation: Imitation of aspects of internal processes of a system and their results; usually to visualise, statistically compare with, or predict real-world occurrences.

[128] Symbiosis: A close interdependence or association (in the literal sense, of animals or plants of different species) often of mutual benefit.

the causes that resulted in the observed effects. An overall model of the interacting systems is therefore required to place their local expression and individual observations in context.

The conventional means of reducing the complexity (and a starting point for scientific explanation) is classification, in which the numerous characteristics of an object result in it being placed in a single class, possibly with added descriptors highlighting particular subclasses to which the observed object belongs. An example might be 'sandstone, light grey, fine-grained, probably aeolian.' The observer is using the remarkable powers of the human brain to recognise significant recurrent patterns and to relate them to hypotheses of their origin (see Invariance and processes 56). The observations are put into the context of a wider understanding of the solid Earth, and the multitude of stimuli reaching the geologists' brain are condensed into one category relevant to the immediate objectives, such as completing a geological map. Classifications can be standardised, but the classes of, say, traditional stratigraphy may be inadequate from a systems geology viewpoint (see Grain, set and patch 251). The mechanisms of the Semantic Web and Grid 50 can assist in coordinating information seen from different viewpoints and clarifying their incompatibilities. Human background knowledge, however, remains essential to extract information in existing records for an unforeseen application (see Reconciliation 186).

An aim of systems geology would be to extend interoperability[129] of the multitude of observations and concepts that share the same background relevance, by bringing them into a formal framework of systems and ontologies, sharing an explicit expression of The surveyor's holistic view 60. Geological survey organisations have achieved considerable uniformity in depicting regional geology, and geology as a whole has adopted consistent terminology in many fields. The geological framework model 105 aims to provide a more comprehensive structure to correlate and compare the findings of geological investigation. However, it is clear that most geological projects, for reasons of efficiency, focus on specific aspects of the systems model at the expense of wider compatibility (see Diverse objectives and products 150). Different projects are likely to use different approaches to sampling, capturing and recording their data, and this must be taken into account. Although the framework should enable users to identify a wider range of possibly relevant material, the background human knowledge of the users, assisted by metadata and ontologies, is required to interpret it correctly.

The sEsm framework provides a means of linking observations of geological entities more explicitly to the environment and geological processes that formed them. Viewing aspects of geology as a system (see The systems approach to Earth science 65) makes it easier to determine where like can be compared with like, and explore relationships with more informative and rigorous statistical methods. Viewing geology as one coherent component of the total knowledge system introduces a more comprehensive approach to the science and broadens its relevance. Methods of systems documentation inevitably differ from those of conventional scientific publication, and are described in The geological infrastructure model 117.

---

[129] Interoperability: Interoperability of information is the ability of concepts, terms or models from various sources to work together, by meeting standards that enable sharing and reuse of information.

Geological surveying is inevitably involved with prediction: filling the spaces on the map between the fragments of evidence, building on local detailed observations with the geologists' wider background knowledge and experience, aiming where possible to ensure that interpretations and prediction conform to global conventions and standards. Broader conclusions are frequently drawn from local interpretations (rather than local observations). Like the geological map, predictions from the sEsm may reflect local circumstances, and careful assessment of the underlying human reasoning is essential. Nevertheless, the advances in communication and quantification in the cyberinfrastructure are reflected in the ability to establish global observation-based interpretations (particularly in geophysics, see Fowler, 2005[130]) and to reconcile interpretations across a wide range of scales. Rafols et al. (2005[131]), suggest that 'particularly good generalisations will result from representing the state of the world in terms of predictions about possible future experience' in the context of reinforcement learning (described in Geological surveying as reinforcement learning 79).

Within a global cyberinfrastructure able to handle large data volumes, the sEsm (thought of as a predictive machine) could help users to describe, integrate and understand Earth systems and their history, from microscopic to planetary scale. It could transform the scope, relevance, accuracy, predictive value and timely delivery of geological knowledge for diverse applications, many of which are vital for human welfare.

Geological surveying as reinforcement learning

<<The solid Earth systems model (sEsm) 71

The use of prior knowledge (that which is already known) is a feature of geological surveying (Forward and inverse models 154), and its links to the cyberinfrastructure have been described in individual investigations (see Curtis and Wood, 2004[132]), without necessarily considering how they fit into the overall picture. In systems geology, however, prior knowledge is an integral part of The solid Earth systems model (sEsm) 71. It may therefore be helpful to look for ideas on this topic shared (but expressed differently) by geologists and by e-scientists concerned with comprehensive knowledge systems. The computational approach to reinforcement learning described by Sutton and Barto (1998[133]) seems appropriate. This section builds on their frequently cited introductory chapter 1, summarises some of their ideas, and explores possible relationships to geology.

[130] Fowler, C.M.R., 2005. *The solid Earth: an introduction to global geophysics*, 2nd ed. Cambridge University Press, Cambridge, UK. 685pp. ISBN 0 521 89307 0

[131] Rafols, E.J., Ring, M.B., Sutton, R.S., Tanner, B., 2005. Using predictive representations to improve generalization in reinforcement learning. *Proceedings of the 19th international joint conference on Artificial Intelligence*, 835-840. Morgan Kaufmann, San Francisco, CA. http://www.ijcai.org/papers/1650.pdf

[132] Curtis, A., Wood, R. (editors), 2004. *Geological prior information: informing science and engineering.* Geological Society, London, Special Publications, **239**.

[133] Sutton, R.S. and Barto, A.G., 1998. Reinforcement learning: an introduction. MIT Press, Cambridge, MA. http://webdocs.cs.ualberta.ca/~sutton/book/ebook/the-book.html and chapter 1 at http://webdocs.cs.ualberta.ca/~sutton/book/chapter1.pdf

They define 'reinforcement learning' not by a particular set of methods but by characterising a learning problem. Geological surveying, as described at various points in The future geological map 125 and summarised in The surveyor's holistic view 60, can be characterised as a learning problem, in which existing geological knowledge is confirmed, corrected, extended, and amended. It could therefore be brought into the wider framework of reinforcement learning. This might ease the transition from conventional geology to systems geology, not by displacing human investigators, but by enhancing their computer support. We therefore consider how conventional surveying procedures could be recast in their terms and concepts.

The concepts of reinforcement learning, like most theories of learning and intelligence, are based on learning from interaction with the environment. The learning is done by an autonomous entity known as an agent. Reinforcement learning "involves interaction between an active decision-making agent and its environment, within which the agent seeks to achieve its goal despite uncertainty about the environment. The agent's actions... affect the future state of the environment... thereby affecting the options and opportunities available to the agent at later times. Correct choice requires taking into account indirect, delayed consequences of actions, and thus may require foresight or planning" (page 7).

In computer science, the agent is likely to be a piece of software. In conventional geological surveying the 'autonomous agent' is a human being – a geological surveyor with or without computer support. The 'environment' in this context would presumably not refer to the real solid Earth, which is external to (and unaffected by) the knowledge system. Rather, the 'environment' would be part of the knowledge system, perhaps represented as a series of geological maps, or (in this systems geology scenario) as The solid Earth systems model (sEsm) 71, but probably held for the most part in the minds of the surveyors. Alternatively, it may be that the 'environment' should be regarded as the solid Earth together with the knowledge system. In any case, the goal of the agents (surveyors) is to improve their geological model (match it more closely to the relevant aspects of the solid Earth) by making observations of real entities (rocks) and recording the resulting object descriptions and interpretations as part of the model. Improvements in the knowledge system may simply add local detail to a map, but the surveyors are also concerned with the wider implications, or 'delayed consequences', of what they record.

"Another key feature of reinforcement learning is that it explicitly considers the *whole* problem of a goal-directed agent interacting with an uncertain environment" (page 5). Geological surveying corresponds to this, as described here in The surveyor's holistic view 60: ...it builds on a top-down view or gestalt (an analysis working down from the structure of the whole to its relations with its constituent parts and their characteristics). At each stage, geoscientists look carefully at everything they know so far, and imagine how the situation might be in its entirety. Analysis of items in isolation cannot provide that understanding, for their significance depends on their place, role and function in the whole.

"Beyond the agent and the environment, one can identify four main sub-elements of a reinforcement learning system: a *policy*, a *reward function*, a *value function*, and, optionally, a *model* of the environment. A *policy* defines the learning agent's way of behaving at a given time. Roughly speaking, a policy is a mapping from perceived states of the environment to actions to be taken when in those states" (page 7). The geological surveyors' policy might indicate how they would react to particular situations. For example, if the goal was to document the fossil fauna, an action might be to exclude (from the study) rocks which the model indicated were of igneous origin.

"A *reward function* defines the goal in a reinforcement learning problem. Roughly speaking, it maps each perceived state (or state-action pair) of the environment to a single number, a *reward*, indicating the intrinsic desirability of that state. A reinforcement agent's sole objective is to maximize the total reward it receives in the long run. The reward function defines what are the good and bad events for the agent" (page 8). Surveyors would not normally assign a numerical value to the satisfaction of making an observation or developing an interpretation, but the actions of doing so and preferring one action to another suggest that each possibility is being informally weighed up to guide the decisions in the surveyors' minds. Costs (negative rewards) are also taken into account ('it will take too long to reach that outcrop, I will look here instead').

"Whereas a reward function indicates what is good in an immediate sense, a *value function* specifies what is good in the long run. Roughly speaking, the *value* of a state is the total amount of reward an agent can expect to accumulate over the future, starting from that state. Whereas rewards determine the immediate, intrinsic desirability of environmental states, values indicate the *long-term* desirability of states after taking into account the states that are likely to follow, and the rewards available for these states... To make a human analogy, rewards are like pleasure (if high) and pain (if low), whereas values correspond to a more refined and farsighted judgment of how pleased or displeased we are that our environment is in a particular state" (page 8). Value functions 'formalise a basic and familiar idea', and in the context of the solid Earth systems model are primarily the concern of The geological business model 93. For example, sets of observations that have the knock-on effect of increasing the predictive power of a significant part of the model (such as detailed mapping of a thin tuff bed that is a good time marker) are of higher value than those with a purely local effect. They could be assessed in terms of the goals of their business model defined in terms of the sEsm. Geological survey organisations and academic studies are particularly relevant, in providing a general-purpose body of geological knowledge that could increase the overall value (predictive power) of the sEsm as a predictive model[134].

"The fourth and final element of some reinforcement learning systems is a *model* of the environment. This is something that mimics the behavior of the environment... Models are used for *planning*, by which we mean any way of deciding on a course of action by considering possible future situations before they are actually experienced" (page 9). Planning models are again a feature of the geological business model, not to be confused

---

[134] Prediction: Drawing conclusions from incomplete evidence. Predictions can result from reasoning about a hypothesis (a suggested explanation of a phenomenon), and are 'useful' (throw light on the likely truth of the hypothesis) if they can be tested by observation or experiment. Also, predictions of as yet unobserved phenomena can stem from a theory, in the sense of a comprehensive explanation supported by facts gathered over time.

with models of solid Earth systems (in Sutton and Barto's terminology the 'environment'). For example, an oil geologist might plan to continue drilling beyond a particular depth only if the drill had by then reached the base of the Cretaceous, for if not the value (the likelihood of striking oil in that well) would be greatly diminished.

"The agent has to *exploit* what it already knows in order to obtain reward, but it also has to *explore* in order to make better action selections in the future... The agent must try a variety of actions *and* progressively favor those that appear to be best." The field geologist might look primarily for one kind of fossil known to have value in stratigraphical correlation, but would also examine others in the same strata, in case they gave more consistent results.

"A *predictive* representation is one that describes the world in terms of predictions about future observations." Rafols et al (2005[135]) argue that such representations are particularly good for generalisation in reinforcement learning, capturing regularities of the environment that allow the agent to increase its cumulative reward. The ability to generalise, as described in Scale-space 255 and Multiresolution survey 259, is an essential feature of The sEsm as a predictive machine 76.

The match of geological surveying to reinforcement learning suggests that (as in many areas of application of geoinformatics) a small-scale multi-disciplinary pilot study would be worthwhile, as a way to explore its value in systems geology. In particular, the explicit consideration of evaluation (in a scientific rather than a monetary sense) could be helpful in directing information searches and in linking and guiding the business and investigation models. Evaluation criteria could emerge from the business model in terms of the relative importance of specific aspects of the geology, could be passed through the sEsm to determine correlated (and therefore relevant) aspects, and implemented in the investigation model. The mathematical framework set out by Sutton and Barto (2005[136]) is likely to be of longer-term value, but the issues it raises, such as optimal control of Markov decision processes, can perhaps be left to subsequent studies.

[135] Rafols, E.J., Ring, M.B., Sutton, R.S., Tanner, B., 2005. Using predictive representations to improve generalization in reinforcement learning. *Proceedings of the 19th international joint conference on Artificial Intelligence*, 835-840. Morgan Kaufmann, San Francisco, CA. http://www.ijcai.org/papers/1650.pdf

[136] Sutton, R.S. and Barto, A.G., 1998. Reinforcement learning: an introduction. MIT Press, Cambridge, MA. http://webdocs.cs.ualberta.ca/~sutton/book/ebook/the-book.html and chapter 1 at http://webdocs.cs.ualberta.ca/~sutton/book/chapter1.pdf

Design requirements for the sEsm

<<The solid Earth systems model (sEsm) 71

The system design should organise relevant information in a widely accepted and understood structure that matches the ways in which geologists think about their science (the geological paradigm[137]). It should provide a coherent structure of subsystems that enable the various components, including tacit background knowledge[138], to work together (see The geological framework model 105), and to cope with sparse and fragmentary information (see The sEsm as a predictive machine 76). It should be able to handle diverse viewpoints reflected in the provenance[139] of the various information sources and user requirements (see Linking beyond the sEsm 115, Projects and information communities 167, Reconciliation 186). It should meet wider industry standards and protocols. It should link to a straightforward user interface to access, process and use the information (see The geological cyberenvironment (gce) 85).

The system design must accommodate the following:

1. existing geological and related information in its original form and at various levels of enhancement for computer access (such as digitised maps) as well as more powerful, cyber-based implementations (see Mapping geology into the knowledge system 282)
2. representations of geological findings in terms of a system, subsystems, objects, attributes, properties, granularity, ontologies, metadata, processes, events, history, and relationships (see The multifaceted model 297, Stages of concept development (cyberinfrastructure) 45)
3. a framework to structure multi-dimensional links to related areas, rock types, stratigraphical position, modes of formation or deformation, granularity (scale), and provenance (see The geological framework model 105)
4. a range of surveying or investigational models, such as measured sections, dynamic and static stratigraphical and other spatial models, various forward and inverse models, complex models, threads of reasoning (see The multifaceted model 297)
5. an appropriate range of modes of representation (see Integrating information types 62)
6. a database of modules, such as minimum revisable units, each with metadata providing information such as:
    6.1. constraints on interpretation and modes of analysis (see The geological infrastructure model 117)
    6.2. ontological control of terminology and semantic links (see The system framework 310, The geological framework model , Other geological initiatives 343 – GeoSciML 345)
    6.3. version control, with access by default to the current approved version of minimum revisable units of the system, or optionally to an archive of superseded approved

---

[137] Paradigm: The set of common beliefs and agreements shared between scientists about how problems should be understood and addressed (Kuhn, 1962).

[138] Tacit knowledge: Knowledge which is acquired through practice and is not or cannot be articulated explicitly.

[139] Provenance (of information): The source, origin or derivation of items of information, which might be formalised in terms of, for example, project, originator, date, place, collection method, archive or database identifier, authorisation.

versions to examine the knock-on effects of changes to areas surveyed at different times (see From document orientation to systems orientation 119)

6.4. links to and from the module (see From document orientation to systems orientation 119)

6.5. provenance, including a business profile and workflow indicating who did what, with which, when and why (see The geological business model 93), methods of evaluating and tracking opinions, interpretations, and data (see The geological investigation model 98)

6.6. access (see The geological cyberenvironment (gce) 85) to quantitative methods of analysis, generalisation, interpolation, simulation and visualisation, that can work together with one another and with the geologists visual assessment, emphasising properties invariant under geological processes (see The geological investigation model 98)

6.7. standard methods for defining and interpolating boundaries (see Boundaries: discontinuities and zero-crossings 264 and The geometry of interpolation 243)

7. a means of distinguishing between knowledge of existence, location and form of located objects (see Ambiguity and map representation 148)
8. methods of reconciling conflicting information (see Reconciliation 186)
9. predictive mechanisms relating reasoning and evidence, including predictive reinforcement learning and generalisation, and multi-resolution interpolation (see The geological investigation model 98, Geological surveying as reinforcement learning 79, The sEsm as a predictive machine 76)
10. data-intensive procedures, such as simulation, and analysis of remotely sensed data (see The sEsm as a predictive machine 76)
11. the means to measure the power and accuracy of prediction, generalisation, and relevance of responses to service requests (see The geological business model 93, Geological surveying as reinforcement learning 79)
12. the means to ensure that all relevant information can be brought to bear on each requirement (see The geological infrastructure model 117)

The geological cyberenvironment (gce)

<<Table of contents 1
<<The emerging geoscience knowledge system 42
<<The solid Earth systems model (sEsm) 71
The geological cyberenvironment (gce) 85
Objectives of the geological cyberenvironment 86
Scope of the gce 87
Structure of the geological cyberenvironment 88
Design requirements for the gce 90
>>The geological business model 93

Cyberenvironments (*Figure 15*) are "a means of enabling research communities to exploit the resources available on the internet... providing an integrated set of hardware, software tools, and services needed to marshal information resources and analyze, visualize, and

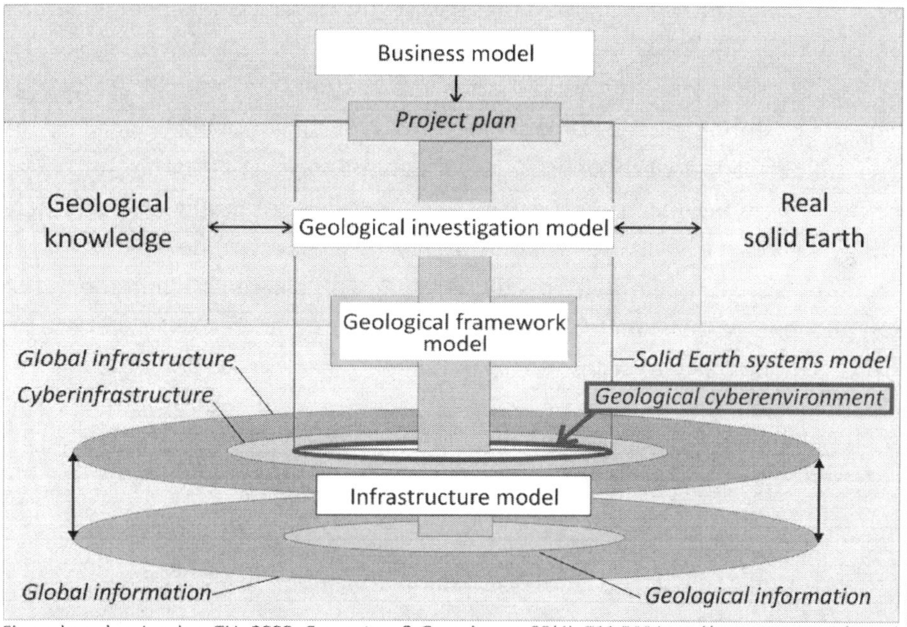

Figure based on Loudon, T.V., 2009. *Computers & Geosciences*, **35**(4), 700-705 http://nora.nerc.ac.uk/7258/

Figure 15: Four models in the geoscience knowledge system (duplicate of Figure 6*). The geological cyberenvironment is outlined in red.*

model phenomena of interest" (US National Center for Supercomputing Applications, 2008)[140]. The geological cyberenvironment (gce) would provide an integrated view of the relevant infrastructure accessed through a user interface that matches the users' working practices, and the familiar concepts and methods of geology as described in The solid Earth

[140] US National Center for Supercomputing Applications, 2008. NCSA 2010: The future of NCSA. http://www.ncsa.uiuc.edu/AboutUs/NCSA_2010.pdf

systems model (sEsm) 71. It should ease the task of making sense of a great diversity of relevant information. The cyberenvironment should assemble aspects of the infrastructure to provide end-to-end support in geological investigations; emphasise integration, support and automation of the working practices of geological investigators rather than the standardisation of software components; and present the knowledge base and associated tools to users as if centred on their own current interests.

An advanced cyberinfrastructure for geological applications requires a shared framework (see The solid Earth systems metamodel (sEsmm) 110) to structure relevant information as a component of a more comprehensive knowledge system (see The geological infrastructure model 117) and provide support corresponding to the cycle of Phases of investigational activity 103.

Objectives of the geological cyberenvironment

<<The geological cyberenvironment (gce) 85

The cyberenvironment provides application services, acting as a gateway between the investigator and the infrastructure, providing the mechanisms that assist geologists in collecting, processing, representing and sharing information from geological investigations. The focus of representing the results of geological surveying is moving from documents to objects and systems, and is becoming less dependent on formal partitioning into map sheets, map explanations, papers in specialist journals, field notes and datasets. Instead, the results can be based on a structured framework linking inclusive, wide-ranging representations, such as: hypermedia accounts of geological objects[141] and their attributes, properties, and relationships; metadata[142] and ontologies[143]; processes[144]; workflows[145]; algorithms[146]; simulations[147]; visualisations[148]; models[149]; and indexes.

The cyberenvironment aims to focus on information of specific interest to the individual user. The new framework should appear simpler and more powerful than the old, because it

---

[141] Objects: Representations of real-world or conceptual things or entities of interest in a particular context.
[142] Metadata is a description of data that is structured to assist the user or computer to find, manage, control and understand the data.
[143] Ontology: A formal representation and shared vocabulary describing concepts, entities and relationships in a domain of knowledge, typically providing a more detailed and rigorous machine-readable specification than a thesaurus or taxonomy.
[144] Process: A particular course of action intended to achieve a result, or a series of natural occurrences that bring about change.
[145] Workflow: The representation of a process or procedure in terms of a sequence of operations to be carried out to complete a task.
[146] Algorithm: A formal set of rules or instructions that can be followed to solve a problem or perform a specific task, such as the instructions of a computer program.
[147] Simulation: Imitation of aspects of internal processes of a system and their results; usually to visualise, statistically compare with, or predict real-world occurrences.
[148] Visualisation: Transforming quantitative data (including the results of interpolation) into sensory information – images that the eye and brain can interpret and visualise.
[149] Model: A formalised representation giving a simplified view of aspects of the real (or of an imaginary) world relevant to the purposes in hand.

can represent geological thinking more exactly; the mechanical tasks can be handled by the computer; and much of the complexity can be hidden from the user. Unfortunately, the simplicity may not be obvious for some time as we learn to design, build and use a radically new system (see The solid Earth systems model (sEsm) 71). The technology must support systems where distributed information sources work together (interoperate) to contribute to a broader and more informative view, facilitating collaboration among topics, organisations and disciplines. The bandwidth of distributed services, which connect knowledge users and creators with hypermedia repositories, must be sufficient to support data-intensive and simulation-based activities (see Remodelling the map 74).

Scope of the gce

<<The geological cyberenvironment (gce) 85

In contrast to the comprehensive scope of technology and science, the inevitably limited experience and brain power of an individual or a team of scientists restrict their knowledge, and call for specialisation (*Figure 16*). Individuals and information communities therefore

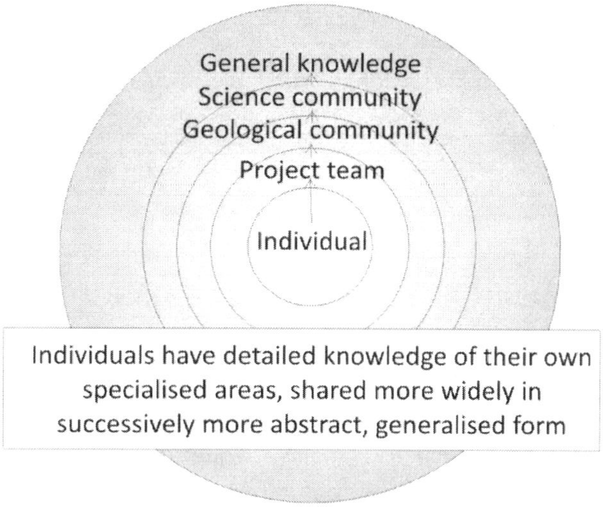

Figure 16: Supporting specialised knowledge.

require focused access to information. Cyberenvironments can help to reconcile the limitations of the human mind with the unbounded Web of Data (see Workflows, collaborative networks and Linked Data 53). Each cyberenvironment is designed to assemble aspects of the cyberinfrastructure relevant to a particular field of enquiry, aiming to maintain global compatibility while providing access through interfaces that match users' working practices. It could portray the knowledge base and associated tools to each group of users as though it were centred on their own current interests. It emphasises integration, and end-to-end support of the workflow of processes of investigation rather than standardisation of software components. Nevertheless, interfaces between the system components must be carefully defined to achieve interoperability.

The US National Center for Supercomputing Applications (2008[150]) describes cyberenvironments as a means of enabling research communities to exploit the resources available on the internet. "Cyberenvironments will provide a broad range of capabilities to scientists and engineers, from executing, monitoring, and analyzing simulations to searching distributed databases to extracting features and analyzing data from sensor arrays—all the while providing the means to interact with colleagues around the world and access the relevant literature and databases. They will be tailored to allow researchers and educators to interact with the cyberinfrastructure using concepts and approaches familiar to their specific scientific or engineering discipline." A relevant example is the Water and Environmental Research Systems Network (WATERS) (see Finholt and van Briesen 2007[151]). One role of the cyberenvironment is to act as an interface connecting two major systems: the knowledge system studied by information scientists and e-scientists, and the systems of the solid Earth studied by geologists (see The geological infrastructure model 117 and The solid Earth systems model (sEsm) 71), requiring collaboration between specialist groups.

Structure of the geological cyberenvironment

<<The geological cyberenvironment (gce) 85

A typical project might conform to standard practice while repeatedly traversing some or all of the following steps (*Figure 17*), mirroring those in geological investigation (see Overview of the geological investigation model 34, Phases of investigational activity 103):

Retrieve: select and retrieve existing relevant information
Assemble: organise, combine and reconcile it with background knowledge
Predict: on the basis of the assembled knowledge, clarify expectations concerning the properties of interest, as predictions based on interpolation and simulation
Acquire new knowledge: observe the solid Earth by eye and instruments, to test, reinforce, amend and extend the expectations
Interpret: classify observations, visualise, analyse, integrate, reason, and explain the conclusions
Share: evaluate, reconcile[152] with standards, codify[153], review, record, and communicate the results
Conform: all the steps must conform to the standards, framework, and ontologies.

---

[150] US National Center for Supercomputing Applications, 2008. NCSA 2010: The future of NCSA. http://www.ncsa.uiuc.edu/AboutUs/NCSA_2010.pdf

[151] Finholt,T., Van Briesen, J., 2007. WATERS network cyberinfrastructure plan. http://www.watersnet.org/docs/CyberinfrastructurePlan.pdf

[152] Reconciliation: Kent (1978, pp. 202-203) points out that people have different views of reality, and that these change with time. But the views overlap and so can be *reconciled* with varying degrees of success to serve different purposes. "By reconciliation, I mean a state in which the parties involved have negligible differences in that portion of their world views which is relevant to the purpose at hand."

[153] Codify: Create a representation or record of something in a form appropriate to the organised system of which it becomes a part.

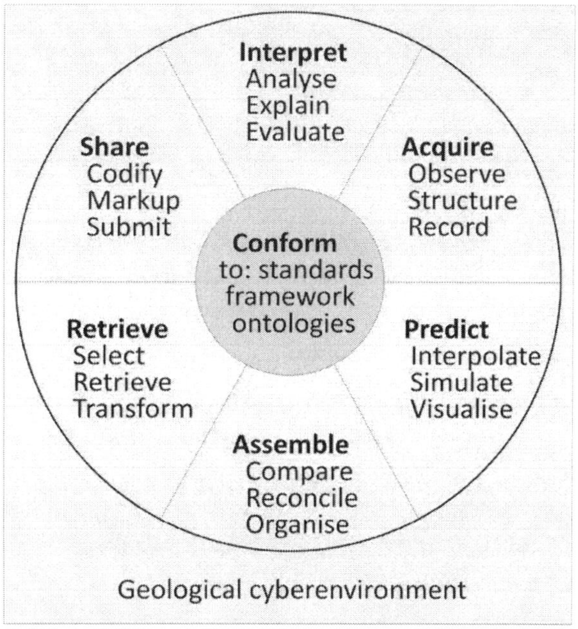

Figure 17: Stages of investigation, repeated frequently in whole or in part during a project

Each stage of the cycle makes its own demands, and the cyberenvironment must respond with appropriate facilities. A systems approach with well-defined and widely accepted subsystems and interfaces can assist efficient development and wider sharing of software. The significance of each stage depends on the nature of the investigation. The positioning of information in a holistic systems context must reflect the workflow throughout the stages of investigation.

Information may be shared at various stages of an investigation. For example, oil exploration emphasises exchange of geoscience data, such as seismic records and downhole logs, moving direct from observation to sharing, interpretation being the responsibility of the recipient. Similarly, geochemical analyses (observations) are exchanged through databases. The results of geological field survey, on the other hand are largely shared at the level of a completed interpretation (as maps and text explanations). The user of shared information must be able to determine its position in the project workflows of investigation and reasoning.

A useful first step in clarifying the structure of the cyberenvironment would be to compile a list of requirements for a future geological cyberenvironment at each stage of geological investigation (in the context of a model of solid Earth systems). This could help to clarify and integrate the network of existing computer applications and conventional methods (including interpretations and databases), and to specify future developments. It would give an appropriate basis for developing and assembling cyberenvironment software. Duplication of effort would be reduced by defining responsibilities for maintaining each subsystem and its interfaces.

Design requirements for the gce

<<The geological cyberenvironment (gce) 85

An achievement of geologists has been to depict the near-surface spatial distribution of stratigraphical units and structural features on maps at various scales throughout the world, and provide text descriptions of the material, its properties and its geological history, together with the underlying reasoning, interpretation, evidence and practical implications. Their training enabled them to take for granted the infrastructure (of instruments, methods, concepts, assumptions, representations, standards, businesses, procedures and mechanisms for collecting, storing and sharing information) that supports their science and defines the environment in which it operates. Advances in the cyberinfrastructure are introducing additional methods of investigation and representation of the results. These are listed at the end of this section and discussed in The future geological map 125.

The cyberenvironment support is determined by the investigators' potential needs (see Design requirements for the investigation model 104), and the background knowledge of the investigator is the basis for understanding and communicating with the geological cyberenvironment. A cycle of six phases of investigation is shown in *Figure 18*, and described in the Structure of the geological cyberenvironment 88. The phases marked *share*, *retrieve* and *assemble* centre on the knowledge system. The phases marked *predict*, *acquire* and *interpret* centre on solid Earth systems. The core labelled *conform* refers to the standards, framework and ontologies shared by both systems, to achieve interoperability between them.

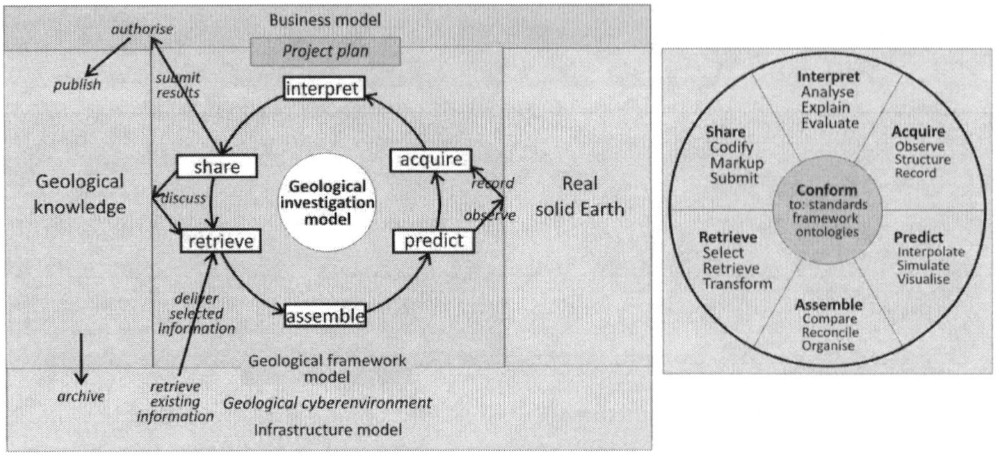

Figure 18: The cyberenvironment aims to provide end-to-end support for the user through the cycle of phases of investigational activity. (Duplicates of **Figure 22** *and* **Figure 17**).

Users are likely to carry out their initial search for information with widely used search engines. Searching spatial and relational data, however, may require additional specialised searching based on GIS and DBMS facilities. Within a systems model, such as The solid Earth systems model (sEsm) 71, these are augmented by hyperlinks within the model, which should provide access to authorised records (in both conventional and hypertext format) relevant to the geology of the area under investigation and related topics. Obtaining the relevant information is subject to availability, pay-walls, and confidentiality. Transforming existing information for a specific application, by generalising and summarising text, spatial or quantitative data, or converting it to a new frame of reference, requires specialist facilities within the geological cyberenvironment.

Most existing information is organised conventionally (whether digitised or on paper), and where relevant, it could be indexed within the sEsm and referred to by users in the light of their own background knowledge. Many geological surveying organisations have extracted and transformed information from conventional records, for example from digitised geological maps and map explanations, as specific geological objects that could be represented directly in the sEsm. Although it is desirable to bring such information into conformance with standard (preferably generic) ontologies, semantic techniques (see Semantic Web and Grid 50) make it possible to link to specific ontologies, for example, in other languages, where other conventions have been followed (see Other geological initiatives 343 – GEON 345) or for obsolete terminology used in old records.

Although the investigation model must link to information in conventional representations, it should also introduce methods that enable surveyors to represent their knowledge more exactly during the field survey process (see Kessler et al., 2009b[154]) and to represent their concepts, procedures and findings more comprehensively. It should address the needs set out by Kessler et al. (2009a[155]) for common software standards in the development of an environmental modelling platform (see Some related initiatives 341 - OpenMI 353) to support a comprehensive subsurface management system. Aspects to be considered in the design of the geological cyberenvironment include:

1. provide access from the sEsm through ontologies and indexes to conventional as well as hypertext records relevant to the geology of the area under investigation (see From document orientation to systems orientation 119)
2. where appropriate, structure the information as object-oriented views of instances, classes and inheritance (see An object-oriented approach 178)
3. establish procedures to evaluate and authorise information and assess the value of other lines of investigation (see The geological business model 93)
4. handle information in three dimensions at all levels of detail, including the ability to extend, modify and refine the existing view of the geology in response to new data or simulations of the surveyor's alternative interpretations, and render and visualise the effects of simulated smaller-scale processes (see A wish list for integrated geometry 221)
5. link semantic, algorithmic, narrative, spatial and tabular information at all scales to and from the objects to which they refer (see Integrating information types 62)
6. link hypertext representations of narrative threads of evidence, description, reasoning and interpretation through the chains of objects to which they refer (see

[154] Kessler, H., Mathers, S., Sobisch, H.-G., 2009b. The capture and dissemination of integrated 3D geospatial knowledge at the British Geological Survey using GSI3D software and methodology. *Computers & Geosciences,* **35**, 6, pp. 1311-1321. http://nora.nerc.ac.uk/7207/1/Kessler_CG_GSI3D_article_final.pdf

[155] Kessler, H., Campbell, D., Ford, J., Giles, J., Hughes, A., Jackson, I., Peach, D., Price, S., Sobisch, H-G., Terrington, R., Wood, B., 2009a. Building on geological models : the vision of an environmental modelling platform. In: *Geological Society of America Annual Meeting 2009, Illinois, USA, 18-21 Oct 2009*. Illinois, USA, Geological Society of America, pp. 24-30. http://nora.nerc.ac.uk/8423/

A framework for the reasoning 135, Microdocuments and the threads of reasoning 190)

7. record the provenance, design and workflow of the investigation as it proceeds, including workflow-based recording of observations and the methods and reasoning that link them to the interpretation (see Workflows, collaborative networks and Linked Data 53, Broadening the framework 291)
8. provide the field surveyor with facilities for electronic surveying, visualisation, calculation, statistical analysis, data-basing, interpolation, GPS, GIS and text handling (see Benefits of an object-oriented system 199)
9. visualise the interpretation of the near-surface geology superimposed on the landscape visible to the surveyor, as in augmented reality (see The importance of space and visualisation 169)
10. provide links from the geological interpretation to annotated, remotely sensed images that provide or clarify evidence (such as landforms, outcrops) (see Object-oriented survey 195, The geometry of interpolation 243, The field survey model 300)
11. relate the observed configuration of the geology to historical configurations and processes (see The field survey model 291)
12. identify the ranges of resolution in scale-space at which observations were made, processes operate and object classes exist (see Zoom 247, Scale-space 255)
13. provide more natural continuous scale change with Gaussian filters (see Scale-space 255)
14. select object boundaries at reproducible zero-crossings (see Boundaries: discontinuities and zero-crossings 264)
15. fill space by justifiable and reproducible procedures, such as process-based interpolation methods and deformable inhomogeneous spatial models (see The role of the dynamic model 287, Broadening the framework 291)
16. for objects and configurations, record spatial relationships and shape statistics with invariant properties suited to regional comparisons and guiding interpolation (see Shape 266, Morphometrics 268)
17. use interpolation algorithms that clarify and justify the visualisation, in preference to unspecified rules of thumb for drawing lines on a map (see The geometry of interpolation 243)
18. separately superimpose knowledge of existence and form on likely position (see Diverse objectives and products 150, Spatial variation and uncertainty 241)
19. at all stages of the investigation, record levels and sources of uncertainty, distinguish where possible between observation, interpolation and interpretation, and where appropriate use Bayesian statistics to track opinions and evaluations of accuracy (see At the interface 133, The field survey model 300)

The geological business model

<<Table of contents 1

<<The emerging geoscience knowledge system 42

<<The geological cyberenvironment (gce) 85

The geological business model 93

Objectives of the business model 94

Scope of the business model 95

Design requirements for the business model 96

>>The geological investigation model 98

The management of a geological project (*Figure 19*) may be undertaken by a professional manager, or may simply be one of the activities in which the geological investigator is

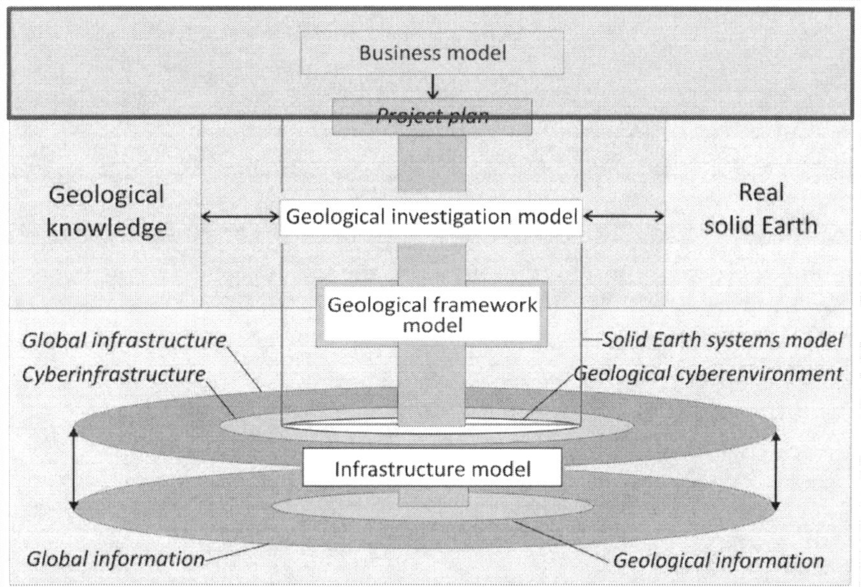

Figure based on Loudon, T.V., 2009. *Computers & Geosciences,* **35**(4), 700-705 http://nora.nerc.ac.uk/7258/

Figure 19: Four models in the geoscience knowledge system (duplicate of Figure 6*). The business model is outlined in red.*

engaged. It may or may not follow a project management methodology, such as PRINCE2 (2009[156]), used in the British Geological Survey. Regardless of who undertakes the task and how, it is likely to influence the success or failure of the investigation. Its success may be measured against its contribution to a business strategy (a coordinated plan of action deploying resources to achieve long-term objectives). Many businesses, with diverse objectives, view and assess geological information in widely different ways. Because

---

[156] PRINCE2, 2009. Projects in Controlled Environments. The Office of Government Commerce (London, UK). http://www.prince-officialsite.com/

geological knowledge can be seen as a national asset, geological surveying organisations are funded to obtain and supply widely relevant and reliable geological information. And to avoid loss or fragmentation of hard-won knowledge, regulators may require oil companies, for example, to collect and deposit (in a shared archive) a defined suite of downhole logs for wells they drill in an area.

Project profiles (see The management view 15) can be explicitly related to regions of The solid Earth systems metamodel (sEsmm) 110, such as geographical areas, classes of geological objects, stratigraphy, properties, processes, relationships and level of detail, guiding the investigational process and helping to evaluate its relevance to other applications. They are the basis for a project plan – the interface linking the business model to The geological investigation model 98. A business objective of many geological investigations (particularly in the academic and surveying areas) is to share results as part of the body of communicated knowledge. The fast changing methods of scientific publication must be considered at an early stage (see From document orientation to systems orientation 119).

Objectives of the business model

<<The geological business model 93

The business model (*Figure 20*) aims to define the objectives of a geological investigation or project, and the resources, priorities and methods for achieving them. In each project, the

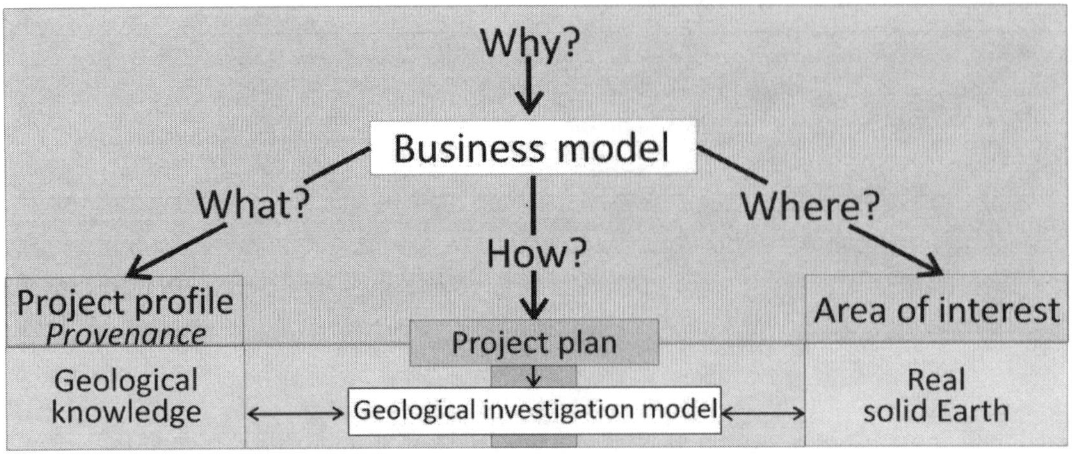

Figure 20: Business model.

business model determines the project plan: why an investigation is undertaken (the objectives), and who carries it out, how and where, thus clarifying for end-users the provenance of the results (their source, derivation and reliability). The project plan aims for efficiency (getting the most value from projects at least cost) and appropriate dissemination and application of the results. It may also address legal issues such as mandatory deposition of information, and intellectual property exploitation and protection. At a scientific level, the business model is concerned with establishing and maintaining the quality of information in the infrastructure and the integrity of the system as a whole.

The project plan is an interface between geological management and geological investigation (see Three views of systems geology 15). It is also concerned with managing the preparation, maintenance, quality and availability of appropriate documentation (see From document orientation to systems orientation 119). The project plan guides the investigational process by clarifying the relative importance to the project of topics, in terms of geographical areas, geological objects, properties, processes and relationships (which can be defined in terms of The geological framework model 105). The project profile subsequently helps to evaluate the relevance of the results to other investigations by indicating shared regions of interest, and identifying the provenance (source, derivation and reliability). In due course, the business model may be seen as a component of predictive reinforcement learning that helps field geologists to organise their procedures during an investigation to optimise results in terms of the objectives (see Geological surveying as reinforcement learning 79).

Scope of the business model

<<The geological business model 93

In collaboration with the geological surveyors, geological management defines the business model for a project (*Figure 20*). It is thus the joint responsibility of the surveyors and their managers (the same individuals may fill both roles). The investigators are responsible to management for the results of the project, and the managers must ensure that the requirements of the project have been fulfilled, the results appropriately documented, evaluated and distributed. A task of the business model is to evaluate results.

In the case of a geological surveying organisation, the aim is to supply basic geological information, rooted in an understanding of the nature, distribution, history and configuration of the rock types, to support a wide range of commercial, regulatory and research activities. It can provide a quality-assessed core of geological knowledge on which many internal and external applications can build (thus avoiding duplication of effort), and to which they can relate their results (thus adding value for all users). Evaluation of the results of geological surveying must take into account the context, the generality of their findings, and the consequent increase in predictive power in specifiable topics, resulting from the regularities they reveal in the geological record.

Unlike the published map, an on-line spatial model could in principle be brought up to date whenever new information became available. In practice, it may be preferable to limit the availability of modifications until the appropriate experts can assess the full implications of change, and the procedures of quality assessment and evaluation are complete. The development of spatial models in the field calls for the same top-down view and feedback process as traditional field mapping. It requires access to the same wide range of information, of which the relevant subset should preferably be available during fieldwork. As the cyberinfrastructure advances, it could enable geoscientists to identify sources and record decisions and reasoning while surveying, as well as visualising the consequences and implications of new evidence. The business model thus relies on a shared understanding among managers, investigators and e-scientists.

Design requirements for the business model

<<The geological business model 93

The business model will require an evolving standards framework for successful migration of authoritative geological information to a systems geology base. This is likely to affect geological survey organisations, geological departments in major oil and mineral companies and their regulatory organisations, but will also affect scientific publishers and aspects of academic geology, notably in the evaluation and publication of their findings. Agreement on standards primarily involves e-scientists, but extends to a wide range of disciplines, including geology. Geological managers, therefore, must ensure that their own specific requirements are taken into account as the standards develop. A lengthy process of experimentation is inevitable, which can be made more productive by wide collaboration (see Workflows, collaborative networks and Linked Data 53).

It may be helpful here to list some possible design requirements:

1. As documentation moves from a conventional to a systems structure, a flexible architecture for handling the various types of object is desirable (see From document orientation to systems orientation 119).
2. A project management methodology (such as PRINCE2, 2009[157]) is likely to be helpful where many separate projects must be coordinated.
3. The major components of a project profile, including provenance and metadata for revisable units in the systems documentation (analogous to bibliographical metadata) should evolve towards global standards.
4. In considering the global standards for documenting solid Earth systems, some long-established nomenclature may have to be reconsidered. For example, linking the definition of stratigraphical units to map scale imposes an unnecessary constraint on systems geology investigations (see Multiresolution survey 259).
5. A standard framework (see The geological framework model 105) and standard ontologies (see Other geological initiatives – GeoSciML 345) should be considered and adopted where appropriate.
6. In general, a large number of possible standards have been proposed in geology and in related fields (see Some related initiatives 341), and will require rationalisation as the technology matures.

[157] PRINCE2, 2009. Projects in Controlled Environments. The Office of Government Commerce (London, UK). http://www.prince-officialsite.com/

7. In the long run, the solid Earth systems metamodel[158] may assist in defining a project plan by identifying regions of interest to the particular application. The predictive model may assist by identifying what is already known, what new information is likely to be most helpful, and subsequently evaluating the significance of the results (see The sEsm as a predictive machine 76).
8. Management procedures for maintaining scientific standards by evaluation within organisations and more generally in peer review of presentation and publication of scientific findings (Heap, 2004[159], Scott, 2006[160]) require continuing review in the light of the advancing cyberinfrastructure. This will involve some aspects specific to modelling Earth systems (see From document orientation to systems orientation 119).
9. Obviously, geological management have a critical role in determining, not only how developments in systems geology will affect their own work, but also in determining its future structure and development. It is hoped that this scenario can assist.

[158] Metamodel: A metamodel is a description of the organisation and function of a model, to assist the user or computer to find, manage, control and understand its contents.

[159] Heap, B. (Chairman), et al. 2004. Peer review and the acceptance of new scientific ideas. Sense about Science, London. ISBN 0-9547974-0-X http://www.senseaboutscience.org.uk/pdf/PeerReview.pdf

[160] Scott, A., 2006. Peer review and the relevance of science. University of Sussex, SPRU Electronic Working Paper Series, No. 145.

The geological investigation model

<<Table of contents 1
<<The emerging geoscience knowledge system 42
<<The geological business model 93
The geological investigation model 98
Objectives of the geological investigation model 99
Scope of the geological investigation model 100
Remodelling geological investigation 101
Phases of investigational activity 103
Design requirements for the investigation model 104
>>The geological framework model 105

This model (*Figure 21*) considers the procedures and methods of geological investigation, as deployed to develop, test, amend and extend The solid Earth systems model (sEsm) 71. It is concerned with observing and interpreting aspects of the real solid Earth against the

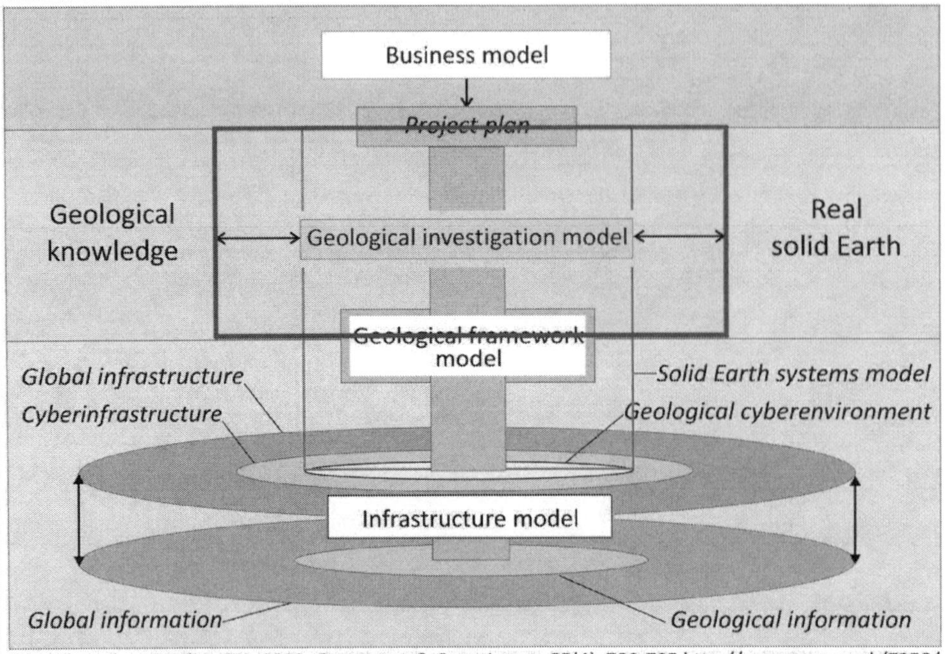

Figure based on Loudon, T.V., 2009. *Computers & Geosciences*, **35**(4), 700-705 http://nora.nerc.ac.uk/7258/

Figure 21: Four models in the geoscience knowledge system (duplicate of ***Figure 6****). The geological investigation model is outlined in red.*

background of existing geological knowledge, and recording the results. In an informal study, or in exploring geoinformatics methods in a new context, the model may lurk unexpressed in the investigator's mind and formal constraints may be inappropriate. However, the focus here is on the comprehensive longer term development of systems geology and its relevance to major projects, in particular the authoritative account of regional geology supplied by geological survey organisations. This scenario therefore emphasises the long-term potential of a model that can bring greater rigour and integration to the science, based on a systems approach that unleashes the synergy of mutually reinforcing techniques from geoinformatics. The methods and rationale are considered at length from a geological viewpoint in The future geological map 125.

A geological investigation is guided by a project plan (its interface with The geological business model 93). The project plan is likely to be flexible, as many geological investigations must proceed step by step, changing course as more is learned. The selected methods of investigation, closely tied to the standards in The geological framework model 105, should be supported by The geological cyberenvironment (gce) 85.

Objectives of the geological investigation model

<<The geological investigation model 98

The geological investigation model aims to bring greater scientific rigour throughout the cycle of phases of geological investigation (*Figure 22*), integrating information from many sources to provide a more coherent and comprehensive view. This scenario emphasises how

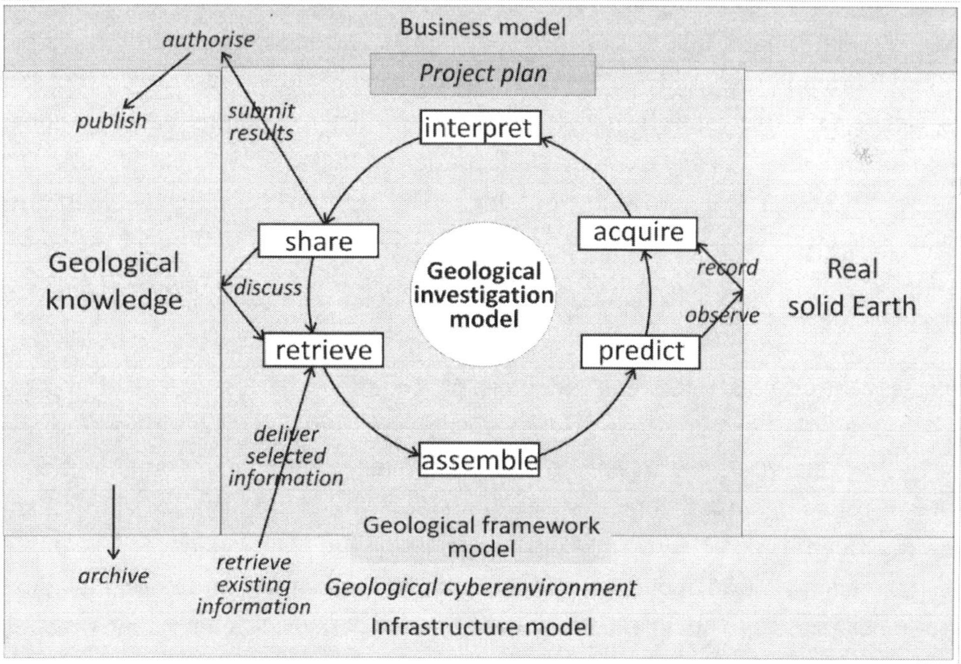

A cycle of project phases is repeated many times in whole or in part, in accordance with the project plan and geological framework model. Arrows show direction of information flow.

Figure 22: The geological investigation model.

more powerful surveying methods, described in detail in 'The future geological map 125', might be supported by the The geological infrastructure model 117 and coordinated through

The geological framework model 105. Investigators can potentially find the required tools, techniques and procedures from interoperable modules assembled in The geological cyberenvironment (gce) 85.

Scope of the geological investigation model

<<The geological investigation model 98

Systems geology extends the scope of conventional geological investigations. It can provide a framework and methods of collecting and representing information that bring fragmentary information into the context of The surveyor's holistic view 60. Unlike conventional representations, digital information can represent many interconnected items in their object-oriented, multi-resolution, three-dimensional, geological-time setting. Information types (see Integrating information types 62), such as text narrative, images, spatial representations, tabular data, algorithmic processes, statistical relationships and human judgment can be recorded and integrated for individual objects at any scale and level of detail. Observation can be separated from interpretation to achieve more rigorous analysis and handling of alternative hypotheses. For example, interpolation methods can be explicitly defined and differentiated from the data on which they operate. The methods can reflect knowledge of scale-sensitive geological processes, rather than obscure rules of thumb, and can extend to deformable inhomogeneous spatial models (see Seeking shared concepts 247) and links with remotely sensed imagery. Description can be separated from presentation for greater flexibility, for example, by developing and recording one digital spatial model, which can then be examined in many different presentations, ranging from multiple static visualisations (such as maps, cross-sections and block diagrams) to immersive reality and augmented reality. The user can follow or develop various threads of thought through interconnected items within and beyond geoscience (see Mechanisms 45, Workflows, collaborative networks and Linked Data 53).

Geologists study processes and their outcomes on a continuum from molecular to planetary scale. The cyberinfrastructure frees survey investigation from the constraint of traditional map scales. An appreciation of Scale-space 255 improves conformity of boundaries by encouraging more rigorous techniques (see Boundaries: discontinuities and zero-crossings 264) for their identification (zero-crossings), establishing the scale ranges of specific processes, relating objects at different levels of detail (ontological hierarchies), generalising (filtering visualisations to mimic the effect of viewing at various distances), and predicting and simulating the style of small-scale objects (visualised as draped over larger objects to give a realistic view that clarifies the distribution of the smaller-scale properties). Workflow recording can provide detailed provenance[161] (see Workflows, collaborative networks and

[161] Provenance (of information): The source, origin or derivation of items of information, which might be formalised in terms of, for example, project, originator, date, place, collection method, archive or database identifier, authorisation.

Linked Data 53). Opinions and evaluations of accuracy can be tracked using Bayesian statistics (see At the interface 133).

The model (see Design requirements for the investigation model 104) should match human thought processes more exactly than conventional methods. As most information is held conventionally, old and new methods must work together. Most of the newer methods have been developed in other fields, and their use in geology has largely been in exploratory academic studies. Extending their application will require collaboration among research workers in various disciplines.

Remodelling geological investigation

<<The geological investigation model 98

Cyber-technology is transforming the knowledge system as a whole, and much has been published about the potential consequences (for example, Foster and Kesselman, 2003[162]). In contrast, most accounts of geoinformatics are largely concerned with specific applications. But gaining the full benefits calls for a comprehensive review of assumptions about the methodology and role of geoscience in the overall knowledge system. The system redesign must be based on geologists' understanding of the subject matter (see Reasoning, models and reality 127 and The future geological map 125). Future developments must operate in harmony with legacy systems and human thought processes, matching geoscience knowledge and procedures to the emerging opportunities (see Mapping geology into the knowledge system 282). For example, long-established methods of mapping are an important facet of the work of many geoscientists. Automating the process of making maps has provided an insight into their significance and is an essential basis for future work.

The map is a means of illustrating the geoscientists' spatial model and linking it to the real world, not an end in itself. The spatial model makes it possible to represent the geology of part of the solid Earth, reduced to a size where its spatial configuration can readily be visualised, studied and adjusted. Digital cartography provides a more flexible representation of the geological objects and their relationships. Three-dimensional modelling tools, such as GSI3D (Kessler et al., 2009a[163]), provide computer assistance to assist geologists in applying their logical reasoning and intuitive understanding to the 3D depiction (Royse, 2010[164]). A benefit of this approach is that the unrivalled skills and knowledge of the geologist dominate, although they are assisted and supplemented by mechanical computation (see

---

[162] Foster, I., Kesselman, C. (eds), 2003. *The Grid: Blueprint for a new computing infrastructure,* 2nd ed. Morgan Kaufmann, San Francisco. 748pp.

[163] Kessler, H., Campbell, D., Ford, J., Giles, J., Hughes, A., Jackson, I., Peach, D., Price, S., Sobisch, H-G., Terrington, R., Wood, B., 2009a. Building on geological models : the vision of an environmental modelling platform. In: *Geological Society of America Annual Meeting 2009, Illinois, USA, 18-21 Oct 2009.* Illinois, USA, Geological Society of America, pp. 24-30. http://nora.nerc.ac.uk/8423/

[164] Royse, K.R., 2010. Combining numerical and cognitive 3D modelling approaches in order to determine the structure of the Chalk in the London Basin. *Computers & Geosciences,* **36**, 500-511.

Unexpressed knowledge 63). Quantitative knowledge derived from observation, theory and experiment might also be included in the longer run, but under the control of an experienced geologist with understanding of its relevance, limitations, and wider consequences.

To guide applications of fast-changing technology, geoscience surveyors must also look beyond the map – at their underlying objectives, methods and models, all of which could be taken for granted in a more settled system. Some existing computer applications that were formalised in isolation may have to be reviewed as components of a more general system (see Seeking shared concepts 247). As providers of an authoritative view of regional geology, geological surveying organisations have a key role in this aspect of the transition to systems geology.

Geological surveying is based on available existing knowledge. It is a process of selectively observing the properties and relationships of located geological objects, bringing appropriate representations of them into a broader knowledge structure, and testing the consequences (see The surveyor's holistic view 60). The surveying process operates at the interface between the real world (or at least that part of the solid Earth accessible to observation by eye or instrument) and the system of geoscience knowledge (which guides the process and is enriched by its results). The procedures of surveying involve Abstracting from reality to model 131 (to reduce the volume of information while retaining the salient points), prediction (to fill the gaps between observations, see The geometry of the spatial model 203), and validation (to test the results, see At the interface 133). The procedures are repeated many times in various sequences in the course of an investigation (see Phases of investigational activity 103 and Figure 22).

The chain of understanding is more complete in the minds of geologists than in its representations, for they see and know more than they can record. Much tacit knowledge can be shared only by demonstration and not as formal records, but may be a vital part of procedures such as stratigraphical correlation. The results are constrained by the fragmentary nature of the geological record, the need to infer the process from the product, the complications of historical geology, and the inherent unpredictability of complex systems (see The imperfect model 156). Nevertheless, the field geologist can successfully fill gaps between observations and interpolate boundary lines on the map to reflect a particular interpretation.

Individual geoscience projects follow their own procedures for collecting and abstracting information, choosing the salient features that meet their specific objectives. However, rather than each project starting from scratch, it may be more efficient to build on the shared foundation that information communities, such as Geological Surveys, can provide. The Survey's products therefore aim to provide a consistent and coherent base of knowledge of optimal value in a wide range of applications. Systems geology will fundamentally change the nature of both the knowledge base and the applications. It offers the prospect of more powerful methods of geological investigation, and the representation of the geologists' conclusions as an integral part of a wider cyber-based knowledge system (see From document orientation to systems orientation 119)

Phases of investigational activity

<<The geological investigation model 98

It is convenient for ease of reference to group the phases of investigation as a cycle, such as that illustrated in *Figure 23*. This can also be the basis for locating support for individual tasks in The geological cyberenvironment (gce) 85 (see *Figure 6*). The investigation model (defining the geological requirements) and the cyberenvironment model (providing the supporting infrastructure) must work in close conjunction throughout the cycle of investigation phases (see Design requirements for the gce 90, Structure of the geological cyberenvironment 88).

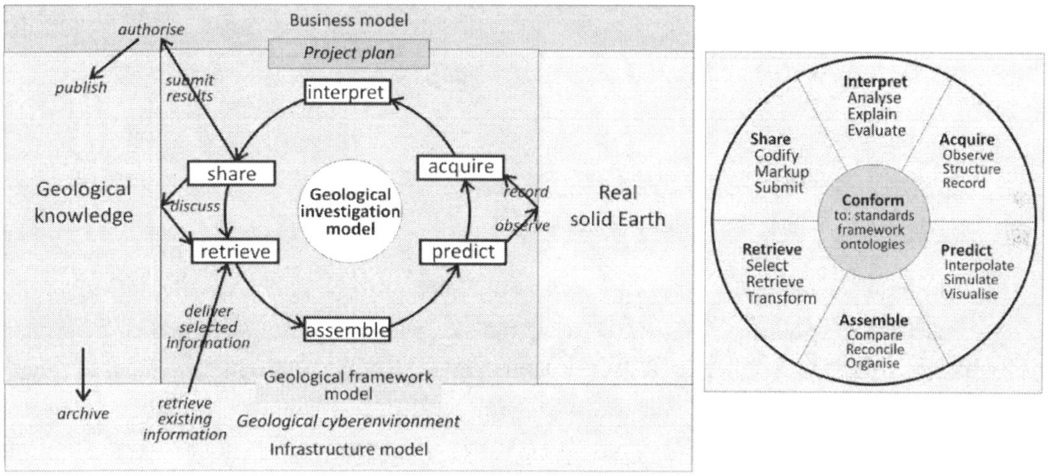

*Figure 23: Duplicates of **Figure 22** and **Figure 17**. The cyberenvironment aims to provide end-to-end support for the individual user through the cycle of phases of investigational activity.*

Geologists might start their investigation with the phase of **retrieving** existing information (lower left in *Figure 17*), selecting what may be relevant, obtaining it from available sources, and transforming it if necessary to meet their specific needs (for example, by re-plotting data gathered from various websites). They might then **assemble** the retrieved information, organising it as required for their own needs, comparing different accounts, and trying to reconcile different views. On the basis of this material, they could form a view **predicting** what they expect to find on the ground. In their minds or in rough sketches, they might interpolate[165] from fragmentary information, simulate[166] and visualise[167] the results.

In the field, with these expectations in mind, they might seek to confirm, refute, amend or extend them by **acquiring** more information by observation – observing, structuring and recording what they saw (see Reasoning, models and reality 127). They might **interpret** the

[165] Interpolation: The estimation of values, for example at a point or along a line or surface, in order to predict a value or complete a visualisation.

[166] Simulation: Imitation of aspects of internal processes of a system and their results; usually to visualise, statistically compare with, or predict real-world occurrences.

[167] Visualisation: Transforming quantitative data (including the results of interpolation) into sensory information – images that the eye and brain can interpret and visualise.

results in the light of their background knowledge, analysing, explaining and evaluating their conclusions. They might **share** what they had learned, initially by discussion with colleagues working on the project. Parts or all of this cycle of activities are repeated many times at different levels of detail in the course of investigation (see Design requirements for the gce 90). In due course, the information could be more formally codified to record the results in appropriate representations, and marked up to record hypertext links. The authors might then submit the results for evaluation by management who might authorise their publication, archived as a component of global information (see *Figure 22* and The geological business model 93). In a formal cyber-based investigation, for example in a geological survey organisation, all phases would be expected to conform (where appropriate) to the standard framework and ontologies. The geological cyberenvironment (gce) 85 should provide the required cyberinfrastructure support.

Design requirements for the investigation model

<<The geological investigation model 98

Systems geology can provide a more powerful approach to geological investigation by integrating concepts and methods within a formal system. The geological framework model 105 proposes a tentative structure (The solid Earth systems metamodel (sEsmm) 110) where they can work together within a uniform pattern of systems documentation (see From document orientation to systems orientation 119) to create The solid Earth systems model (sEsm) 71.

A formal geological survey investigation might begin by retrieving existing relevant information. This systems geology scenario suggests that this should then be assembled within a framework appropriate to the project plan, recorded as the start of the workflow. The assembled interpretation cannot at any stage be regarded as a factual account of all aspects of the selected fragment of the solid Earth. Rather, it is an interpretation based on what is known, and predicting what was not observed. The surveying process can be regarded as a reinforcement learning process, which modifies, amends and extends the current view or prediction as the investigation proceeds (see Geological surveying as reinforcement learning 79, The sEsm as a predictive machine 76). The future geological map 125 suggests that the surveyor might in future map the spatial concepts, not onto two-dimensional sheets of paper, but into a multi-dimensional, multi-resolution framework. The power of Unexpressed knowledge 63 suggests that this must be undertaken by an experienced geologist who is fully in control of the process and aware of the implications.

Some potential developments in the investigation model are listed in the Design requirements for the gce 90. They should also be kept in mind during the investigational design.

The geological framework model

<<Table of contents 1
<<The emerging geoscience knowledge system 42
<<The geological investigation model 98
The geological framework model 105
Objectives of the geological framework model 106
Scope of the geological framework model 107
Remodelling the systems framework 108
The solid Earth systems metamodel (sEsmm) 110
Linking beyond the sEsm 115
>>The geological infrastructure model 117

The framework model (*Figure 24*) is a structure that depicts and clarifies the principal relationships in The solid Earth systems model (sEsm) 71 in order to organise and assemble dispersed information relevant to geology. It is a multidimensional map connecting geological thinking and its computer representation. It should help to locate and integrate

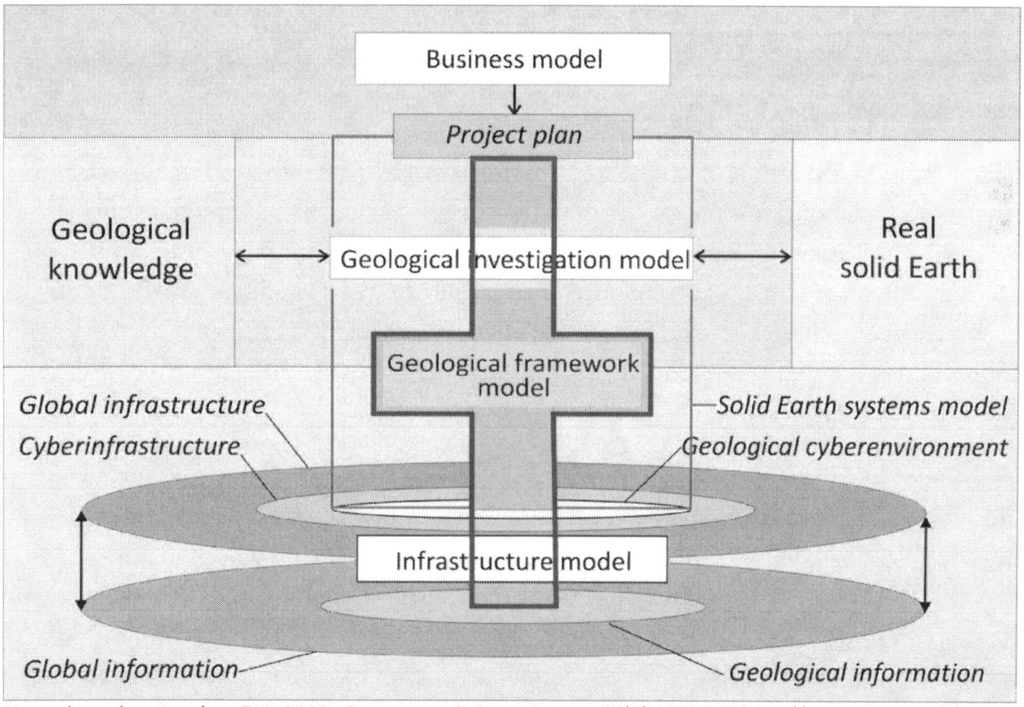

Figure based on Loudon, T.V., 2009. *Computers & Geosciences*, **35**(4), 700-705 http://nora.nerc.ac.uk/7258/

Figure 24: Four models in the geoscience knowledge system (duplicate of Figure 6*). The geological framework model is outlined in red.*

ideas, concepts, workflows of investigation, and threads of reasoning that are shared among the various models of the geoscience knowledge system. Aspects of the framework refer to the geographical disposition and configuration of the geological objects in the solid Earth systems model; their observed and interpreted properties, composition and relationships, at all scales; geological processes and their interaction with pre-existing configurations of objects, and the resulting events and historical changes through geological time.

The framework[168] model should ensure that all aspects of The solid Earth systems model (sEsm) 71 and its supporting software can work together through shared standards. The geological framework model interfaces with The geological business model 93 through the project plan. It interfaces with all aspects of The geological investigation model 98 that conform to the standard framework and ontologies (see *Figure 6*). It interfaces with The geological infrastructure model 117 through The geological cyberenvironment (gce) 85.

Objectives of the geological framework model

<<The geological framework model 105

The framework model aims to provide a shared, explicit structure to link geological thinking with its computer representations. Its intended purpose is to provide a map of content and terminology so that users and systems developers know where to put information, where to find it, and how to handle and process it. Important objectives are: to enable the business, investigation and infrastructure models to work together; to simplify and coordinate the interface to applications in The geological cyberenvironment (gce) 85; and to strengthen the support for quantitative analysis.

To meet these objectives, it attempts to formalise and clarify stable and uncontested elements and relationships among the findings of geology. It reflects aspects of the scientific paradigm,[169] described by Kuhn (1962[170]) as: 'universally recognised scientific achievements that for a time provide model problems and solutions for a community of practitioners... a map whose details are elucidated by mature scientific research, and since nature is too complex and varied to be explored at random, that map is as essential as observations and experiment to science's continuing development.'

As a result of their training and experience, geologists have some such structure in their minds. It is given concrete form by decisions on what to investigate in the solid Earth and how, which standards to follow (as in taxonomies and indexes) and where to publish (as in geological map series and specialist journals). The radical changes to the infrastructure now call for a more extensive, explicit framework. The framework should aid collaboration and shared understanding by reinforcing convergence on a set of underpinning ideas which

[168] Framework: A logical structure and guidelines giving a broad overview for classifying and organizing complex information, within which detail can be added as required.

[169] Paradigm: The set of common beliefs and agreements shared between scientists about how problems should be understood and addressed.

[170] Kuhn, T.S., 1962. The structure of scientific revolutions. The University of Chicago Press, Chicago, 172 pp.

constrain and define the system. It should assist in Reconciliation 186 of information from a wide range of dispersed sources and encourage standards to support interoperability.

Scope of the geological framework model

<<The geological framework model 105

A framework should reflect a shared paradigm, providing a solid, stable base for normal science (see An overview of systems geology 12), and is necessarily resistant to change. The geological framework model is seen as linking the solid Earth systems metamodel to ontologies[171] and to indexes[172] that connect to appropriate information in distributed information stores. The framework should provide The solid Earth systems model (sEsm) 71 with the potential to link diverse information at all granularities (levels of detail or resolution). The geological cyberenvironment (gce) 85 relates to that framework to provide the user with services based on the cyberinfrastructure. Unlike the stable framework, the cyberenvironment reflects the growing and ever-changing methods of scientific investigation and communication of the results. But, like biology and ecology, the framework and cyberenvironment intertwine and must develop together, with well-defined systems interfaces, and procedures for evaluating and accepting change.

The framework and associated ontologies must reflect a distilled consensus of expert views, in the spirit of the Stratigraphic Guide: "...agreement on stratigraphic principles, terminology, and classificatory procedure is essential to attaining a common language of stratigraphy that will serve geologists worldwide. It will allow their efforts to be concentrated effectively on the many real scientific problems of stratigraphy, rather than being wastefully dissipated in futile argument and fruitless controversy arising because of discrepant basic principles, divergent usage of terms, and other unnecessary impediments to mutual understanding" (Hedberg, 1976[173], page v). The framework is thus the concern of the geological community as a whole and, to some extent, of associated disciplines.

---

[171] Ontology: A formal representation and shared vocabulary describing concepts, entities and relationships in a domain of knowledge, typically providing a more detailed and rigorous machine-readable specification than a thesaurus or taxonomy.

[172] Indexing: The intellectual analysis of the subject matter of a document to identify the concepts represented in the document and the allocation of descriptors to allow these concepts to be retrieved.

[173] Hedberg, H.D. (editor), 1976. International stratigraphic guide: a guide to stratigraphic classification, terminology and procedure. Wiley-Interscience, New York.

Remodelling the systems framework

<<The geological framework model 105

Most geological knowledge is held in the minds of geologists and most geological information is held in conventional publications. The reader of a conventional publication is assumed to have background knowledge of general aspects of the science, and of the specialist area to which the publication refers. Relevant information from other sources may be repeated to establish context, probably citing and possibly quoting from existing publications. Information added to the formal literature is thus embedded in a sequence of previously existing information, and its significance may be lost if it is taken out of context (see A framework for the reasoning 135). In effect, readers of a scientific paper are each being guided by the author in building in their own minds small sub-systems of knowledge, influenced by their own interests, requirements, background knowledge and prejudices. Each reader thus has to identify the nuggets that interest them in a multitude of documents, and extract them from the various contexts of the authors' interests.

As Licklider (1960[174]) pointed out: "Any communication between people about the same thing is a common revelatory experience about models of that thing" (see Models and frameworks 47). He suggested that the clerical and intellectual tasks of locating, extracting and digesting information from many sources could be helped by a symbiotic partnership between human brains and computers, mediated by a shared model accessible to both the human user and the machine. The solid Earth systems model (sEsm) 71 is proposed as a shared model for viewing the solid Earth as a system (see Objects, ontologies and systems 48).

Geological knowledge is inevitably imperfect, exploration of diverse views is essential for evolution of the science, and acceptance of alternative ontologies[175] is necessary, as information has been assembled from many sources over many decades. However, geological survey organisations throughout the world have long been using their local knowledge to assess all available sources and test them in the field to provide a coherent, authoritative, widely useful, standardised view of the regional geology. This body of geological information is over-printed on topographical maps, giving a direct link to the real-world entities to which the map refers. In several geological surveying organisations they have been supplemented by three-dimensional spatial models, object-oriented views of the map content, and digital information recording in the field. Techniques from geoinformatics are individually applied to assist in analysing and presenting the information.

The results of geological surveying are central to the systems geology scenario, and could provide an essential core of information as a starting point for a solid Earth systems model. But they remain in a pre-digital framework that cannot fully reflect the holistic, systems view of the geologist. Whereas the map positions spatial information (separated from explanatory text narrative) in a two-dimensional framework, the sEsm can integrate

---

[174] Licklider, J.C.R., 1960. Man-computer symbiosis. *IRE Transactions on Human Factors in Electronics,* vol HFE-1, p. 4–11 (March 1960). http://memex.org/licklider.pdf

[175] Ontology: A formal representation describing concepts, entities and relationships in a domain of knowledge, typically providing a more detailed and rigorous machine-readable specification than a thesaurus or taxonomy.

information types[176] and hierarchies of objects in a three-dimensional, geological time, multi-resolution framework. The construction of an sEsm aims to bring geological investigation into a more comprehensive setting where:

- the emphasis is on geological objectives and obtaining more powerful, rigorous and efficient results by means of a systems view supported by the cyberinfrastructure
- mutually supporting geoinformatics techniques are assembled to work together in parallel with the user's thinking, as one accessible, coherent, synergistic system
- detailed accounts of local geology can be developed in their wider context of global geological observations, processes and history

Inevitably, however, geological investigations are fragmentary, matching the geographical areas and topics of interest defined by the business plans of their originators. The diversity of sub-systems reflects the diversity of applications. For example, the aspects of the Earth that are relevant in oil exploration differ (with significant overlap) from those of interest in civil engineering or in the history of effects of climate change in the geological past. Sharing the information requires knowledge of its provenance[177], as well as conformance with agreed protocols and standards. Organising the information requires a structure (such as The solid Earth systems metamodel (sEsmm) 110) that defines the primary sub-systems and their relationships, developed by the geological community within the broader framework of a whole-Earth model.

The core ontology, structure and framework of the sEsm must follow well-defined and widely accepted standards, in order to make sense of the diversity of information. It must also be extensible to accommodate relevant (but unforeseen) types of geological objects, attributes, relationships, properties, processes, events and processing techniques, enabling geologists to supplement and refurbish existing knowledge and carry it forward by migrating in step with The geological cyberenvironment (gce) 85 and The geological infrastructure model 117. It must build on conventional representations and existing concepts.

A federated structure can reach out to share information in collaborative networks linking activities in many specialist fields, while maintaining a core of widely relevant material. An sEsm could conform to such a structure through its metamodel[178] (see The solid Earth systems metamodel (sEsmm) 110), a framework that aims to provide a structure for ontologies and indexes, improve machine efficiency, help human understanding, and enable users to access pertinent information held centrally or serviced from archives in the computing 'cloud'. Geological survey organisations are well placed to provide national hubs as a central source of systematic, located, geological information within a federated structure. The task matches their long-standing aim: to record and communicate a comprehensive quality-assessed view of the national or regional geology appropriate for

---

[176] Information type: The manner in which information is represented and processed, for example, as spatial images, narratives, data, algorithms, tacit and background knowledge.

[177] Provenance (of information): The source, origin or derivation of items of information, which might be formalised in terms of, for example, project, originator, date, place, collection method, archive or database identifier, authorisation.

[178] A metamodel is a description of the organisation and function of a model, to assist the user or computer to find, manage, control and understand its contents.

generalised interpretations that underpin a wide range of applications. A structure of national sEsm's, sharing appropriate content, standards and metamodel, would be an appropriate extension of existing organisational structures. It could aim at improving interoperability for worldwide, inter-disciplinary sharing of the current diversity of sources and content of geological information.

The solid Earth systems metamodel (sEsmm)

<<The geological framework model 105

The metamodel, or description of the structure and organisation of the sEsm (as opposed to its content), is a central part of the framework (see also Loudon and Laxton, 2007[179]). It provides a multidimensional map to locate and connect ideas, concepts, workflows of investigation, and threads of reasoning. It refers to the three-dimensional disposition and configuration of the objects that comprise the solid Earth (where things are and how they are arranged); their observed and interpreted properties, composition and relationships, at

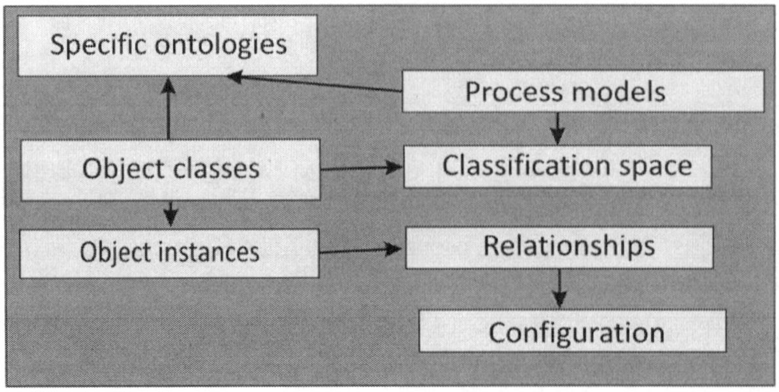

*Figure 25: Extract from metamodel in **Figure 9**.*

all scales; geological processes and the outcome of their interactions with pre-existing configurations of objects; and events and historical changes throughout geological time. Tentative proposals for its contents are described below. The links shown in *Figure 25* illustrate how the configuration[180] of a particular set of object instances might be represented in terms of their space and time relationships. They form the core of a well established descriptive activity, interacting with the more speculative interpretative aspects of the historical processes (see *Figure 26*) that created the observed objects.

---

[179] Loudon, T.V. and Laxton, J.L., 2007. Steps towards Grid-based Geological Survey: suggestions for a systems framework of models, ontologies and workflows. Geosphere, 3 (5), 319-336. http://nora.nerc.ac.uk/1084/

[180] Configuration: The spatial arrangement, pattern, form and shape of objects or their properties. Used by Simpson (1963) in contrasting 'The actual state of the universe or any part of it at a given time, its configuration, is not immanent and is constantly changing' with 'The unchanging properties of matter and energy and the likewise unchanging properties and principles arising therefrom are immanent in the material universe.'

Object classes: Consider first (*Figure 25*) the classes of object, or things of interest, such as rock types or fossils or stratigraphical units.
Classification space: The classes exist in what we might call classification space. Each of its innumerable dimensions represents one property, and the values of the properties of an object class are defined by a point or region in this space.
Specific ontologies provide a controlled vocabulary that identifies the object classes and their properties, relationships and characteristics (see Objects, ontologies and systems 48).
Object instances are actual occurrences of an object class. They might be recorded, for example, by noting the location and extent of a stratigraphical unit on a geological map, a description of a hand specimen, or the chemical analysis of a stream sediment sample.
Relationships among object instances include spatial relationships, such as: one stratigraphical unit lying unconformably on another, or a dyke or a fault intersecting a sequence of sediments; or fossils oriented vertically in the containing sediment; or crystals fitting together in a particular way, as seen on a microscope slide. Spatial relationships may imply time relationships, such as before, after, or during.
The **configuration**, or shape and arrangement of the various objects can be determined by their spatial relationships. In particular their locations may be determined by their relationship to the topography shown by a map or remotely sensed image.
Process models, or the geologist's perception of the geological processes that created, and operate on, the object instances play an essential part in the reasoning that leads the geologist to a particular interpretation of what is observed.

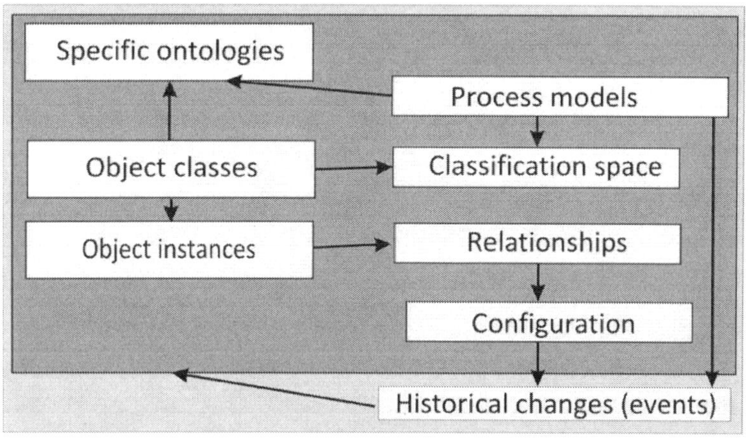

*Figure 26: Extract from metamodel in **Figure 9**.*

Together, these concepts provide a description of the configuration of a set of object instances at a moment in time, their classification, and the processes that operate on them to bring about change. This configuration can be embedded in geological time (*Figure 26*). Geology is a historical science, concerned with deciphering records of evolving configurations of objects and processes throughout geological time. Each configuration therefore has its place at a point in an additional dimension, representing geological time. All this can be considered at any level of granularity (detail or resolution). This requires yet another dimension (that is, another property with its own axis of measurement).

Shared, generic ontologies classify the content, and make it easier for tasks from different disciplines to work together (see Semantic Web and Grid 50). Semantic Web concepts allow links to specific ontologies if need be. Examples of generic ontologies and concepts are shown on the right of *Figure 27*, and widely used examples are published on the internet (see Workflows, collaborative networks and Linked Data 53).

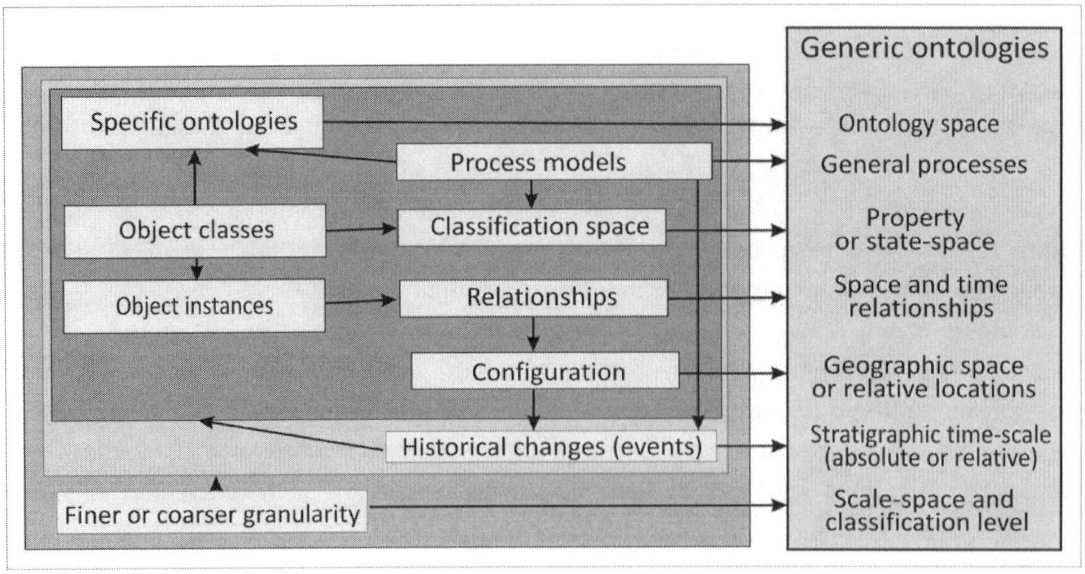

Figure 27: Extract from metamodel in ***Figure 9****.*

Just as metadata are data describing data, so a metamodel is a model describing a model. The solid Earth systems metamodel (sEsmm) 110 describes a model of the systems of the solid Earth which operated throughout geological time, interpreted from their present-day outcome.

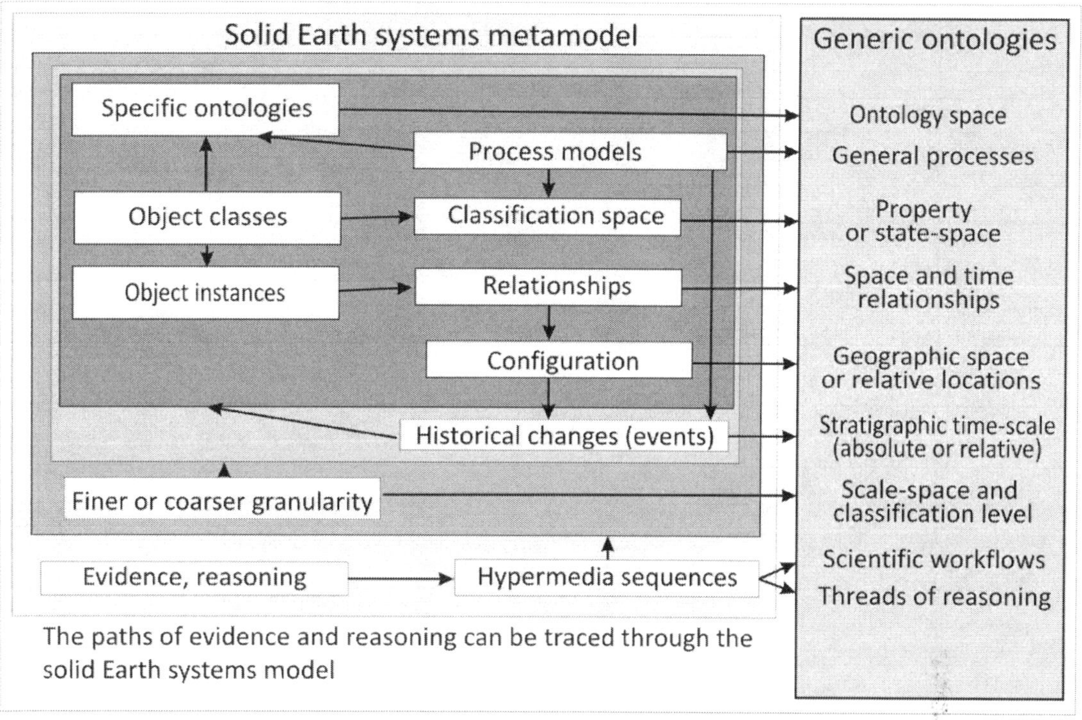

*Figure 28: Extract from metamodel in **Figure 9**.*

Paths of evidence and reasoning can be located and traced through the metamodel (*Figure 28*). The geological framework model can be completed (*Figure 29*) by linking the solid Earth systems metamodel and ontologies to indexes or rdf links (see Workflows, collaborative networks and Linked Data 53), which in turn connect to appropriate information (identified by URI's) in distributed information stores in the computing cloud[181].

The concepts of cyberinfrastructure development suggest that the framework should focus, not on the network or on documents, but rather on the things to which they refer (see Workflows, collaborative networks and Linked Data 53). The metamodel relates each observation through its geographical coordinates to its location on the solid Earth. It can also place an observation or a search for information in its context, relating it to other information, not just in geographical terms but in terms of the innumerable dimensions of the metamodel, including such aspects as geological age and history, lithology, environment of deposition and properties.

[181] Cloud computing: Distributed computing supplying services, such as data and processes, to the desktop or mobile device from the 'cloud' of large, distributed data centres.

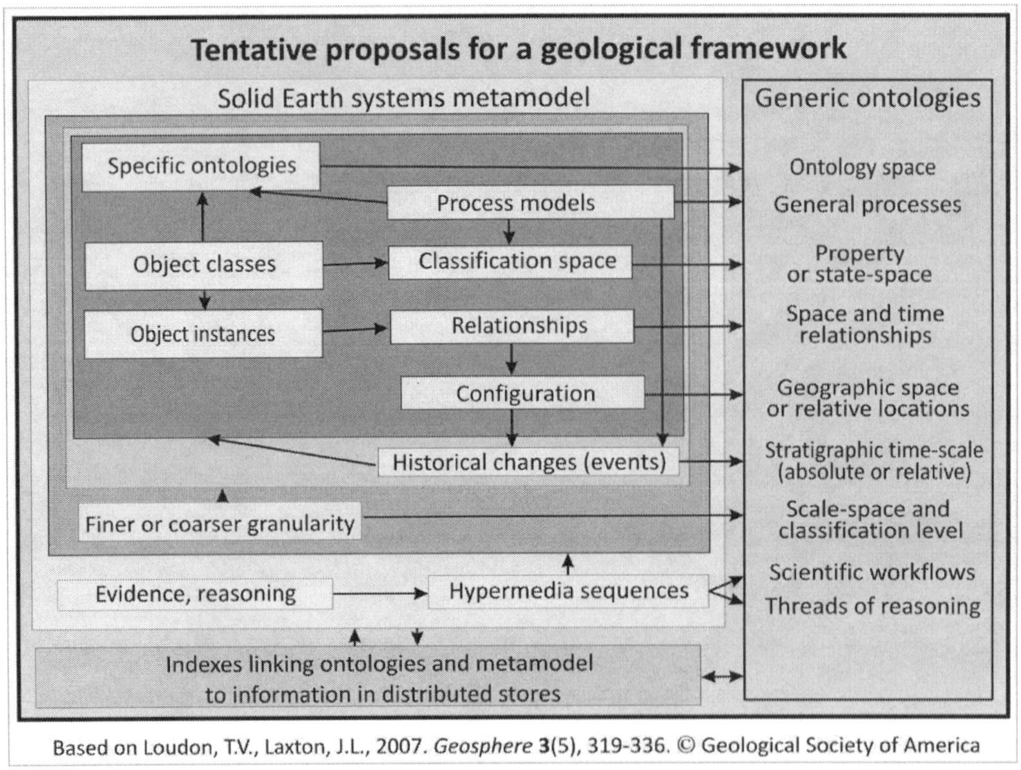

Figure 29: A proposal for a geological framework (duplicate of **Figure 9***)*

An important benefit of the systems approach is its support for quantitative methods. If appropriately applied, statistical methods can bring greater rigour, insight and predictive power to investigation of the solid Earth. They depend on comparing like with like, and the systems framework makes it easier to understand and disentangle the complicated interactions of geological environments, processes and observational bias. Standardised ontologies and defined operational procedures make the shared descriptive language more exact. But the significance of data depends on how they relate to the aspects of the underlying system that they record. The framework model, therefore, positions data within a multi-dimensional system space that clarifies their relationships. The framework must be widely acceptable to geologists, representing a structure they recognise as part of the underlying paradigm. The solid Earth systems metamodel is a tentative scenario for such a framework.

However, observations must also be placed in the context of the project in which they were made. Ideally, the provenance[182] of a project and its implications should be clarified at the level of the business plan and extended to the workflow[183] in which information was collected and interpretations developed. The project design is likely to include a number of

[182] Provenance (of information): The source, origin or derivation of items of information, which might be formalised in terms of, for example, project, originator, date, place, collection method, archive or database identifier, authorisation.

[183] Workflow: The representation of a process or procedure in terms of a sequence of operations to be carried out to complete a task.

sub-models, dealing with say, structure, sediments, igneous intrusions, petrology, and metamorphism. Although the models can be designed to throw light on one another, each, of necessity, brings its own observational bias. Individual items of information must therefore be seen in context. For example, measurements of crystal orientations collected in a study of consolidation processes in an ancient lava flow are unlikely to be directly relevant to a statistical study of microfabric to assess the overall tectonic structure: they are the results of different processes and incompatible sampling schemes. Placing datasets from different sources into a shared framework does not imply that they can be appropriately analysed in the same way. Decisions on the analysis can be assisted by metadata recorded as part of the project documentation, but must ultimately rely on human judgment.

Investigators hold more geological knowledge in their minds than they can record. Many, perhaps most, geological records are based on the tacit knowledge (see Unexpressed knowledge 63) of experts. The systems model must therefore correspond to their patterns of thinking. Furthermore, most representations of geological knowledge are in conventional form, on paper or scanned from paper documents. But an increasing amount, particularly of geological map information, has been reworked into an object-oriented format, and in some cases related to standard ontologies. Information collected specifically for the emerging knowledge system will have only a minor role for some time (see Design requirements for the investigation model 104). The project profile and other metadata are essential for understanding and reinterpreting existing information. The task of the investigators is to use their background knowledge to find, assess and apply the information in the multidimensional framework.

Linking beyond the sEsm

<<The geological framework model 105

Much conventional geoscience literature is entirely relevant to the solid Earth systems model. In geological survey maps, marks and symbols have a standardised significance across wide areas, and can be extracted and studied in a wider framework. Although related information in the conventional literature can often be located and displayed by a search engine, in most cases it requires human interpretation in the context of the document where it appears.

Knowledge of the solid Earth system can be described at various levels

- the geological paradigm of agreed, shared concepts and beliefs
- established standards, definitions and agreed best practices
- (potentially) the solid Earth systems model
- consistent geological maps and spatial models
- authoritative published and evaluated descriptions and interpretations
- informal and unpublished records
- unrecorded knowledge, informal demonstration and discussion

The background specialist knowledge expected from a reader of the scientific literature corresponds more or less to the relevant scientific paradigm as described by Kuhn (1962). It is acquired by training and experience in various levels and areas of specialisation (see *Figure 30*). Textbooks specifically aim to extend the students' knowledge of what is already known and widely accepted. The geological framework model 105 aims to encapsulate an intuitive view of how geologists might think of the structure of a solid Earth systems model. Associated ontologies and references to standards can help to clarify the content.

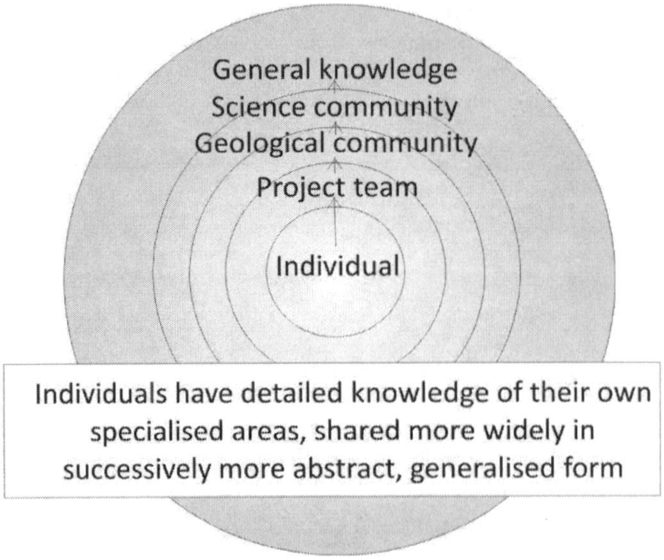

Figure 30: Supporting specialised knowledge (duplicate of **Figure 16***)*

Just as conventional geological information is scattered across many libraries and publishers, the sEsm must access securely archived digital information across widespread organisations and locations. It would aim to place the results into one consistent overall structure, particularly appropriate to systematic results from geological surveying. Although it is not suggested that the model need be restricted to 'authoritative' information, evaluation of the information is essential as a guide to users. A strong core of comprehensive, quality-assessed, surveyed information, maintained by stable, long-term organisations, is needed for a coherent, consistent model of geological systems, as described in The emerging geoscience knowledge system 42.

The geological infrastructure model

<<Table of contents 1
<<The emerging geoscience knowledge system 42
<<The geological framework model 105
The geological infrastructure model 117
Objectives of the infrastructure model 118
Scope of the geological infrastructure model 118
From document orientation to systems orientation 119
>>The future geological map 125

The infrastructure (*Figure 31*) comprises the facilities and mechanisms used in the processes that capture, store, process, share and provide information. The role of the conventional infrastructure is being supplemented, and in some cases replaced, by the cyberinfrastructure, including on-demand services based on Web documents and online systems documentation. It can support the geological infrastructure model, and provide a more comprehensive systems view of geology and the solid Earth in its wider context. The geological infrastructure model must integrate on the one hand with conventional scientific literature and thinking and on the other hand with the concepts of the Semantic Web and Grid 50 and notions such as those of The service-oriented knowledge utility 52 (see From document orientation to systems orientation 119).

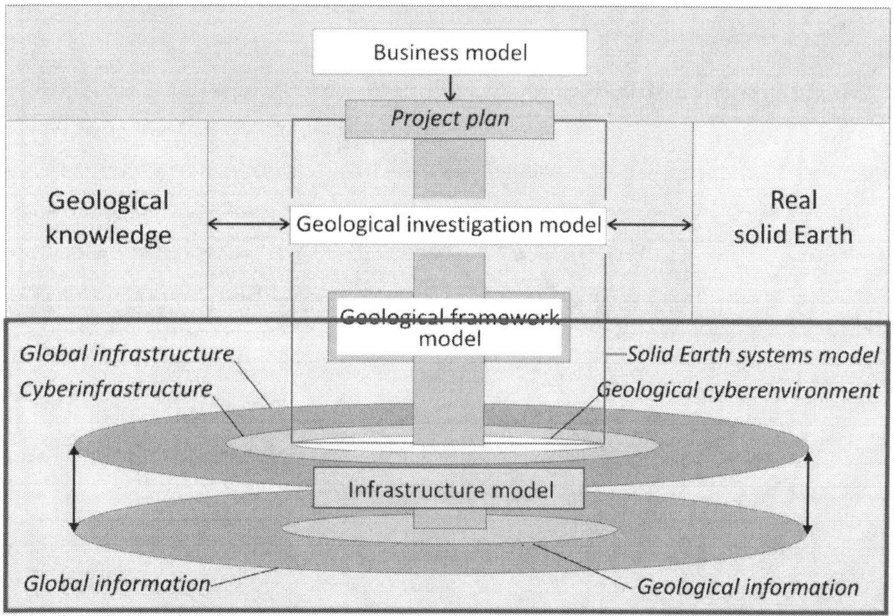

Figure based on Loudon, T.V., 2009. *Computers & Geosciences,* **35**(4), 700-705 http://nora.nerc.ac.uk/7258/

Figure 31: Four models in the geoscience knowledge system (duplicate of Figure 6). The infrastructure model is outlined in red.

The geological cyberenvironment (gce) 85 should provide an interface to the geological infrastructure appropriate to the needs of geologists and users of geological information.

The resulting complex network of services offered by different providers will require global standards, integration of models, means of quality assessment, and a shared conceptual structure (see The geological framework model 105). The geological business model 93 must provide mechanisms to maintain the quality of information in the infrastructure.

Objectives of the infrastructure model

<<The geological framework model 105

The infrastructure model that underpins scientific knowledge is changing, and in response geological survey organisations are gradually replacing products such as printed maps and map explanations. An on-demand knowledge utility will provide services in response to requirements specified by users, and deliver products created from distributed sources of information, computation and expert knowledge. The cyberinfrastructure has the potential to support a systems view of The solid Earth systems model (sEsm) 71 accessed by means of The geological cyberenvironment (gce) 85, bringing scattered fragments of information into a meaningful, rapidly accessible context. As most geological information is in conventional form, the cyberinfrastructure and the conventional structure must evolve in harmony.

Scope of the geological infrastructure model

<<The geological framework model 105

The infrastructure model will be designed and implemented primarily by e-scientists. But geologists have a major responsibility for the objectives and design of the geological infrastructure, which must relate closely to the unstated knowledge of geologists and to the conventional scientific literature. The cyberinfrastructure for geology must relate to other fields with a strong spatial emphasis (see Yang et al., 2010[184]). The focus here is on trusted products from geological survey organisations, seen as a potential integral component of the wider semantic Grid (see The service-oriented knowledge utility 52, and Workflows, collaborative networks and Linked Data 53).

[184] Yang, C., Raskin, R., Goodchild, M., Gahegin, M. 2010. Geospatial Cyberinfrastructure: past, present and future. Computers, Environment and Urban Systems. **34**, 264-277

From document orientation to systems orientation

<<The geological framework model 105

The geological infrastructure[185] is obviously affected by its historical development. It contains both conventional and digital documentation, which must work together. The forms of conventional, published, geological literature include books, scientific papers (generally brought together by topic in specialist journals), and maps (which may be part of a series possibly covering a wide geographical area). Other formats include short notes (brief scientific papers), specialist encyclopaedias and glossaries, monographs, standards manuals, informal reports, datasets and logs. Recent publications, although conventionally organised, are generally available also in digitised form that can be accessed by search engines and selectively displayed and printed. They may refer to databases, geographical information systems (GIS)[186], models, and other parts of the infrastructure which are likely to be in digital form with their own specific structures. However, this scenario emphasises the importance of comprehensive systems documentation, and therefore it is the significant differences between mainstream conventional and systems documentation that are significant here. Their structures are therefore summarised in *Figure 32*.

Geological books and scientific papers have a linear structure. They begin at the beginning, and follow a line of thinking through to the end, typically describing the background understanding of the subject, followed by new observations or interpretations, leading to discussion and conclusions. The document provides its own context for discussing the various issues involved, together with references to other sources and an expectation of the reader's background understanding. Points in the document can be identified by page or chapter number. The book is significantly longer than the paper, and normally covers a wider range of ideas. It may be addressed to students of a particular topic, whereas the paper typically introduces new observations and ideas to those already familiar with the topic. On publication, both are archived in many libraries including copyright libraries. Some books may be entirely revised as distinct new editions, usually after several years. Scientific papers are seldom revised after publication, but may be superseded by later papers.

Geological maps within series, such as those published by geological survey organisations, are (like books) published, archived, and may be revised and republished as new editions. Their structure, however, is two-dimensional rather than linear. They display a two-dimensional view of the geology, with topographical contours and cross-sections to help the user visualise the vertical dimension. Points in the map can be identified by grid reference or latitude and longitude, and can be related directly to the overlaid topographical map, and thence to the real world. In the margin, the key may indicate adjacent map sheets in the same series, and relevant maps at other scales, extending the two-dimensional structure and introducing the scale dimension. Map explanations, reports and related books may have

[185] Infrastructure: The basic facilities, services and installations needed for a system to function.

[186] Geographic Information System (GIS): An integrated system for the capture, storage, management, retrieval, analysis, manipulation and display of geographically referenced spatial data and its attributes.

Medium	*Referenced by*	*Guides to content*	*Structure*
Book	Author Title Edition Date Publisher, city ISBN	Table of Contents List of references Index Glossary	Linear sequence (Chapter 1 Section 1 ... Chapter 2 ...)

Internal references by chapter, section, and/or page number.
Updated by new edition if any.

Medium	*Referenced by*	*Guides to content*	*Structure*
Scientific paper	Author Title Date Journal name Volume, part Pages ISSN	Abstract List of references	Linear sequence (Section 1 Section 2 ...)

Internal references by page or section number.
Updated piecemeal by subsequent papers.

Keywords may be available, search engines can retrieve by full-text indexing of both of above.

Medium	*Referenced by*	*Guides to content*	*Structure*
Geological map	Title Map Series Sheet number, edition Publishing agency Dates of survey Date of publication Surveyors ISSN	Marginalia may include map description, references, stratigraphic key, cross-sections, map legend, illustrations, etc.	Two geographical dimensions. Keyed to other scales and related maps

Overprinted on base map, allowing internal references by coordinates (lat/long, National Grid, etc.) or geographical feature. Colour coded link from stratigraphic table to areas on map.
Updating by new edition of sheet.

Above publications are archived by copies in multiple libraries, including copyright libraries, and may also be held as electronic copies in archival web sites.

Medium	*Referenced by*	*Guides to content*	*Structure*
Web document	Web address	Search engines, search history	Network

Internal and external references by hyperlinks

Web documents generally regarded as ephemeral, unless in explicitly archival website.

Medium	*Referenced by*	*Guides to content*	*Structure*
System documentation	Web address Provenance	Internal documentation by organisation responsible	Network

Generally archived by organisation responsible

Figure 32: Comparison of documentation structures

links to or from a specific set of map sheets (or their GIS equivalents), and to points within them.

Books, papers and maps (digital or not) can represent only a small part of a geologist's thinking. A geologist considers manifold properties of the rocks, their form and configurations at all scales, their historical evolution and the physical, chemical and biological processes that created them: a web of hierarchies and networks in a multitude of dimensions (see The geological investigation model 98). Furthermore, a geological investigation involves all information types (see Integrating information types 62), which cannot be represented in one style of paper document.

The cyberinfrastructure can offer more comprehensive representations than the conventional linear or two-dimensional structures. The author of a Web document places it in a hierarchy by specifying its address, with levels separated by the / symbol, and can identify specific points in the document by the # symbol. Many connections in Web and GIS representations can be made within and between documents, by links that can be followed rapidly by the user (including those between text narrative and spatial information), thus representing a network structure. The ability to display multiple windows on the screen and view several dimensions and information types side by side (see Integrating information types 62), and to support the multimedia hypertext threads and the associative trails envisaged by Bush (Mechanisms 45), provide a more comprehensive and flexible means for geologists to represent their ideas. Web documents are typically more ephemeral than books or papers, but are much more rapidly disseminated at lower cost than conventional publication. Websites that index and access archived material, such as those of publishers and research organisations (for example, the NERC Open Access Research Archive[187]) and the development of e-books, open the prospect of information of enduring significance benefitting from the flexible structures of the Web and the rapid access of the internet.

In the case of systems geology, The systems approach to Earth science 65 and The geological cyberenvironment (gce) 85 are possible candidates for such an environment (see Remodelling the map 74). A systems model should be able to give a more complete record of geological thinking. However, documenting a system brings its own challenges, and calls for a flexible architecture for the documentation (see for example, Other relevant fields - DITA 355). The systems approach regards a set of scientific phenomena as a single, coherent set of interacting parts that function as a whole, comprising, in this case, the solid Earth (see The systems approach to Earth science 65). The aim is to share concepts to achieve a fuller understanding of the interactions among the components, and unify individual systems as subsystems of the larger whole. By bringing the geologist's holistic view of the science into an explicit representation, correlations among the components can be explored more readily, supporting Geological surveying as reinforcement learning 79 and The sEsm as a predictive machine 76. It is particularly important, therefore, to maintain a well-defined ontological framework (see The geological framework model 105 and Some related initiatives - GeoSciML 345) and a disciplined approach to adding to the core of geological information (see The geological business model 93). Geological survey organisations have

[187] NORA, 2010. NERC Open Research Archive. http://nora.nerc.ac.uk

long experience of maintaining a consistent system for geological mapping, and should be well qualified to respond to the task of migrating to comprehensive, digital, systems documentation.

In general, unlike much conventional publication, systems documentation is likely to be controlled and undertaken by (or on behalf of) the organisation responsible for the system. Access to systems is generally restricted, much of it concerned with back-office activities and only a small part made visible to users by information delivery (*Figure 33*). For example, online users of a banking system may be restricted to a small part of the system, such as information on customer products and on their own financial transactions. Systems may be designed for rapid updating from several sources and formats (see, for example, OmniMark, 2010[188]), so that, for example, new observations and corrections to a geological map can be recorded and shared among the investigators while a survey is in progress. But, like back-office activities, it is inappropriate to make them more widely available until the project is complete. Geological survey organisations aim to provide an authoritative view of the geology, and therefore are likely to make information (derived from the archive) generally available only after authorisation and approval. Nevertheless, the flexibility of the Web structure makes it possible for detail and the workflow of surveying to be archived as appropriate for future reference.

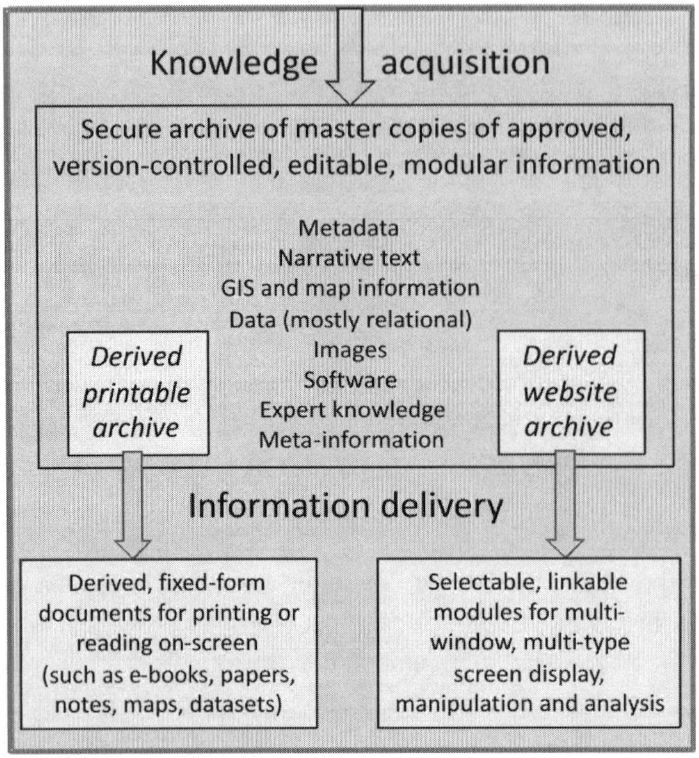

Figure 33: Flow of information in a modular information system

---

[188] OmniMark, 2010. A high-performance content processing platform. http://www.stilo.com/Products/OmniMark/tabid/57/Default.aspx

Subdividing geological information by map sheet is being superseded by geographical information systems, but the information nevertheless develops as modules[189], probably resulting from work over a limited period by an individual or small group, and varying in size from description of a single outcrop to a regional analysis. Each module may have relevance in various contexts, and can therefore be linked in as a component of the systems network. It must be sufficiently self-contained that it makes sense in relevant contexts. Updating could in principle be immediate, for example if an error or new evidence was found. In practice, maintaining a coherent system requires careful management and authorisation of changes and assessment of any knock-on effects (see The geological business model 93). The basic module of system documentation could therefore be a self-contained micro-document or minimum revisable unit (mru[190]). It should include metadata[191] on its provenance, such as authors, authorised by, date of acceptance, version number and lists of references from (and citations to) the micro-document.

Changes within a module would result in it being replaced by a new version. The latest approved version would normally be the one visible to the user. Earlier versions of the module should be archived, so that historical reconstruction of an interpretation is possible, including knowledge of the context of other micro-documents in which the work was carried out. Approval of a new version would have to consider knock-on effects within the system, and deal with these appropriately, including revision of modules at a higher level if need be. The change would not be visible to the user until the revised module was approved, although changes in progress might be flagged.

A system, such as The solid Earth systems model (sEsm) 71, requires A framework for the reasoning 135 to organise the content, enable users to find the appropriate components, and ensure that modules are appropriately linked to maintain the integrity of the system as a whole (see The geological framework model 105). Published documents and e-documents, such as books, maps, web-sites and papers could be generated from system modules, and held in their own archives, as shown in *Figure 33*. The framework could also be used as a structure for referencing and indexing conventional documents. Referencing between systems documents and conventional documents is a routine procedure. As always, readers must use their expert knowledge to assess the relevance to their specific needs.

The Web structure supports a flexible response to the users' requirements. However, the user who has located useful information, and wishes to study it in detail, is likely to prefer a printed version. And internal references in a paper document require page or sequential section numbers, which a networked structure cannot readily supply. On the other hand, the extracts could fit naturally into the structure of an e-book (provided the reading device can

[189] Module: A subdivision of a system that can be combined with others in various ways to perform different functions, can be independently replaced or upgraded, or can be plugged into the system to extend its functionality.

[190] Minimum revisable unit: A self-contained subset of information in a documentation system: a component of the system designed to be revised as necessary without endangering the integrity of the system as a whole

[191] Metadata: Metadata is a description of data (often used in a broad sense of representations of information or knowledge) and its context. It is structured to assist the user or computer to find, manage, control and understand the data.

handle hypertext references) thereby retaining the modularity and providing the flexible information trails foreseen by Bush (see Mechanisms 45). Users could be offered predetermined sequences of modules, covering areas of general interest, along with the option of printing or editing the sequence to correspond more closely to their interests. Approved, authorised and edited extracts might also be assembled manually from the sEsm to fill the role of conventional books, maps or scientific papers, visible to appropriate search engines and suitable for printing on demand.

The future geological map

<<Table of contents 1
<<Introduction and overview 1
<<Overview of the future geological map 38
<<The emerging geoscience knowledge system 42
The future geological map 125
Reasoning, models and reality 127
From map to digital model 145
Reconfiguration 165
An object-oriented approach 178
The geometry of the spatial model 203
Transforming space 224
Seeking shared concepts 247
Mapping geology into the knowledge system 282
>>References 314
>>Glossary 328
>>Index 359

Summary: Geoscientists have the deepest knowledge of their subject and must control the development of systems geology[192]. It is to them that The future geological map 125 is primarily addressed. The advancing infrastructure can extend the meaning of 'geological mapping'[193] – looking beyond the depiction of geology on a paper map to model it as a component of a digital knowledge base. This calls for reconsideration of the underlying objectives, methods and models (see Reasoning, models and reality 127) that are taken for granted in a more settled system. Geoscience maps provide shared, located, spatial information central to the science. Future users will expect delivery of comprehensive spatial information to their digital desktop or field notebook (see From map to digital model 145), with flexible visualisation and hypertext links within a comprehensive cyberinfrastructure[194].

---

[192] Systems geology: A view of geology re-based on the developing cyberinfrastructure and regarded as a system (a set of interacting parts that function as a whole) embedded in the wider knowledge system.
[193] Mapping: Conventionally, geological mapping leads to a graphical depiction, usually on a flat surface, of spatial relationships and forms of geological features or properties in a selected area of the Earth's surface or subsurface. In the mathematical definition, mapping relates the elements of one set to those of another. A broader definition of geological mapping could be 'relating elements of geological observation or interpretation of the solid Earth to corresponding elements in appropriate models in the geoscience knowledge system'.
[194] Cyberinfrastructure: An integrated assemblage of computing, information and communication facilities, deploying the combined capacity of multiple sites to provide a framework to underpin research and discovery, typically with broad access and end-to-end coordination.

Geoscience models (that simplify reality by selectively condensing information) can be brought together (see Reconfiguration 165) as the core of a system that provides an integrated interpretation interfaced with the real world. It potentially distinguishes between observations, deductions, assumptions and interpretations; integrates information types; supports new investigational techniques and multiresolution analysis; and clarifies the reasoning with links to geological process models, stratigraphical and ontological[195] frameworks and a broad knowledge base[196]. When feasible, visualisations and explanations should overcome the ambiguities of the conventional map and should offer user options to select area, content, scale and form of presentation, with reprocessing to meet specific needs. However, the models are based on fragmentary geoscience knowledge and an incomplete general geoscience model.

It is proposed that the diverse models should be integrated within a comprehensive systems view, based on geological reasoning. Object-oriented methods (see An object-oriented approach 178) are appropriate for this purpose, and can integrate information from initial survey to final presentation. They provide support for recording and reconciling information that comes from many sources and can be filtered and visualised for a range of applications. Computer representation of the geometry (see The geometry of the spatial model 203The geometry of the spatial model) can provide a shared mathematical framework for analysis, simulation and visualisation. The static view can be extended with dynamic models (see Transforming space 224). Methods (such as zero-crossings, patch dynamics, scale-space analysis and multiresolution and deformable models) have been developed in other fields but are relevant also in geology (see Seeking shared concepts 247). This scenario concludes that the systems view and its associated methods are an appropriate basis for Mapping geology into the knowledge system 282. A structure for this purpose, focusing on the core aspects of regional geology, is described in the section on The emerging geoscience knowledge system 42 as The solid Earth systems model (sEsm) 71[197], supported, from field survey to end-use, by The geological cyberenvironment (gce) 85.

---

[195] Ontology: A formal representation describing concepts, entities and relationships in a domain of knowledge, typically providing a more detailed and rigorous machine-readable specification than a thesaurus or taxonomy.
[196] Knowledge base: A dynamic repository for information and methods for accessing and processing it. It is generally machine-readable and online, and may include the means to access expert knowledge.
[197] Solid Earth systems model (sEsm): An approach to structuring distributed knowledge of the science of geology to provide an integrated view in the context of sciences of the solid Earth as a whole. A model of the systems of the solid Earth, organised within a framework or metamodel that depicts and clarifies the principal relationships among the findings of geology, providing a multidimensional map to locate and connect ideas, concepts, workflows of investigation and threads of reasoning.

Reasoning, models and reality

<<Table of contents 1
<<The future geological map 125
Reasoning, models and reality 127
The need to look again 127
The dialectic model 129
Abstracting from reality to model 131
At the interface 133
A framework for the reasoning 135
The stratigraphical framework 139
Stratigraphical units in space and time 141
>>From map to digital model 145

Abstract: Models are simplified views of reality. Spatial models can separate observations, deductions, assumptions and interpretations more effectively than a map. The geological spatial model can condense information from numerous observations as a generic interpretation in harmony with scientific understanding of the historical geology. The interpretation is confronted with reality by locating it in space through surveying procedures based on human knowledge. Short notes in an indexed object store can record the reasoning, linkable as explanations for user-defined areas and topics. The global framework of space and time-sequenced stratigraphical units connects field correlations with historical geology. Top-down surveying extends existing knowledge with directed observations and interpretations, potentially gaining from digital access to related material and rapid updating.

The need to look again

<<Reasoning, models and reality 127

Summary: *The advanced infrastructure enables us not only to automate existing procedures but also to tackle underlying requirements in new ways. For example, the British Geological Survey has implemented a digital geoscience spatial model as an integrated computer-based representation of the results of geoscience survey. This scenario focuses on concepts and rationale rather than implementation, and points to a more fundamental change, viewing geology as a major component in a solid Earth systems model.*

The cyberinfrastructure is changing the procedures by which we abstract information from the real world, and visualise, generalise, record and share it. The major benefits will stem, not so much from automating things we already do, as from enabling us to tackle the underlying requirements in new ways. This requires geologists to reconsider their thought processes, communication and working practices, and introduces insights from mathematics supported by powerful tools in the emerging infrastructure.

Collaboration is needed among at least three groups of specialists: experts in e-science, mathematicians, and geoscientists. The human mind has its limitations and none of these groups can have a complete understanding of the other disciplines. The mathematician can look at what geologists do, and e-scientists can implement their solutions. But that just embeds current procedures in a new setting. There is therefore a need to explore more comprehensive scenarios where the concepts can work together to provide a new perspective on geoscience surveying. Geoscientists, to whom this is primarily addressed, know their own business best and must therefore understand and control its development. But the complexity and scale demand a collaborative framework.

Geological maps and sections illustrate our view or spatial model of the anatomy of the Earth. They show the position, spatial patterns and spatial relationships of its components and their properties at many scales, from global summary to intricate local detail. Automating the process of making maps and bringing their content into a geographical information systems[198] environment provide an insight into their significance (USGS, 2007[199], USGS, 2010[200]). It is an essential basis for future work, but on its own is an incomplete response to future needs. The map is a means of illustrating the geoscientists' spatial model, not an end in itself.

"Spatial modelling is a new name for an old concept. It refers to what has long been a core activity of a geological survey, namely, piecing together a picture of the geometrical configuration and disposition of sequences of strata or other rocks, their constituent materials, characteristics, and properties, and relating that picture to ideas of their history and origin. The novelty lies in basing the model on computer methods rather than the conventional reports, maps and cross-sections. This is not to suggest that the interpretation and ideas can come from anywhere other than the geologist. But it is suggested that a computer model could give a better medium for the geologist to express his ideas; build up his interpretations; organise, analyse, summarize and share his observational data; explore the consequences of his hypotheses; reconcile information from diverse sources with his expectations and background knowledge; display the results of his work and transmit them to the users" (Loudon, 1982[201]). The concepts have been developing for some time, and are now being implemented more extensively, thanks to current developments in the

---

[198] Geographic Information System (GIS): An integrated system for the capture, storage, management, retrieval, analysis, manipulation and display of geographically referenced spatial data and its attributes.

[199] USGS, 2007, (Committee on Research Priorities for the USGS Center of Excellence for Geospatial Information Science, Mapping Science Committee, National Research Council). A Research Agenda for Geographic Information Science at the United States Geological Survey. National Academies Press, 156 pp. ISBN-10: 0-309-11154-4 http://books.nap.edu/catalog.php?record_id=12004

[200] USGS, 2010. National Geospatial Program http://www.usgs.gov/ngpo/

[201] Loudon, T.V., 1982. The case for computer-based spatial models in geology. *BGS Project 22A, Progress Report.*

infrastructure. The British Geological Survey embarked on a major project to implement a digital geoscience spatial model in 1999 (Smith, 2005[202], Howard et al., 2009[203]).

In contrast, this scenario (an outline of one possible course of future development) largely dodges the transient, practical problems of implementation and is intended to encourage fresh thinking on underlying concepts. It addresses the question 'Where do we want to go, and why?' rather than the detail of 'How do we get there?' It focuses on longer-term issues and a rationale for modelling the results of geological surveying as a potential part of a comprehensive solid Earth systems model that includes subsurface geology and other branches of geoscience. Nevertheless, only implementation of new methods can improve the process of geoscience survey, hence the insertion of short notes on that subject. Short-term benefits can justify the early developments, but to keep moving in the right direction we must keep longer-term goals in mind. To guide applications of fast-changing technology, geoscience surveyors must look beyond the conventional map – at their underlying objectives, methods and models (see The dialectic model 129), all of which could be taken for granted in a more settled system.

Implementation note: An inexpensive but essential aspect of systems planning is thinking ahead, and the main purpose of this scenario is to encourage such thoughts. These implementation notes are added as an aside to suggest how thoughts might be converted to real benefits. Actual implementation of a development phase might start with pilot studies to clarify the practical aspects, including a review of other work, an assessment of available technology and standards, and a cost-benefit analysis. But pilot studies seldom scale directly to full implementation. Work must proceed step by step, securing each foothold before moving to the next.

The dialectic model

<<Reasoning, models and reality 127

Summary: *Conceptual* models *(simplified and formalised views of aspects of reality) are fundamental to understanding, but misleading if inappropriately used. The map reflects a discourse between arguments from the model and from reality, but amalgamates observations, deductions, assumptions and interpretations in a single view. Spatial models have the potential to separate these, identify and reconcile multiple viewpoints, and record the reasoning.*

In order to guide their observations and make sense of them, scientists develop *conceptual models* – formalised mental images giving a simplified view of aspects of the real world relevant to the purposes in hand. There are many types of model in geoscience, from the

[202] Smith, I.F. (editor), 2005. Digital Geoscience Spatial Model, Final Report. British Geological Survey Occasional Publication No. 9, 56pp. http://www.bgs.ac.uk/downloads/start.cfm?id=535

[203] Howard A.S., Hatton B., Reitsma, F., Lawrie, K.I.G., 2009. Developing a geoscience knowledge framework for a national geological survey organisation. Computers & Geosciences, **35**, 820–835. doi:10.1016/j.cageo.2008.06.004

general models representing geologists' background knowledge, to comprehensive spatial models and the hierarchy of more specific models that contribute to them. Models are fundamental to scientific understanding, but are misleading if inappropriately used. By definition, models differ from reality, whether or not they are implemented on the computer. Different models may refer to different aspects of the real world: different objectives may call for different models.

Selecting the model appropriate to the purpose is crucial. A well-known example involves the question discussed by Mandelbrot (1982[204]): How long is the coastline of Britain? Take a map as your model, and the answer seems to depend on the amount of detail as determined by the scale. Take the real world as your guide, and you realise that the map model is inappropriate to the question and the answer is more interesting than the incalculable numerical estimate. This raises issues about how far the map or model depends on observations of the real world, how far it depends on the underlying theory, and how far the interpretations are consistent and appropriate. The map can reflect only a view of reality, although that view can be selected to serve a broad range of needs and its limitations can be recognised. "A good geological map is much more than an objective presentation of the distribution of rock units, their structure and their relations; it is also a subjective presentation of interpretations based on a multitude of observations and, to a greater or lesser degree, based on theories and prejudices held at the time the map was made." (Harrison, 1963[205])

Trifonov (1984[206]), in the USSR, discussed the mechanism of geological mapping fitted to an explicit worldview (his only two references are to works by Lenin and Karl Marx). A current worldview is more likely to be influenced by the business objectives, the paradigms of geoscience, the disciplines of an information community, and the global imperatives of the internet. Nevertheless, Trifonov offers valuable insights into the role of the geological map in the *dialectic* between model and reality (in the sense of a discourse juxtaposing arguments from the two sources and resolving their contradictions). The map, based on the geologists' observations and interpretation, depicts the form and position of geological objects: "an incomplete, approximate reproduction of the real structure... As a result of this kind of abstraction, the object appears on the map much simpler than it is in reality." But "the map is much more than a simple descriptive system storing information about the object, because it may serve as an instrument for the forecasting of still unobserved phenomena."

Maps, rather than the real world, may be studied to arrive at broader explanations. A hierarchy of ideas, where broad theories are built from narrower hypotheses, is a recognised feature of the scientific method. But the geological map gives only a snapshot of the dialectic process, and what is observed, what is assumed and what is deduced are not clearly distinguished. In consequence, a map might be studied without realising that by taking, say,

---

[204] Mandelbrot, B.B., 1982. *The fractal geometry of nature*. Freeman, San Francisco, 460pp.

[205] Harrison, J.M., 1963. Nature and significance of geological maps, in Albritton, C.C. (editor), 1963. *The fabric of geology*. Freeman, Cooper & Co, Stanford, pages 225-232.

[206] Trifonov, G.F., 1984. Maps as stages of the cognition process in geology. *In*: Dudich, E. (editor), 1984. *Contributions to the history of geological mapping*. Proceedings of the Xth INHIGEO Symposium 1982, Budapest, Hungary. Akademiai Kiado, Budapest, 47-53.

the probable configuration of a surface as its only model, it misses a less likely (but entirely possible) configuration suggesting an oil prospect. There is also a risk of unwittingly introducing invalid circular arguments. Explanations based on study of maps might appear to be validated against the real world, when in fact they merely reflect the models and assumptions on which the maps were based. On the other hand, a spatial model with integrated information types should be able to clarify and record the reasoning. This might in turn identify which parts of the model are valid in particular applications (see Diverse objectives and products 150). Before considering how this might be done, we need a clearer view of the reasoning processes involved in Abstracting from reality to model 131.

Abstracting from reality to model

<<Reasoning, models and reality 127

Summary: Abstraction *(reducing the volume of information while retaining salient features) pervades geoscience survey, progressively reducing the complexity of reality by selective observation and recording, interpretation, explanation, and reworking into more condensed forms. What is 'salient' depends on the surveyors' worldview – their broad model, or paradigm, of the science, tuned to its business setting. The interpretations reflect geology as a historical science – a configuration of objects in space that changes through time, echoing a sequence of real, individual but related events driven by processes controlled by immanent*[207] *laws (like the law of gravity, independent of time or place). The interpretations should lead to a coherent view consistent with observations of the real world.*

The related processes of summarising, generalising and abstracting reduce the volume of a body of information while retaining its salient features, as described for map generalisation by Buttenfield and McMaster (1991[208]). In a sense, much of the process of geoscience survey is one of abstraction. Abstraction (see The geological investigation model 98) repeatedly filters information from the real world, reducing its volume by selective observation, selective description, measurement, representation, recording, classification, analysis, interpretation, explanation, and generalisation. At each step, the procedures are guided by feedback from general concepts derived earlier. From the immense detail of the real world, salient points are observed and the more important are selectively recorded. The records are reworked to provide a coherent account in reports and maps. These findings may be summarised in turn as review articles and smaller-scale maps. Subsequently, more general interpretations and explanations reduce the detail to an explanatory framework that can predict significant pattern but not specific occurrences. Abstracts, titles and keywords are provided. At each step, the volume of information is reduced. But this definition of abstraction begs the question of how we determine what is salient and what is important. The answer must lie in the geologists' business model and *worldview*, that is, their broader

---

[207] Immanent: Something naturally inherent and intrinsic within and throughout its domain.

[208] Buttenfield, B.B. and McMaster, R.B. (editors), 1991. *Map generalization: making rules for knowledge representation*. Wiley, New York. 245pp.

model of the science (itself the result of abstraction). Here, geologists assemble the exemplars that form their shared *paradigm*, taking into account knowledge from other sciences and aiming at consistency throughout the hierarchy of explanation. For a fuller account, see Loudon (2000[209], part K).

The broad model is tuned to the business setting of the investigation (see Overview of the geological business model 93). Thus, oil geologists would have different objectives and procedures from academic geologists whose business was research and education, or engineering geologists looking for geological hazards, or amateurs satisfying their curiosity, or employees of a Geological Survey trying to maintain a knowledge base for a wide range of applications. Each business setting and project has its broad model that determines what is important, what is observed and what is recorded. Metadata[210] referring to this business model are essential for users to understand the results and their provenance. The metadata may be formally recorded as such, but at this level are more likely to be regarded as part of the paradigm and discussed in reports or assumed to be part of the users' background knowledge. Results obtained for one purpose can be reworked and at least partially reused for others. The differences lie in the emphasis and the detail rather than in the general understanding, and information is readily exchanged between projects and between information communities.

Geology is a historical science describing the development of the present distribution of rocks through a hypothetical sequence of configurational[211] changes that were the outcome of processes and events in the geological past. Frodeman (1995[212]) suggested that consequently there may be no sharp distinction between interpretation and data (see Invariance and processes 56). We can picture the geological model as focusing on specific tasks but without rigid boundaries. It reflects, and is a part of, a hierarchy of models, from a broad global worldview to detail such as an explanation of a ripple mark. Some of the models are tied to observations and measurements of the real world. Others reflect processes following scientific laws that are immanent (like the law of gravity, not unique to a particular situation) but which operate on particular configurations of objects and events resulting from the flows and accidents of historical geology (see Complex and emergent systems 159). In the course of abstraction, many analytical models come into play, shedding light on one another. Statistical, spatial and quantitative models coexist with broad qualitative interpretations. They may be expressed in computer programs, images and narrative records, or may remain only as perceptions and concepts in a geologist's mind.

---

[209] Loudon, T.V., 2000. Geoscience after IT: a view of the present and future impact of information technology on geoscience. Elsevier, Oxford. 142 pp. Also available as *Computers & Geosciences*, Special Issue, **26** (3A), A1-A142. (Part K at http://nora.nerc.ac.uk/2406/)

[210] Metadata: Metadata is a description of data (often used in a broad sense of representations of information or knowledge) and its context. It is structured to assist the user or computer to find, manage, control and understand the data.

[211] Configuration: The spatial arrangement, pattern, form and shape of objects or their properties. Used by Simpson (1963) in contrasting 'The actual state of the universe or any part of it at a given time, its configuration, is not immanent and is constantly changing' with 'The unchanging properties of matter and energy and the likewise unchanging properties and principles arising therefrom are immanent in the material universe.'

[212] Frodeman, R., 1995. Geological reasoning: geology as an interpretive and historical science. *GSA Bulletin*, **107**(8), 960-968

They must lead to a coherent view, consistent with the supporting evidence, and yielding predictions that can be tested against further observations of the ultimate arbiter – the real world.

You may recall that Tennyson's (1842[213]) Lady of Shalott had a problem (apart from her silly name) familiar to some theoreticians and philosophers. Her view of the world was confined to images on her screen (the 'magic mirror') and her work on the web. Her attempt to extend the dialectic by direct observation of the outside world was fatal: "Out flew the web and floated wide; /The mirror crack'd from side to side; /'The curse is come upon me,' cried /The Lady of Shalott." It can be argued (see At the interface 133) that a vital function of geological surveying is to bridge this accursed interface between the worlds of geological concepts (expressed in images and models) and external reality.

At the interface

<<Reasoning, models and reality 127

Summary: *Like the map, the spatial model bridges the* interface *(the shared boundary) between model and reality. The scientist's concepts and ideas are adjusted to match observations, and tested, corrected, and refined by feedback. Rules and conventions determine the procedures that convert observations and interpretations to records in the model or marks on the map. The procedures might potentially be automated as an expert system, but at present the rules are elusive, dependent on experience and on understanding the thought processes of colleagues.*

The spatial model, like the geological map, lies at an *interface* (the shared boundary between parts of a system, or the means of interaction between them). On one side is background knowledge of the geological setting; agreed classifications, terms and definitions; a hierarchy of immanent scientific laws, explanatory theories, hypotheses and models; and configurational ideas of the local historical geology within its regional context. All are mental constructs or their recorded representations. On the other side is the real world, observed, measured and studied in the light of that background knowledge. The model or map records the results of the dialectic process between the two sides. The flexible configuration of interpreted spatial patterns and relationships is imagined, depicted as cartoon models (see Stratigraphical units in space and time 141), or derived from process-response models. It is adjusted and warped (morphed or rubber-sheeted) to fit the rigid geometry and fixed observations representing the real world. The consequences are fed back, adjusting and refining the models in a continuing cycle, to visualise and predict the configuration and disposition of the local characteristics and properties of the rocks.

One can judge the validity of the interpretation by the consistency and credibility of the arguments and their sources. One can assess its accuracy by checking it against observational records, and testing its predictions against additional observations of the real

[213] Tennyson, A., 1842. The Lady of Shalott. http://charon.sfsu.edu/tennyson/tennlady.html

world. This emphasises the value of the rigid geometry of the map or spatial model. It establishes a one-to-one relationship between points and areas on the ground and their counterparts on the map, matching the configurational details of the observations with those of the models. The map illustrates spatial models of real objects, retaining form and relationships while reducing the scale for ease of visualisation. At the interface, observations can be recorded, the consequences of hypotheses explored, and ideas developed, tested and corrected by reference to reality.

To cross the interface, rules and conventions must exist for transforming the relevant parts of the geoscientists' concepts, ideas and observations into marks on the map (or records in the model) and vice versa (but see The imperfect map 145). Furthermore, both the creators and users of the map must know these conventions, whether they are aware of it or not. The surveyors must know them, to decide what to record on the map as a result of what they see and what they know. The users must know them, in order to 'read' the map, appreciate the local configuration, build a picture of the local geology, and know what to expect on the ground. The map is thus also the interface where information passes between the surveyor and the user, who view the knowledge from different perspectives (see Reconciliation 186). If the rules and conventions could be stated formally, they might be embedded in an expert system.

In a computer-based expert system, techniques of knowledge representation might structure the geologists' ideas within a knowledge base as hierarchies of objects, properties, processes, events and states (see, for example, Sowa, 2001[214], Sowa, 2000[215], Mennis, Peuquet and Qian, 2000[216], and The solid Earth systems model (sEsm) 71). Inference engines might then represent in computer software the geologists' procedures for drawing conclusions from the knowledge base and aligning them with observations for any local area. For example, interpolation procedures based on knowledge of the geological setting might complete lines and surfaces adjusted to fit the scattered data. Bayesian analysis and belief networks (see for example Howson and Urbach, 2006[217]) offer a mathematical approach to combining new data with existing knowledge, seeing observations as a means of modifying opinions rather than determining absolute truth.

We return to this topic (see The geometry of the spatial model 203, Broadening the framework 291, Object-oriented survey 195) to offer a more robust view. Here, we may note that the rules and conventions are elusive, dependent on understanding the training and thought processes of colleagues. Much geological interpretation is based on intuition rather than inference. It relies on analogies that are seldom identified or articulated, from a vast legacy of information that may never reach a computer knowledge base. Expert systems

---

[214] Sowa, J.F., 2001. *Processes and causality.* http://www.bestweb.net/~sowa/ontology/causal.htm

[215] Sowa, J.F., 2000. *Knowledge representation: logical, philosophical, and computational foundations.* Brooks/Cole, Pacific Grove. 694pp.

[216] Mennis, J.L., Peuquet, D.J., Qian, L., 2000. A conceptual framework for incorporating cognitive principles into geographical database representation. *International Journal of Geographical Information Science*, **14** (6), 501-520.

[217] Howson, C., Urbach, P., 2006. *Scientific reasoning: the Bayesian approach*. 3rd ed. Open Court Publishing Company, Chicago. 352pp. ISBN 0-8126-9578-X

thus offer no immediate solution. The methodologies are nevertheless relevant now, for they can be applied informally to clarify ideas about the framework and to point towards future solutions. For the foreseeable future, the geological model is dependent on human insight. It must focus initially on interactive visualisations of spatial patterns and relationships, supported by explanatory text, dominated by legacy information and unrecorded experience. Nevertheless the spatial model, unlike the static map, allows us to record the dynamic interplay of model and reality, not just the conclusions but also the dialectic process, creating fertile ground for the growth of a diversity of new ideas. Evolution of these ideas or memes requires their evaluation, reconciliation and selection (not just by the host community). They must be recorded, catalogued and communicated, to provide A framework for the reasoning 135, with a mechanism to ensure that consistent sets of favoured ideas prevail.

Implementation note: The conventions followed on a map or model should be documented as metadata, using global standards where appropriate. Developments in the representation of reasoning and knowledge (such as Geological surveying as reinforcement learning 79) deserve at least a watching brief, in particular for their relevance to the framework and metadata as related to the processes and concepts of geoscience.

A framework for the reasoning

<<Reasoning, models and reality 127

Summary: *Unlike the map, the model can record the interplay of interpretation and observation, encouraging the evolution of new ideas. It should record the evidence and reasoning for checking by others, to identify modifications required by new evidence or new interpretations, assist reconciliation of ideas and establish the validity of applications. Conventional map explanations are illustrated narrative reports. The spatial elements and narrative framework can be more closely linked in a model, free of sheet boundaries in an* object store *(computer library) of short notes or microdocuments and segmented spatial models. They should be fully indexed and marked-up for generating a variety of more complete and coherent accounts.*

To fulfil its role as an interface between the geoscience knowledge base and the real world (see At the interface 133), the spatial model must include an account of the underlying interpretation and reasoning, establishing the relationships between the information in the knowledge base[218] and the evidence in the real world. The reasoning should also be recorded (see The geological business model 93) for at least four other reasons:

1. In a scientific product, it should be possible to assess and verify conclusions by tracing them back to the evidence. Scientists should be able to explain their procedures, and their colleagues should be able to follow their reasoning and reproduce the results. "All work

[218] Knowledge base: A dynamic repository for information and methods for accessing and processing it. It is generally machine-readable and online, and may include the means to access expert knowledge.

should be documented... All completed [map] work needs to be supported by all the raw data, whether obtained from external services or internally generated." (Tearpock and Bischke, 2003[219], page 6).

2. It should be possible to track the amendments to the model needed to respond to small changes in the interpretation or observations. For instance, erosional features might be discovered at a supposedly conformable junction, or a supposed outcrop might turn out to be a large boulder. The design must ensure that limited adjustments of this kind avoid needless disruption to the model as a whole, but that where necessary their knock-on effects can be propagated through appropriate levels of interpretation (see Geological survey documentation 6). Equally, two-way hyperlinks should help human experts to identify the consequences of a change to the metadata[220] (such as a revised definition of a stratigraphical unit), or help to explore the outcome of alternative hypotheses. Ideally, therefore, it should be possible to trace the process of abstraction from the final summary back to the details on which it was based, and vice versa.

3. The Reconciliation 186 of ideas that conflict and overlap depends on understanding their background. For example, explaining the significance of discrepancies between biostratigraphical and lithostratigraphical correlations requires insight into the underlying reasoning.

4. The validity of each part of the model for particular applications may depend on the underlying procedures. For example, a formation boundary drawn as a smooth line on the map because of poor exposure is inappropriate for determining the absence of small-scale folding.

As suggested earlier, coherent visualisations[221] (including standard maps and sections), assembled from their graphical elements, take the central role when obtaining, comparing and correlating spatial information. Supporting explanations are communicated by links to narrative text. The spatial elements need this narrative framework (Loudon, 2000, part J[222]) just as a map needs an explanatory report. But in clarifying the reasoning and tracing the process of abstraction, the roles change. A coherent narrative account (addressing the user's episodic memory) takes the lead, and the maps or visualisations (addressing spatial memory) act as accompanying support (see Integrating information types 62). The spatial and narrative frameworks, their mutual dependencies and two-way links must be designed to make sense in both situations.

---

[219] Tearpock, D.J. and Bischke, R.E., 2003. *Applied Subsurface Geological Mapping*. 2nd ed. Prentice-Hall, Upper Saddle River, NJ. 822pp.

[220] Metadata: Metadata is a description of data (often used in a broad sense of representations of information or knowledge) and its context. It is structured to assist the user or computer to find, manage, control and understand the data.

[221] Visualisation: Transforming quantitative data (including the results of interpolation) into sensory information – images that the eye and brain can interpret and visualise.

[222] Loudon, T.V., 2000. Geoscience after IT: a view of the present and future impact of information technology on geoscience. Elsevier, Oxford. 142 pp. Also available as *Computers & Geosciences*, Special Issue, **26** (3A), A1-A142 (Part J at http://nora.nerc.ac.uk/2405/).

In the long run, map-sheet boundaries are as irrelevant to the text narrative as they are to the visualisations. Spatial objects extend beyond map sheets. The explanatory documents may therefore evolve from sheet-based reports to an object store of *microdocuments* (a computer library of short notes, perhaps corresponding to the memes[223] mentioned in Abstracting from reality to model 131). A database management system might help to index the microdocuments for convenient searching under various headings, provided their contents are recognised as interpretative objects, inherently redundant and immutable. They are not normalised, their contents may overlap and they can be superseded but not altered or destroyed (see Geological survey documentation 6, From document orientation to systems orientation 119). The structure of the object store should encourage and support complete, consistent, maintainable and retrievable records and metadata. It should reflect the structure of the reasoning and areas of responsibility. It must be possible to link together the microdocuments (just as graphical elements can be linked) to give a coherent account of a specified topic and area. Links between microdocuments and the sets of associated graphical elements can establish the detailed reasoning behind the depictions on the map or spatial model.

The headings for searching the object store must include location and stratigraphical position. Searching might also be possible on criteria mentioned earlier including: provenance[224]; geoscience topics; business settings; project descriptions; analogous cases; background theory; and metadata defining objects, processes, events, procedures, operations, activities, properties and their relationships. However, the threads of reasoning through the microdocument library are likely to be fashioned, not of database references, but of hypertext linkages. Each microdocument must therefore have a uniform resource identifier (URI). Mark-up[225] languages can identify and relate the layers of reasoning and can connect them to and from spatial features. Links can cite sources to identify the business models, specific metadata and background ideas. Explanations can be tied to spatial objects, thus making it possible to visualise consequences of the reasoning, relating observations to interpretations. The linkages from the microdocuments, like those from the spatial entities, help to clarify the implications and the knock-on effects when information is updated.

The process of geological survey is described in Abstracting from reality to model 131 as a hierarchy of abstraction from detailed observations through to a more generalised explanatory model and framework (see The solid Earth systems model (sEsm) 71, The geological framework model 105). As they develop, they inform subsequent observations and define classifications and nomenclature. There is constant feedback at all levels, refining

[223] Meme: According to Dawkins (The Selfish Gene, 1976): "a unit of cultural inheritance, hypothesized as analogous to the particulate gene and as naturally selected by virtue of its 'phenotypic' consequences on its own survival and replication in the cultural environment." Examples are ideas and concepts passed from mind to mind by imitation or explanation, evolving through variation, selection and heredity.

[224] Provenance (of information): The source, origin or derivation of items of information, which might be formalised in terms of, for example, project, originator, date, place, collection method, archive or database identifier, authorisation.

[225] Mark-up: Symbols inserted in a document in a mark-up language such as SGML, HTML or XML, to tag the beginning and end of character sequences that can be interpreted by machine, and can be omitted for displaying to the user.

the model as we learn more, and adjusting subsequent investigation in response to the evolving ideas. The framework (see The geological framework model 105) should become more stable as ideas mature, and standards based on the framework can then be more widely accepted. Standards, including those for the text explanations, would ideally be aligned across all topics. In practice, however, the spatial model must relate to contributions from many projects with differing business settings, topics, instrumentation, investigational procedures, and explanatory frameworks that inevitably influence their results. A project map may have its own specifications and internal reasoning (see Semantic Web and Grid 50). For example, a gravity model of a particular date might be mapped to its own standards, consistent over a large but bounded area. As pointed out elsewhere (see The importance of space and visualisation 169), its individual measurements cannot be related to those in other models, but the finished product is a spatial object that could be compared, in the light of metadata, with completed models from other topics, such as geochemistry, and reconciled (see Reconciliation 186) with the overall model.

A map or spatial model may reflect the distribution in space of a *property* of an object. For example, a gravity map might record the distribution of measured values reflecting the densities of nearby zones of the Earth's crust. Or a map might reflect the *composition* of objects in terms of, say, geochemistry, lithology or sand/shale ratio. But a map showing the *disposition* (locations) and *configuration* (pattern, arrangement and spatial relationships) of stratigraphical objects and other rock bodies, as affected by the structural geology, throws most light on the genesis and subsequent history of the rocks. It is the key to geoscience explanation, and therefore the core of the overall model. It depends on stratigraphical classification and nomenclature. Geoscience surveys normally refer to limited areas, but the stratigraphical framework, being based on geological time, correlates rock sequences world-wide. Just as a map ties spatial objects to geographical co-ordinates and thus to the real world, so stratigraphy ties conceptual events of many kinds at many locations to the single shared axis of geological time. The shared framework of space and geological time helps geoscientists to identify conflicts of observation or interpretation, to resolve or account for them, and to relate the various models in one coherent and consistent story (see The stratigraphical framework 139).

The spatial model should offer a hospitable environment for records from new surveying tools (for example, Xu et al., 2000[226]), which conventional mapping procedures cannot provide. But we cannot start from a clean sheet. A vast amount of existing information is available as maps and reports. The system must be designed to accommodate past work as well as new methods whose spatial coverage is inevitably incomplete. The new methods should not be constrained by the older technology, but must nevertheless share a framework with the legacy data.

Implementation note: A store of spatial objects, of any information type and including microdocuments, must handle their differing locations and spatial extents without confining

---

[226] Xu, X., Aiken, C.L.V., Bhattacharya, J.P., Corbeanu, R.M., Nielsen, K.C., McMechan, G.A., and Abdelsalam, M.G., 2000. Creating virtual 3-D outcrop. *The Leading Edge*, 19 (2), 197-202.

them within arbitrary map sheet boundaries. It depends on effective GIS-based[227] browsers, and indexing with database techniques and headings. Authors can build coherent accounts of larger areas and broader topics by generalising from individual local objects, recording links for the reader to drill down to greater detail as required. Reuse of the spatial objects in different contexts avoids unnecessary repetition. Step-by-step pilot studies are needed in the slow process of surmounting the barriers to a new approach. For example, surveyors can store field notes and observations as spatial objects and insert links to them from a digitised map and map explanation (see Some related initiatives – SIGMA 343).

The stratigraphical framework

<<Reasoning, models and reality 127

Summary: *The shared framework of space and geological time helps to manage conflicting ideas and relate the various models as one coherent story. This standard, global, geographical and stratigraphical framework is as essential for spatial models as it is for maps. Stratigraphical units seem well matched to the human thought processes on which the overall model depends. The units are themselves models and in a computer environment can be seen not just as objects on a map but also as involved in processes and relationships.*

The ability to compare and correlate a diversity of maps requires complex global metadata[228]. To quote Hedberg (1976[229], page 7): "It is possible to classify stratified rocks according to any of their properties: lithology, fossil content, magnetic polarity, electrical properties, seismic response, chemical or mineralogical composition, and many others. Rock strata can also be classified according to such attributes as their time of origin or their environment of genesis. ... units based on one property do not generally coincide with units based on another, and their boundaries not uncommonly cut across each other. ... a different set of units is needed for each. ... all involved intricately in achieving the same major goals of stratigraphy – to improve our knowledge and understanding of the Earth's strata and from this to outline the nature of past events, processes, and life on Earth."

International committees maintain a standard stratigraphical framework to which geoscientists in general can work. "All of our classifications and terminologies of natural bodies are no more than an attempted ordering contrived by human beings for the purpose of aiding our own imperfect conception and understanding of the infinite complexities of nature; and as such they have all the weaknesses of the human minds in which they originated. Classification and terminology of rock strata are no exception. ... [However,]

---

227 Geographic Information System (GIS): An integrated system for the capture, storage, management, retrieval, analysis, manipulation and display of geographically referenced spatial data and its attributes.

228 Metadata: Metadata is a description of data (often used in a broad sense of representations of information or knowledge) and its context. It is structured to assist the user or computer to find, manage, control and understand the data.

229 Hedberg, H.D. (editor), 1976. International stratigraphic guide: a guide to stratigraphic classification, terminology and procedure. Wiley-Interscience, New York.

agreement on stratigraphic principles, terminology, and classificatory procedure is essential to attaining a common language of stratigraphy that will serve geologists worldwide. It will allow their efforts to be concentrated effectively on the many real scientific problems of stratigraphy, rather than being wastefully dissipated in futile argument and fruitless controversy arising because of discrepant basic principles, divergent usage of terms, and other unnecessary impediments to mutual understanding." (Hedberg, 1976, page v).

"*Stratigraphic procedures* and principles ... are applicable to all earth materials, not solely to strata. They promote systematic and rigorous study of the composition, geometry, sequence, history, and genesis of rocks and unconsolidated materials. They provide the framework within which time and space relations among rock bodies that constitute the Earth are ordered systematically. ... [and] define the distribution and geometry of some commodities needed by society ... *Stratigraphic classification* systematically arranges and partitions bodies of rock or unconsolidated materials of the Earth's crust into units based on their inherent properties or attributes. ... A *stratigraphic code* or guide ... provides the basis for formalization of the language used to denote rock units and their spatial and temporal relations." (NACSN, 1983[230], page 847).

A stratigraphical framework is as essential for geological spatial models as it is for maps. Spatial relationships seen in the field or on the map face, together with reasoning from historical geology, establish the sequences of time relationships shown on map keys. The stratigraphical units can coexist with quantitative representations of other properties and seem well matched to the background knowledge and subtle intuition of the human thought processes on which the geological model depends (Loudon, 2000[231], pages A93-96). However, information technology may modify our view of the behaviour of stratigraphical units. They are conceptual constructs showing a simplified view of reality. In other words, stratigraphical units are models[232]. Within a computer environment they can be viewed not just as objects on a map but also in terms of the processes and relationships in which they take part.

The conventional geological map portrays formations as static objects. Look at the map key to see the stratigraphical sequence, note the colour representing the formation of interest, and look for that colour on the map face to see its spatial distribution. In the spatial model, however, we could point to the key to see stratigraphical sequences and relationships, and click on an item to invoke its model within the user-selected area. If the input was legacy map data, the result may be no more than the usual pattern of colour and symbols in two dimensions. But if the field data were collected with a digital model in mind, surveyors could record their exploration of alternative models and identify the reasons for selecting a particular interpretation. Users could select the mode of visualisation appropriate to their

---

[230] NACSM (North American Commission on Stratigraphic Nomenclature), 1983. North American Stratigraphic Code. *The American Association of Petroleum Geologists Bulletin*, **67**(5), 841-875.

[231] Loudon, T.V., 2000. Geoscience after IT: a view of the present and future impact of information technology on geoscience. Elsevier, Oxford. 142 pp. Also available as *Computers & Geosciences*, Special Issue, **26** (3A), A1-A142 http://nora.nerc.ac.uk/2405/

[232] Model: A formalised representation giving a simplified view of aspects of the real (or of an imaginary) world relevant to the purposes in hand.

needs. Those with access to the full data (not just the conclusions) might conceivably select from a range of options to control the geometry and interactions with structural geology, according to their own objectives, interpretations and opinions. Another benefit of thinking of the stratigraphical units as models may be the greater flexibility of relating them beyond the map to depositional, structural and other models, thus connecting the current configuration more closely to the past events and processes of dynamic stratigraphy (see Stratigraphical units in space and time 141).

Stratigraphical units in space and time

<<Reasoning, models and reality 127

Summary: *International bodies maintain the stratigraphical framework. Stratigraphical correlation in the field feeds information on the spatial extent of the stratigraphical units into the map or model. Dynamic stratigraphy relates the units to the events and processes of historical geology and their outcome. The framework, spatial model, and dynamic stratigraphy can be regarded as three subsystems within the current system of geological investigation. But a geological map, with stratigraphical key and generalised vertical sections, shares elements from all three. Good subsystem interfacing will be crucial to successful visualisation of all these aspects in a spatial model without sheet boundaries.*

Stratigraphical terminology and classifications define hierarchies of stratigraphical units and arrange them systematically in sequences based on inferred relationships in geological time. The units are referenced to type sections, giving precedence to earlier definitions where practicable. One function of the map is to record the spatial aspects of stratigraphical units, extending the stratigraphical sequence from inferred time into measured space. Spatial extension of the units leads to spatial models of two types.

The first type of spatial model is a *cartoon model* of the events and processes of historical geology and their results, configuring the spatial objects relative to one another rather than to geographical coordinates. The word 'cartoon' is not used in any derogatory sense, but indicates that the model is a preparatory design or sketch. Here, it sketches a historical configuration based on what is known of its present-day outcome and arguments from immanent[233] laws that do not depend on position. This preliminary sketch is characteristic of studies of historical geology. It may guide and be guided by interaction with the second type of spatial model, which places the spatial objects in absolute rather than relative positions.

The second type of spatial model extends the stratigraphical units in space by stratigraphical correlation in the field, tracing their occurrences by correspondence in character and stratigraphical position (see Hedberg, 1976[234], page 14). It is based on comparisons with occurrences of the unit elsewhere and ultimately with the type section. In turn, their extension on the map sets a precedent for further correlation. One mode of classification of

[233] Immanent: Something naturally inherent and intrinsic within and throughout its domain.

[234] Hedberg, H.D. (editor), 1976. *International stratigraphic guide: a guide to stratigraphic classification, terminology and procedure*. Wiley-Interscience, New York.

the stratigraphical sequence is into formations, explicitly selected for showing on a map. "The proposal of a new formation must be based on tested mappability ... No formation is considered valid that cannot be delineated at the scale of geologic mapping practiced in the region when the formation is proposed." "Because the surface expression of lithostratigraphic units is an important aid to mapping, it is commonly advisable, where other factors do not countervail, to define lithostratigraphic boundaries so as to coincide with lithic changes that are expressed in topography." (NACSM, 1983[235], pages 858 and 856). "The geographic extent of lithostratigraphic units is controlled entirely by the continuity and extent of their diagnostic lithologic features" (Hedberg, 1976, page 31). Surveying the unit boundaries ties them to geographical coordinates. This second type of spatial model is characteristic of the geological map.

We can identify (1) the stratigraphical framework, (2) geological maps and spatial models, and (3) dynamic stratigraphy as the subjects of separate subsystems within the current system of geological investigation. But the close links between spatial and stratigraphical models raise again the question of whether long-established subsystem boundaries remain appropriate to the advanced infrastructure (see Reconfiguring the system 172). At first sight, information technology seems to reinforce the boundaries between the subsystems. Separate organisations, with separate publication procedures, are responsible for the three subsystems, each with its own methods, characteristics and infrastructure requirements.

The stratigraphical framework requires an international organisation to maintain the metadata, involving collaboration among many scientists with primary allegiances elsewhere. Time, unlike space, proceeds in only one direction, and the parallel sequences of the stratigraphical framework and its complex time relationships suggest that directed graphs provide the appropriate mathematical structure for this information in the cyberenvironment. Robust software is available as Critical Path Analysis programs, more usually applied to studying manufacturing processes, or the graphs might be handled as modified Petri Nets in UML (Sowa, 2000[236]). They appropriately show time relationships (such as before, after, during) and correlate and compare paths through parallel sequences of activities and events. Because of its unique features, the stratigraphical framework could be regarded as a separate subsystem.

Mapping and spatial modelling can be regarded as a separate subsystem, because it is regional or local, and likely to be the responsibility of state-funded organisations or commercial organisations with specific terms of reference to prepare and maintain a coherent spatial view. It is likely to require software based on cartographical, surface modelling, visualisation and geographical information systems.

The third subsystem has the task of portraying the historical geology or dynamic stratigraphy, providing an account of past processes and a general description of their outcome. It has its own objectives, perspectives, deadlines and modes of communication,

---

[235] NACSM (North American Commission on Stratigraphic Nomenclature), 1983. North American Stratigraphic Code. *The American Association of Petroleum Geologists Bulletin*, **67**(5), 841-875.

[236] Sowa, J.F., 2000. *Knowledge representation: logical, philosophical, and computational foundations.* Brooks/Cole, Pacific Grove. 694pp.

often within an academic setting. It is not tied to any specific area or to any formal mapping programme. Here, the cyberinfrastructure can contribute statistical descriptions and process-response models for elucidating geological processes and their consequences. Again it seems to reinforce the existing division into subsystems.

This seems to suggest clearer boundaries for The geological investigation model 98, equating it with the second subsystem. But look again at the geological map. The map face illustrates the spatial model. The map key lists the sequence of stratigraphical units, thus drawing on the stratigraphical framework. A generalised vertical section may summarise some properties and relationships of the units (notably thickness variation and relationships with adjacent units) that throw light on the historical geology. The map face and marginalia thus share elements from all three subsystems. The spatial model aims to offer a seamless picture unbroken by map sheet boundaries, allowing users to select their own scale and items and areas of interest. But the marginalia (and microdocuments mentioned in A framework for the reasoning 135) must then be assembled to match the specific content of each visualisation. Software must draw information for a designated area or spatial object from all subsystems, and be able to trace a change in one subsystem to its knock-on effects in other subsystems. In a cyberenvironment, an interface to the stratigraphical metadata as a whole (such as BGS, 2010c[237]) should enable workers in many fields to communicate with the same stratigraphical framework. Equally, models of the properties of stratigraphical units and of their origin and geological history could supplement and assist visualisation of the local spatial model.

At this early stage of implementation, we must build on the legacy of map sheet information. However, the evolving system should take advantage of developing opportunities, and must be seen in a global context (see Reconsidering geological mapping 277). The benefits from an integrated view across the subsystem boundaries are considered in The surveyor's holistic view 60 and The need to harmonise the geometry 203. However, the immediate problems for a cyber-based system may lie, not in changing the subsystem boundaries, but in defining their interfaces, ensuring that they can work well together, and managing the timing of parallel developments. For example, a regional survey might accept responsibility within its region for maintaining a view of the spatial extent of the stratigraphical units, and of the local expression of diachronous relationships. It has to make complex decisions about whether to base stratigraphical keys on those of existing map sheets or to embed them in new tiling structures for seamless display. If new structures are developed, this could be done in isolation, or could await international agreement on interfaces to other subsystems or the availability of mainstream software at an indefinite future time. On the basis of local circumstances, a feel for the herd instinct, and taking a view on the inevitable unknowns, managers must make uncomfortable decisions on when to move ahead and on how broad a front, when to backtrack to conform to other initiatives, and when to lag behind and risk getting lost.

The process of abstraction or generalisation has been a theme of this section. Salient information is assembled from observations and measurements and interpreted with

[237] BGS, 2010c. The BGS rock classification scheme. http://www.bgs.ac.uk/bgsrcs/home.html

increasing generality, extending the integrated model representing the geoscientists' paradigm of accepted ideas, explanations and exemplars, and tying it to a framework of space and geological time. This is in no way dependent on the infrastructure, but placing the activity in a computer environment should, in the longer run, lead to more efficient, rigorous and effective survey procedures (see From map to digital model 145). In the section on The imperfect map 145, we look at some of the shortcomings of traditional surface and subsurface geological maps (related of course to limitations of the medium, not the practitioners). Overcoming these is one of the challenges in developing spatial models. We consider the wide range of map uses and various approaches to modelling, leading to suggestions in An object-oriented approach 178 for a framework to accommodate the diversity.

Implementation note: An information system thrives only if both contributors and users gain benefits from it. Acknowledging the authors of map sheet and monitoring document sales are inappropriate in a seamless model. It is not difficult to devise a replacement charging and reward system for microdocuments, but general acceptance of a stable system by authors, users and management may remain an active issue for many decades. Meantime, parts of the system can already be implemented. Experimentation with new methods also calls for a long time-scale, and may be driven forward by the early-adopters for whom innovation is its own reward. As the aim here is to outline a conceptual scenario, the questions of motivation and driving forces are mentioned only briefly. They nevertheless represent a key issue in system implementation.

From map to digital model

<<Table of contents 1
<<The future geological map 125
<<Reasoning, models and reality 127
From map to digital model 145
The imperfect map 145
Ambiguity and map representation 148
Diverse objectives and products 150
Forward and inverse models 154
The imperfect model 156
Complex and emergent systems 159
>>Reconfiguration 165

Abstract: Looking ahead, spatial models can be more rigorous and flexible than the map, avoiding its information overload and meeting a wider range of objectives. Standard maps for wide-ranging spatial comparisons will be joined by spatial models offering visualisations and thematic presentations with user selection of area, content, scale and form of presentation. As an inverse model, the overall geological model must call on diverse approaches to resolve multiple hypotheses. Explanations relate to a general model of Earth history that is complex and fragmentary, involving self-organising processes that restrict deterministic models. Nevertheless, a broad understanding is achieved through hierarchical classification of objects and processes.

The imperfect map

<<From map to digital model 145

Summary: *To evolve efficiently, current developments must be informed by looking ahead at longer-term trends. In due course, advanced systems will comprehensively address geoscience survey, making full use of mathematical and systems concepts to overcome earlier constraints. They will employ a battery of surveying techniques, and provide more flexible visualisation and thematic presentations for a wide range of applications. They will help to overcome rigidities of scales, arbitrary and inconsistent interpretation, and inflexible techniques.*

The main theme in Reasoning, models and reality 127 was the extension of existing procedures to take advantage of the flexibility of computer-based systems. A secondary theme has hinted that this is a first step to more advanced systems, which may influence the future of geoscience survey by overcoming significant limitations of conventional procedures. To establish this second theme, we look at the possible future evolution of the system of geoscience survey, from map-based concepts to spatial models that are integrated from initial investigation to end use and make greater use of mathematical and systems concepts. The aim is to suggest how current developments can be aligned with future trends determined by underlying user requirements.

The geoscientist may think in terms of conceptual spatial models, but is influenced by their representation as graphical models – maps and cross-sections that are projections of three-dimensional objects on a flat sheet of paper. They worked well for depicting surface geology for well-defined and limited objectives and methods. But these standard graphical models are less appropriate for today's wide range of applications, user demands and techniques (geophysical, geochemical, subsurface and so on). Sets of thematic maps for the same area can address specific topics. For example, Edwards et al. (1987[238]) provided a range of thematic maps for the Southampton area of the UK. They included solid geology, drift geology, drift thickness, rockhead contours, sand and gravel resources and end-use analysis, clay resources, worked ground, aquifer distribution, engineering geology, slope stability, landfill and waste disposal, boreholes, and Sites of Special Scientific Interest. In this case, the thematic maps were professionally drafted from material selected from a computer database and printed in small numbers. Although an expensive and inflexible solution, within the database it identified and related spatial objects that can be visualised in various contexts. It did not address the issues (mentioned in Representing spatial information and relationships 173 and A framework for the reasoning 135) of map sheet constraints, inefficient and inflexible presentation, laborious updating, and weak object-level linkage.

The fixed scales of mapping led to an emphasis on spatial objects of a particular size range. The geological significance of the objects, however, does not necessarily depend on their size. For example, the mineralogy or the spatial relationships of microscopic grains seen in a thin section may provide crucial evidence for a regional interpretation. As another example, permeability of a reservoir might depend on spatial continuity of minute interstitial cavities, their development controlled by processes operating on scales from grain-to-grain contact and sedimentary structures to major faulting. Decisions about fracturing techniques might reflect estimates of the scale distribution of permeability. An understanding of how this relates to lithologies and formative processes might be gained from studies of extensive surface exposures. But these examples do not fit readily into the map scale framework. Geoscience explanations and applications refer to objects and processes of size and resolution unrelated to map scales. Constant map scales are needed for regional and thematic comparisons, but they imply arbitrary and undesirable spatial filters on data collection, storage and interpretation.

Geoscience data are sparse and incomplete. Interpolation[239] models are needed to fill the gaps and display our view of the geology, unobscured by the pattern of data collection. In mapping subsurface geology, three-dimensional surfaces such as formation tops may be represented by contours. The gaps between data points can be contoured by hand,

---

[238] Edwards, R.A., Scrivener, R.C., Forster, A., 1987. Applied geological mapping, Southampton area. *BGS Research report ICSO/87/2*. British Geological Survey, Keyworth. 69pp.

[239] Interpolation: The estimation of values, for example at a point or along a line or surface, in order to predict a value or complete a visualisation.

following one of several predetermined approaches, with interestingly varied results, or mechanically by computer, following rules expressed as software processes (Tearpock and Bischke, 2003[240], chapter 2). More often, contouring involves geological interpretation. Explaining the procedures is hard work – the textbook just mentioned devotes 822 closely packed pages to an account of applied subsurface geological mapping. As its authors point out (page 16) different geologists inevitably produce different maps from the same data. "The differences in the finished maps may be the result of the geoscientists' educational background, experience levels, interpretive abilities, or other individual factors... the differences can also be the result of the method of contouring..." They suggest (page 21) that: "the specific method chosen for contouring may be dictated by such factors as the number of control points, the areal extent of these points, and the purpose of the map. It is essential to remember that, no matter which method is used in making a subsurface map, *the map is not correct*... What is important is to develop the most *reasonable and realistic interpretation* of the subsurface with the available data, whether the maps are constructed by hand or by computer." Similar remarks could be made about the intersection of formation tops with the land surface, or about the other features shown on maps of the surface geology. The geology should presumably be consistent on and below the ground.

These approaches are not altogether satisfactory. Geologists produce different hand-drawn maps from the same data, and arrive at different solutions at different times. The reasons for contouring in a particular way are seldom fully explained, and may be opaque even to the originator. However, computer contouring also gives different results depending on the methods used, and again the reasons for selecting particular options may be obscure. Links to geological reasoning may be unclear and important geological constraints may be overlooked, for geologists and mathematicians approach interpolation from different backgrounds. More extensive spatial models therefore bring a need to explore further the links between the mathematics and the geology (see The geometry of the spatial model 203, Seeking shared concepts 247).

Traditional methods of geological survey rely on images rather than mathematics. Maps and cross-sections illustrate spatial models conceived by geologists through visualising the form, relationships, behaviour and appearance of geological objects, not by performing mathematical calculations. Geologists presumably manipulate the images in their minds, imagine the consequences of geological processes and depict them graphically. Dots, symbols, annotations, lines, colour and ornament on a traditional map define the graphical expression of the geoscientists' spatial view – a graphical model, illustrating the underlying conceptual spatial model. They are easy to draw using a line-drawing instrument like a pen, pencil or computer stylus or cursor. They are appropriate for a model depicting the locations of observations, boundaries between stratigraphical objects, contours on the top of a formation, or its line of intersection with the land surface.

The formation boundaries are themselves selected for convenient drawing on a map, with well-defined, continuous, single-valued, sharp boundaries, and with contents clearly distinct

---

[240] Tearpock, D.J. and Bischke, R.E., 2003. *Applied Subsurface Geological Mapping*. 2nd ed. Prentice-Hall, Upper Saddle River, NJ. 822pp.

from the material outside and consistent with the lithostratigraphical hierarchy (see Stratigraphical units in space and time 141). The form of representation inevitably influences our thought processes and the way we define, visualise and study the spatial object. This has knock-on effects up the chain of abstraction and reasoning. We might even begin to believe that real formations could possess the properties just listed (see Ambiguity and map representation 148). If the primary objective is making a lithostratigraphical map, other approaches such as biostratigraphy or sequence stratigraphy might be neglected even where they could have thrown more light on past events and processes. When the graphical model is manipulated for cartographical presentation or scale change, particularly by computer, there is a danger that the full implications of the conceptual spatial model may be overlooked.

Ambiguity and map representation

<<From map to digital model 145

Summary: *Using conventional methods, geologists must force a mixture of ambiguous, contradictory, subjective and vague ideas into the precise and rigid geometry of the map. The single image is overloaded with diverse information. The map user has problems in determining the sources, evidence and reasoning; recognising uncertainty, scenarios and multiple hypotheses; identifying the salient points, sampling schemes and design of the investigation; reconciling overlapping maps of related topics; adjusting the map to a new or evolving interpretation; and reinterpreting the map to meet various objectives.*

The map illustrates spatial aspects of the geology in a form that relates to, and can be tested against, the real world. However, it proves difficult to tie down the precise significance of the marks on the map. As mentioned with a dyke swarm in Representing spatial information and relationships 173, the need to complete lines on a map can impose exact and potentially misleading geometrical representations on features whose correct position and spatial relationships are unknown. If, instead, the features were omitted, their absence would be at least as misleading. For example, a fault might be known to cross a map because it occurs in adjoining sheets, but its position might be poorly defined. If it were omitted, the user might draw the false conclusion that it was not there. Placing it in an arbitrary position might be more helpful, but could still mislead. In general, if an object's existence is known but not its position, where should it be placed on the map? Also, if an object's configuration or shape is known, but not its precise positioning, how can the shape be indicated? For instance, we could show the buried meanders of a river channel on a map as sinuous features indicating their shape. But if we did not know where the bends occurred, we would have to place them arbitrarily, making an unsupported prediction of their location. Or a straight channel might be drawn, giving better estimates of the likely lithology at any point, but showing a misleading pattern. Similarly, if a formation top had very different values at two wells some way apart, and the formation was known to be flat-lying with sharp vertical breaks due to steep faults, reef edges or whatever, where should the break be placed on a contour map? It could be shown at any position between the two wells, as all are equally likely, or the surface could be contoured as an even slope between the two data points despite knowing that the configuration was unrealistic. The map can show the existence, location or configuration of spatial objects. It can show known positions or likely ones, known shapes or likely scenarios. But they are different images. They cannot all be shown together and the viewer may not know which ones are represented.

The field evidence for the position of a formation boundary may rest on a combination of observations of outcrops, inferences from topographical features, analogies with the behaviour of the formation elsewhere, extrapolations from the subsurface, and knowledge of the geological setting and history. A smoothly drawn curve might indicate that the boundary was seen to follow that line in the field, or that the feature was thought from other evidence to be gently folded, or it might show a likely position where detail was uncertain. An intricately convoluted line, on the other hand, might mean that the pattern was clearly visible in the field. Or it might mean that it was drawn parallel to a nearby tightly-folded surface, or that it corresponded to a pattern of folding believed to be present throughout the area, or that it was a smooth surface intersecting complex topography. How should one depict the intersection of a smoothed surface known only from sparse boreholes and a land surface known in full detail? The shapes of lines on the map give mixed messages.

The three-dimensional picture may combine lines from a formation boundary (mapped in the field) with contours (from interpolation of borehole data) depicting the top of the same formation below ground. In some areas, the subsurface geology may be better known than the surface geology, in others the reverse may be true. The amount of detail and sources of knowledge about the boundary vary from place to place. Spatial properties measured in the field, such as location, elevation and slope, refer not to points (in a geometric sense) but to areas of ill-defined size and shape, perhaps controlled by accidents of exposure. As an interpretation develops, specific spatial relationships and patterns may emerge as salient features that should be preserved during the process of abstraction. For example, the pattern of slopes of beds in adjacent areas might suggest a particular pattern of folding, or non-parallelism of adjacent beds might suggest an unconformable relationship. These suggestions might be confirmed by detailed field observations that cannot be shown at the scale of the map. The patterns might consequently be exaggerated on the map to emphasise their significance. The map has a uniform scale but varying and unidentified resolution.

Our current interpretation influences what we observe, our interpolation procedures, and the emphasis we place on various features. As the survey proceeds, the initial work may have to be reviewed as more is learned. The evolving ideas of the surveyor make it difficult to maintain a consistent sampling scheme. Consequently records of strike and dip, for example, may be unrepresentative and unable to support statistical estimation of the average slope, amount of folding or shape of the folds. Changes in interpretation and the conventional wisdom of the day can have startling effects on a map (for example, see Harrison, 1963[241], pages 228-229). As mapping is a holistic[242] process (see The surveyor's holistic view 60), a new viewpoint may call for a complete resurvey. But even within a single

[241] Harrison, J.M., 1963. Nature and significance of geological maps, in Albritton, C.C. (editor), 1963. *The fabric of geology*. Freeman, Cooper & Co, Stanford, pages 225-232.

[242] Holistic: A view of a system that emphasises its properties and interrelationships acting as a whole, as opposed to the reductionist approach of studying its components in isolation as distinct entities.

map sheet, the learning process can introduce subtle discrepancies as the investigation proceeds. Sometimes, there may be no grounds for choosing between alternative interpretations, each with its own implications and penumbra of uncertainty, other than the desire to produce one single map.

In summary, geologists must force ambiguous, unclear and uncertain ideas into the precise geometry of the map. Consequently, the representation of marks on the map poses problems in:

- detecting arbitrary choices reflecting inadequate evidence
- determining what graphical decisions were made and why
- distinguishing between representations of existence, location and form
- recognising scenarios, likelihood, uncertainty and causes of uncertainty
- ascertaining specific controls on map content, such as stratigraphical and cartographical conventions, deductions from historical geology, inference, interpolation, analogy, measurement, observation
- distinguishing observed smoothness from uncertainty
- emphasising salient points and retaining them when generalising
- identifying, matching and reconciling features or processes of different scales or levels of detail
- reconciling maps of related topics in the same area
- identifying sources and justifications for items on the map
- determining the sampling scheme and design of the investigations
- handling multiple hypotheses
- adjusting the map to match an evolving interpretation

A particular problem in geological survey is meeting the wide range of possible requirements (see The imperfect map 145). The system design should reflect user requirements, where these are known. This may therefore be an appropriate point to tentatively consider some Diverse objectives and products 150 that can be identified at this stage.

Implementation note: A change in geological surveying practices will be necessary to gain the full benefits of the cyberinfrastructure. This can only be a long-term development but should be considered at an early stage to encourage experimentation and avoid an implementation that locks out new methods.

Diverse objectives and products

<<From map to digital model 145

Summary: *The Geological Survey spatial model should present a coherent, integrated view of a wide range of geoscience information. It should lead to testable predictions, and provide useful information for prospective users with diverse interests. Users' objectives influence their interpretation and preferred form of visualisation. The objectives are varied: establishing and testing scenarios and interpretations; estimating; predicting; generalising; categorising; recognising, detecting and separating patterns; correlating and reconciling* *surfaces; and explaining their origin. User options to select area, level of detail, content, and interpolation and visualisation procedures could be helpful. Standard default views corresponding to current map scales and formats are essential for ease of access and compatibility with older material.*

An aim of geological survey organisations is to build one coherent, integrated view of a wide range of geoscience information – a view that can be tested, verified, evaluated and updated. It should be consistent with observations, the stratigraphical framework, the concepts and processes of historical geology, and with other related models. It should tie in with work in adjacent areas and other topics. It should be possible to see all available detail, but also to generalise the results to visualise larger areas in less detail. Where the available evidence is insufficient, simplifying assumptions are needed to give a complete solution and the consequent limitations should be made clear. The model should be productive or fruitful, in the sense of leading to testable predictions for a wide range of relevant phenomena. It should be useful, in the sense of providing information of value to its potential users. For example, it might give a basis for more detailed studies (such as site investigations); broader studies (a regional search for potential oil basins); or different viewpoints (providing a model of the lithostratigraphy which could clarify a biostratigraphical analysis).

In creating or using a spatial model, the geoscientist might have one or more objectives in mind that influence the way the evidence is interpreted and the results are visualised. Examples are:

- Providing a generic view, as a basis for more specific interpretations to meet explicit objectives. Example: a Geological Survey map.
- Reconciling subjective bias through a neutral view of the evidence. Examples: to suggest and guide exploration of alternative interpretations; to resolve a dispute about, say, allocation of oil reserve estimates between adjacent leases.
- Deciphering the sequence. Example: working out the relative stratigraphical positions of the exposed beds, and establishing the local stratigraphy.
- Testing initial interpretations. Example: a model to demonstrate that identifications and interpretations being developed in the field conform to a plausible geological pattern.
- Estimation and prediction. Examples, to assess: the depth at which a well is likely to reach a particular formation; the gold reserves in a deposit; the amount of folding in an area.
- Testing predictions. Examples: matching simulations of geological processes against the observed geology; testing predictions from other topics, such as geophysics or geochemistry, against the known geology or vice versa; predicting values to compare with new or withheld data.
- Exploring best- or worst-case scenarios. Examples: estimating the largest feasible oil resources before relinquishing a lease; estimating the risk of significant faulting from test boreholes before mining a coal seam; assessing geological threats to a radioactive waste repository.

- Generalisation. Example: to obtain a regional overview by simplifying and combining detailed local models.
- Recognising or detecting pattern. Examples: visualising the spatial characteristics of a surface to throw light on its origin; testing whether a surface with specific characteristics, such as a dendritic pattern of river valleys, could fit the available data; looking for characteristic patterns to detect faults.
- Separating patterns of different scale or type. Examples: examining deviations from the regional structure to identify local anomalies that might indicate, say, data errors or wrong identifications, geological hazards (such as sink holes), or economic opportunities (oil-bearing anticlines, ore deposits); separating the effects of faulting and folding; separating depositional and structural features.
- Categorising types of surface. Examples: establishing classes of strata with similar properties (cluster analysis); characterising surfaces from high-energy and low-energy depositional environments (discriminant function); extending the classification to new areas (discriminatory analysis).
- Comparing, correlating and reconciling surfaces. Examples: comparing surfaces from seismic surveys and well picks and combining information from both sources for a more accurate view of the structure; reconciling gravity and geochemical data with a geological model (see The importance of space and visualisation 169); relating the spatial variation of a property, such as porosity, to the variation of other properties like grain size or position within a basin.
- Explaining the origin of surfaces. Examples: relating the observed surfaces to a conceptual model of their formative processes; examining the pattern of folding to throw light on past stress patterns and their variation through geological time.

There are many business settings that give direction to geoscience surveying, many different objectives, and many models for interpreting the results. Methods of modelling surfaces or mapping land-surface geology must be suited to the requirements. The emphasis placed on different sources of information varies from one application to another. Even data or basic information reflects a particular viewpoint. No map or model can meet all the conflicting objectives, each of which may call for a different approach. Diversity enables ideas to evolve, but forces difficult choices on a Geological Survey attempting to offer a broadly relevant view based on widely accepted procedures that all users can understand.

There is a strong case for providing default options of standard views matching consistent, widely accepted, expert opinions. But the processes of interpolation and generalisation are applied, explicitly or implicitly, at each stage of observation and interpretation from field observation to final product. Reuse of the objects in investigations with other objectives may therefore require reinterpretation. This is likely to be more reliable if it is based on the original observations. A better understanding of interpolation and generalisation (see Seeking shared concepts 247) may lead to a more rigorous approach; the process is nevertheless heavily dependent on human knowledge. Where practicable, a range of options, with additional information and alternative selection, interpolation and visualisation procedures, could help users to meet their more specific requirements. Access to the raw observations, the sampling procedures, and the reasoning underpinning the conclusions would help in reusing information from an investigation with one objective in order to meet another (unforeseen) objective.

The spatial model can offer a sequence of standard views, corresponding to stages of interpretation and generalisation. It was suggested (see Abstracting from reality to model 131) that identification and description are a first stage of interpretation. A second stage of interpretation is interpolation to fit lines and surfaces between points that are known from direct evidence. Further stages of interpretation involve generalisation to look at larger features at a smaller scale and coarser resolution. The stages might broadly correspond to conventional products: field notes, survey-scale maps, generalised smaller-scale maps. Indeed, such a match is necessary to incorporate legacy information in the Geological Survey model at local, regional, national and global levels of detail. Different objectives call for access at different levels of detail, where users might be able to define aspects, such as methods of visualisation, to meet their specific needs.

The range of scales at which Geological Survey maps are produced depends on standard, largely manual procedures to go from scale to scale. Scale reduction requires generalisation rather than interpolation, although both procedures may lead to similar mathematical functions (see Seeking shared concepts 247). To some extent, a spatial model can provide automatic generalisation, but expert opinion may still be needed to choose the salient and important geoscience features and properties that should be preserved on change of scale. Where different geological themes or topics are handled separately, a shared framework of generalisation stages (analogous to standard map scales) would allow the topic models to be compared and the results integrated at various levels of detail (Downs and Mackaness, 2002[243]). When Surveys can present their authoritative view of the geology as digital models, rather than geological maps, a multi-resolution view of the surveying process will lead to a more comprehensive understanding of scale-space (see Scale-space 255).

The top-down view of field mapping (see The surveyor's holistic view 60) must be carried through to an integrated modelling process. This should explicitly record the surveyors' reasoning, which could guide users in forming their own evaluation or new interpretations. The top-down view implies that local interpretations are consistent with wider knowledge. For example, the interpretation of relatively small features, such as isoclinal folds or drumlins, may be influenced by knowledge of their setting in larger-scale processes, such as plate tectonics or climatic variation. The wider knowledge is likely to involve Forward and inverse models 154, which raises additional issues.

Implementation note: Spatial models must be able to provide standard views at various scales, likely to match those of existing maps. Local, regional, national and global levels of detail may be appropriate. Users of the models could benefit from options to select, interpolate and visualise the information to meet their own specific needs.

---

[243] Downs, T.C., Mackaness, W.A., 2002. An integrated approach to the generalisation of geological maps. *Cartographic Journal,* 39(2), 137-152. http://www.geos.ed.ac.uk/homes/wam/DownsCartJournal2002.pdf

Forward and inverse models

<<From map to digital model 145

Summary: *The Geological Survey spatial model looks at the outcome of past events, and tries to work backwards to understand the processes by which it came about, a so-called inverse model. Where the forward model, predicting the outcome of a known process from the initial conditions, is clear, as in some areas of geophysics and even in stratigraphy, it may be possible to compute the inverse solution. Inverse models, however, tend to yield various alternative solutions. Supplementary information from a battery of diverse approaches, including the judgment and prior knowledge of experts, can restrict the multiplicity of hypotheses.*

The Geological Survey spatial model addresses an inverse problem, in the sense that we are not looking directly at a process affecting objects with known properties and composition and predicting the outcome (that would be the *direct problem* of establishing a *forward model*). Rather we are looking at the outcome of past events, and trying to work back to understand the processes by which it came about (an *inverse model*). The interpretation of this geological history is the key to predicting what we have not yet seen of the present situation – essential for preparing maps, cross-sections, and spatial models. The approach has been embedded in geological thinking for some considerable time, but the increasing importance of computer representation and quantitative manipulation call for an explicit re-examination of the rationale and consequences of inverse modelling.

Geologists have for some decades been developing computer-based forward models to simulate sedimentary processes (Harbaugh et al., 1999[244]). An extensive repository of modular numerical models to simulate the evolution of landscapes and sedimentary basins and the transport and accumulation of sediments and solutes is available at the CSDMS Project (Community Surface Dynamics Modeling System) (2008a[245],b[246]). The models incorporate complex relationships, random noise and feedback mechanisms. Sequences of simulations can help to predict the statistical properties of the outcome and identify the variables to which the processes are most sensitive.

Quantitative inverse models for objective evaluation of stratigraphical interpretations, and for assessing the accuracy and uncertainty of stratigraphical predictions were advocated by Cross and Lessenger (2001[247]). "[Stratigraphic] inversion is a systematic process of searching for a forward model solution that best matches observed stratigraphy". They argue that stratigraphical inversion models can provide a scientific basis for objective evaluation of

---

[244] Harbaugh, J.W., Watney, W.L., Rankey, E.C., Slingerland, R., Goldstein, R.H., Franseen, E.K., 1999. *Numerical experiments in stratigraphy: recent advances in stratigraphic and sedimentologic computer simulations.* SEPM (Society for Sedimentary Geology) Special Publication 62. SEPM, Tulsa.

[245] CSDMS, 2008a. Community Surface Dynamics Modeling System Project, Strategic Plan 2008-2013. http://csdms.colorado.edu/mediawiki/images/CSDMS_Strategic_Planv3F-48-op.pdf

[246] CSDMS, 2008b. Community Surface Dynamics Modeling System Project. http://csdms.colorado.edu/wiki/Introduction

[247] Cross, T.A., and Lessenger, M.A., 1999. Construction and application of a stratigraphic inverse model. In Harbaugh, J.W. et al (editors). *Numerical experiments in stratigraphy: recent advances in stratigraphic and sedimentologic computer simulations.* SEPM (Society for Sedimentary Geology) Special Publication No. 62, Tulsa, pages 69-84.

stratigraphical interpretations, and for assessing the accuracy and uncertainty of stratigraphical predictions, but only where adequate forward process-response models are available. Current work on quantitative inverse modelling in geology and many other disciplines suggests that the approach will be of increasing importance in geoscience, and will be relevant to many geological spatial models. It may lead to a clearer view of our methods and of the need to reconcile and build on information from many sources.

Inverse models tend to yield a number of alternative solutions, as Chamberlin (1897[248]) recognised in his method of multiple hypotheses. Gorbachev (1995[249]) provides an example in downhole logging. "...we have to establish the medium characteristics on the basis of field parameters measured in a hole, i.e. we deal with *inverse problems*. The properties of formations beyond a borehole are estimated from measurements in the hole. These measurements are of an integrated nature and are contributed to by several zones: the borehole itself, the formation adjacent to the hole that has been disturbed by drilling, the undisturbed (more distant) zone, and the surrounding beds. The influences of the zones may cancel each other so that quite different models of the medium may correspond to similar values of borehole measurements... To overcome [this] we reduce the number of possible solutions by drawing on supplementary (a priori) information. Most importantly, these are data provided by other geophysical methods based on different physical measurements". As Gorbachev points out, if solutions to a particular problem are obtained by, say, three different procedures, then the solutions form three intersecting sets, and the set of possible models is restricted to the region of their intersection. This is one reason for the availability of more than 50 basic log types based on electrical, electromagnetic, nuclear, acoustic, gravitational, magnetic, thermal, and geochemical methods. It emphasises the benefits of a battery of exploration techniques and of integrating wide-ranging evidence more effectively. Reconciling separate views of the same phenomena (such as those of stratigraphy prefixed by bio, seismo, litho, chrono, and sequence) can help to narrow the range of possibilities. A Geological Survey spatial model therefore aims to integrate information and lines of argument from many sources.

In many geological and geophysical problems, the broader picture has to be established from limited observations of local detail. With particular reference to seismic prospecting, Bleistein et al. (2001[250], page 1) state: "Our goal is to present a theory for determining the characteristics of the interior of a body based only on observations made on some *boundary surface*. In particular, we are interested in finding ways of *imaging* structures inside a body ... [and] actually determining values of certain *material parameters* characteristic to these structures." Such problems arise over a wide range of scales from those of interest in solid-Earth geophysics to those of medical tomography and material science. "The distinguishing feature of the methods in this text that makes them applicable to all these problems in that the wavelengths of the signals in our data are small, in an appropriate sense, compared to

[248] Chamberlin, T.C., 1897. The method of multiple working hypotheses. Journal of Geology. Reprinted in 1995, Journal of Geology **103**, 349-354.
[249] Gorbachev, Y.I., 1995. *Well logging: fundamentals of methods*. John Wiley & Sons Ltd., Chichester. 324pp.
[250] Bleistein,N., Cohen, J.K., Stockwell, J.W., 2001. *Mathematics of multidimensional seismic imaging, migration and inversion*. Springer-Verlag, New York. 510pp.

the length scale of the physical model." Inverse problems of this kind arise in many fields, for example, in geophysics, atmospheric science, oceanography, geophysical tomography, or trying to understand the working of the brain by observing its reactions to stimuli. They have generated an extensive literature on mathematical approaches to the inverse problem.

In a geological context, Wijns et al. (2004[251]) describe an interactive approach to their stratigraphical modelling program, guiding the solution by using expert human knowledge to assess simulations at each stage of a step-by-step refinement of an inverse model. This makes use of the experts' prior knowledge and the formidable processing power of the human brain, not just in evaluating the final result, but also in narrowing down possibilities and directing the search towards the most appropriate solution. Geological prior information is defined by Curtis and Wood (2004[252]) as that which is provided as an *a priori* component of a solution to any problem of interest. "That is, it comprises all information that pre-existed to the collection of any new or current data sets that were designed specifically to help solve the problem. Geological prior information takes many forms, ranging from basic assumptions of physics, chemistry, biology and geology, to the design of the problem to be solved, and to the use of prior experience from previous studies in order to interpret new data and provide a solution." Howard et al. (2009[253]) emphasise the importance of prior knowledge to geological surveying, and the need to record how it is used in the field or office during the inverse-modelling process of geological interpretation. The aim of the inverse approach in geological surveying is likely to be selection of the most appropriate forward models with maximum predictive power, evaluated by the accuracy of representing relevant aspects of real world geology as tested against new or withheld information. Its limitations are considered further in The imperfect model 156.

The imperfect model

<<From map to digital model 145

Summary: *The general model of the systems of the solid Earth refers to the all-embracing process in three spatial dimensions and geological time by which the Earth and its component materials evolved from their birth to the present day. As a direct model, it is complex and poorly understood. Available evidence is inadequate and unrepresentative. The usual strategy for investigation is to simplify by classifying the continuum into hierarchical sets of subsystems, objects, properties, processes, states and events. The overall model*

---

[251] Wijns, C., Poulet, T., Baschetti, F., Dyt, C., Griffiths, C.M., 2004. Interactive inverse methodology applied to stratigraphic forward modelling. *In* Curtis, A. and Wood, R. (eds) 2004. Geological Prior Information: Informing Science and Engineering. Geological Society, London, Special Publications, **239**, 147-156. 186239-171-8/04/$15.00.

[252] Curtis, A., Wood, R. (editors), 2004. *Geological prior information: informing science and engineering.* Geological Society, London, Special Publications, **239**.

[253] Howard A.S., Hatton B., Reitsma, F., Lawrie, K.I.G., 2009. Developing a geoscience knowledge framework for a national geological survey organisation. *Computers & Geosciences*, **35**, 820–835. doi:10.1016/j.cageo.2008.06.004 http://nora.nerc.ac.uk/7128/1/Author_final_Howard2009C%26G.pdf

brings these sets together as an account of Earth history which, by looking for features that imply a particular cause, can be tied to observations of the present-day situation. Within this framework, object classes and their expected properties, form and behaviour can be identified at any level of detail, defined in metadata, depicted on a map, manipulated in a model, and described and discussed as a text narrative.

The obvious difficulty in considering the inverse model[254] (see Forward and inverse models 154) for geoscience as a whole, even in a broad non-mathematical sense, is that it relates to a complex direct model that is only broadly understood. In principle, The solid Earth systems model (sEsm) 71 refers to the all-embracing process, in three spatial dimensions and in geological time, by which the Earth evolved from its birth to the present day. It is a view of Earth systems – systems in the sense of sets of interacting parts that function as a whole. Most of the component sub-systems operated under extreme conditions and over a time scale that cannot be observed or reproduced experimentally. Most of their products have been reworked and no longer exist. Thus they too cannot be observed, although their past configurations determined later developments. Even where present-day analogues of ancient processes are available for study, for example in sedimentology, random variations rule out exact prediction of their outcome. Processes interfere with one another, such as weathering affecting the results of sedimentation, and their consequences cannot always be disentangled. Faced with this complexity, the sparse available evidence seems inadequate and unrepresentative. There is no prospect of capturing the totality of the overall process in a set of equations and solving for an inverse model. Instead, to see how information technology can contribute, we might start by looking at how geoscientists traditionally tackle the daunting task of studying this complicated system and how they are able to draw useful conclusions.

Simplification by classification is an initial strategy for handling complexity, lumping together much detailed variation. The space-time continuum can be classified into sets of discrete subsystems, objects, properties, processes, states and events. Their roles in Earth history provide a broad framework, giving an account of the historical geology. This conforms to Simpson's (1963[255]) distinction between configurations[256] of objects that evolve in geological history in partly unforeseen ways, and the processes (based on immanent laws) that change them (see Abstracting from reality to model 131, Overview of the solid Earth systems model 26). The historical view of objects and processes relates them in space and time to one another (rather than to absolute co-ordinates) because the processes operated regardless of location and to some extent of scale, and there was no Greenwich Meridian and few time-stamps on the ancient Earth. The complexity of the representation, and of our reasoning, can be simplified, where appropriate, by projecting the four-dimensional situation onto

[254] Inverse model: A model addressing the 'inverse problem' of looking at the outcome of past events, and trying to work back to understand the processes by which it came about.

[255] Simpson, G.G., 1963. Historical science, in Albritton, C.C. (editor), 1963. *The fabric of geology*. Freeman, Cooper & Co, Stanford, pages 24- 48.

[256] Configuration: The spatial arrangement, pattern, form and shape of objects or their properties. Used by Simpson (1963) in contrasting 'The actual state of the universe or any part of it at a given time, its configuration, is not immanent and is constantly changing' with 'The unchanging properties of matter and energy and the likewise unchanging properties and principles arising therefrom are immanent in the material universe.'

fewer dimensions. Thus, a stratigraphical table might show only the time dimension, a generalised vertical section might show only depth, and a cross-section or map might be limited to two dimensions.

This simplified view of processes operating on configurations of geological objects enables us to identify significant properties and key features that we can look for as evidence of the sequence and mode of origin of specific rock bodies. In plate tectonics, for example, the plates might be regarded as spatial objects, and related to structural, sedimentary, igneous and metamorphic events. The events are tied to episodes in the history of the plates (collision, subduction, and so on) that created rock bodies permanently imprinted with characteristic features. The broad non-quantitative inverse model can thus be tied to observations located on a map, by looking for features that imply the cause.

The general model provides a consistent basis for more detailed studies. The strategy of classifying the continuum into named object classes (such as deltaic deposits) can be carried down to any level of detail (see The object-oriented perspective 179). It simplifies by bringing together within a single concept many ideas about expected properties, form, and behaviour. Named instances of the object class (such as Recent deposits in the Mississippi Delta) inherit the properties of the class (see The geological framework model 105).They can be depicted spatially in a map, model, or sketch, can be manipulated in a model and discussed in a text account. At these more detailed levels in the hierarchy, the overall system can be split into subsystems that deal with a specific area, stratigraphical range, topic, level of detail, map scale, objective, or means of investigation. This allows us to focus on areas and topics of immediate concern (such as stratigraphical formations near the land surface within a defined area), or on specific objectives (such as estimation of ore reserves). Detailed studies are thus undertaken against the background of existing knowledge and within the broader historical framework, which itself resulted from generalisation of earlier detailed studies and is continually modified as more is learned.

Each individual study simplifies the immense complexity of the real world according to its own unique perspective, viewed from its business setting and objectives (see The geological business model 93). Some projects, notably seismic surveys, apply inverse models referring to the procedures of investigation, in order to place the outcome of the project in a broader context. The outcomes of projects from different sources are generally integrated by visual and statistical examination of spatial patterns and distribution. Reconciliation of results from a diversity of studies with different topics and objectives (see Reconciliation 186) can narrow the range of possible interpretations (see Forward and inverse models 154). Matching the detail from separate projects requires the prior knowledge of expert geoscientists, and possibly interdisciplinary collaboration through interactive computing. There is an inevitable trade-off between the benefits of collecting information according to rigid standards that enable results to be more readily shared, and associated costs that may be unnecessary for the immediate objectives. However, there are problems other than the difficult implementation. Dynamic stratigraphy must deal with Complex and emergent systems 159.

Implementation note: A helpful step towards easier spatial integration in the modelling environment is to establish standard procedures for representation, interpolation and visualisation. The standards could best be developed within a global, forward-looking framework that could also carry legacy information into its future setting. Provision must be made, however, for integrating non-standard information.

Complex and emergent systems

<<From map to digital model 145

Summary: *The reductionist mode of explaining complicated phenomena reduces them to simple parts controlled by mechanical processes governed by the deterministic laws of physical science. However, many geological processes belong mathematically in the realm of complex, self-organising systems. Adjacent parts of these interact according to simple rules, without any central control. But feedback in the process means that effect is not proportional to cause, and the linear equations of physics do not apply. Patterns emerge, and can be described in terms of attractors associated with preferred states. But unpredictably minute changes to the initial conditions can lead to quite different outcomes. The patterns arise over a range of scales, and tend to form hierarchies, just as stratification, say, can occur from microscopic detail to all levels of stratigraphical units. The diversity of models relevant to geological entities calls for a flexible system, and is well suited to the object-oriented approach.*

Mathematical models are effective tools in fields like seismic exploration and downhole logging. The investigators can to some extent control the conditions and express the physical processes in terms of linear equations. Some geological processes, however, may be less amenable to such methods. Mathematically, many belong in the realm of complex, self-organising systems. Readable introductions to this subject include Nicolis and Prigogine (1989[257]), Heylighen (2001[258]), Bar-Yam (2002[259]) and references to the extensive literature can be found in the Usenet Newsgroup (2008[260]). The systems approach offers an alternative to the "*reductionist*" mode of explaining complicated phenomena that reduces them to simple parts controlled by mechanical processes governed by the deterministic laws of physical science. This mode is typical of the forward[261] and inverse[262] models[263] described in Forward and inverse models 154 and The imperfect model 156. Instead, the systems approach looks at characteristics of the system as a whole (see also The surveyor's holistic

---

[257] Nicolis, G., Prigogine, I., 1989. *Exploring uncertainty.* W.H. Freeman, New York. 313pp.

[258] Heylighen, F., 2001. *The science of self-organization and adaptivity*, in: The Encyclopedia of Life Support Systems. EOLSS Publishers Co. Ltd. (in press). http://pespmc1.vub.ac.be/Papers/EOLSS-Self-Organiz.pdf

[259] Bar-Yam, Y., 2002. "Significant points" in the study of complex systems. http://www.necsi.org/projects/yaneer/points.html

[260] Usenet Newsgroup comp.theory.self-org-sys, 2008. Self-Organizing Systems (SOS) FAQ: Frequently asked questions. http://www.calresco.org/sos/sosfaq.htm

[261] Forward model: The model of a process affecting objects with known properties and composition and predicting the outcome.

[262] Inverse model: A model addressing the 'inverse problem' of looking at the outcome of past events, and trying to work back to understand the processes by which it came about.

[263] Model: A formalised representation giving a simplified view of aspects of the real (or of an imaginary) world relevant to the purposes in hand.

view 60). It provides fresh insights in many varied fields, such as biology, ecology, evolution, market theory and thermodynamics, exposing characteristics and structures that they have in common. There are possible applications to the social aspects of information science concerned with representation and communication of geological knowledge, and to the aspects of brain science concerned with understanding it (see Stages of concept development (geological thinking) 56). However, our main interest here is in the light that complex systems may throw on geoscience processes.

A *system* is defined as a set of interacting parts that function as a whole. The investigators decide where to place the boundaries and interfaces that separate the system from its environment (see Boundaries: discontinuities and zero-crossings 264), and that subdivide the system into component *subsystems*. Properties of the system that can change and take on different values are referred to as *variables*. *Phase* or *state space* is a mathematical abstraction in which each variable is regarded as a separate dimension. It can thus represent any conceivable state of the system, an unimaginably vast number of possibilities (see The solid Earth systems metamodel (sEsmm) 110). In reality, the number of possibilities is reduced because natural systems show a high degree of organisation. *Organisation* is defined as the arrangement of parts so as to promote a specific function, in this case to maintain a particular configuration in spite of disturbances that would otherwise disrupt it.

A system may show patterns that appear to arise spontaneously by "*self-organisation,*" that is through the interaction of adjacent parts according to simple rules, without any central control. The granularity of sedimentary rocks, their stratification at any level of detail, sedimentary structures, folding, faulting, volcanic activity, even tectonic plates and convection cells, show patterns that may result from self-organisation. The system in question could be the broad geoscience model that includes the major processes in the Earth's crust, but also those in the deep Earth and in the oceans and atmosphere. More usually, a smaller subsystem and its boundaries are selected to define a more manageable problem. However, even a small sedimentary structure cannot be understood without reference to local interactions between major processes that determine, say, the nature, form and slope of the substrate, the availability of sediment entering and leaving the system, and the operation and strength of currents. The investigators' definition of the system boundaries determines whether the input and output of matter and energy as sediment and currents are part of a closed local system or are seen as crossing the interface from its external environment.

The form of a sedimentary structure has a local effect on the current. For example, the flow of air might develop an eddy in the lee of a sand ripple. The ripple creates the eddy, which deepens the trough of the ripple, which strengthens the eddy, which deepens the trough some more, and so on. This process where effect is fed back to cause (the input and output are connected by a causality loop) is known as *feedback*. It can be *positive*, where the results intensify the process, as in this example. It is *negative* where the feedback brings deviations back towards the average value. As the ripple grows in size, it might reach a height where the eddy tends to break away from the surface. This provides negative feedback that inhibits further deepening of the trough.

One consequence of the feedback effect is that complex systems cannot be explained by the *linear* equations familiar in physics, where effect is proportional to cause. The interaction of nearby components of the system, such as saltation of grains on a depositional surface, can produce surprisingly complex patterns, as you see when walking across a sandy beach on a windy day. Even although each grain interacts only with its neighbours, the pattern or structure covers a much larger area than the individual component grains. The structure interacts with the wind pattern, which again might be seen as the result of local interactions between air masses in the atmosphere operating on a different scale from the sand grains. The properties that emerge from this self-organisation are known as *emergent*. Because of them, we cannot assume that the behaviour of the system as a whole can be explained by building up a picture from the behaviour of the parts. Instead, we must view the system as a single, coherent and organised whole, with properties that cannot be reduced to those of its components.

The tendency of the system to create organised processes and objects can be described in terms of *attractors* – preferred positions in state space, such that if the system starts from another state, it tends to evolve towards an attractor. The pattern of attractors can be visualised in terms of a *fitness function,* which calculates a value measuring the degree of suitability or desirability of a particular state. In three dimensions of state space, two of them representing property variables and the third (vertical) the fitness function, we can imagine a fitness landscape where states are plotted at an elevation that represents their fitness. On a smooth landscape with a single peak, the system will evolve by moving from its current position on the landscape up the line of steepest slope until it reaches that single attractor. A rugged fitness landscape, typically the result of wider interactions between the variables, will have not just a highest point, but also many lower foothills, each with its own highest point. If the system evolves by following the slope upwards from an arbitrary starting point, it is likely to arrive at, and remain at, the top of a foothill, the nearest sub-optimal attractor. Physicists prefer to think of a ball rolling down the landscape with the highest values of the fitness function in the valley bottoms, but their diagrams give the same message upside down (they seem not to worry about their ball trickling away down-river). The huge number of dimensions in state space cannot all be visualised simultaneously, but three-dimensional systems behave like those in more dimensions.

Random elements in the processes can introduce variety and allow the situation to evolve rather than sticking in a stable state associated with one attractor. The random elements allow the system to explore state space by shaking it away from sub-optimal peaks to a new starting point from which it could reach alternative attractors. The valley bottoms represent critical points in the system where small random changes can create large effects, moving the system out of one stable state into the zone of influence of another attractor. Cross a small river, and moving up-slope takes you to the summit of a different hill. "Which of the possible configurations the system will settle in will depend on a chance fluctuation. Since small fluctuations are amplified by positive feedback, this means that the initial fluctuation that led to one outcome rather than another may be so small that it cannot be observed. In practice, given the observable state of the system at the beginning of the process, the outcome is therefore *unpredictable*." (Heylighen, 2001[264], section 3.7). The simple rules of self-organisation between adjacent parts of the system can thus create complex behaviour in the system as a whole. In such a system, quantitative inverse modelling may not be feasible, prediction may be uncertain, and because the system destroys information as it evolves, the complete past history can never be recovered.

The objects, processes and feedback mechanisms operate within systems at different scales or levels of detail, such as individual grains, individual ripples, the rippled bed as a whole, the sequence of beds of which it is a part, facies, formations and so on up the hierarchies of stratigraphy. As larger systems are modelled, their component parts are also larger and more generalised. "...a self-organizing system may settle into a number of relatively autonomous, organizationally closed subsystems, but these subsystems will continue to interact in a more indirect way. These interactions too will tend to settle into self-sufficient, 'closed' configurations, determining subsystems at a higher hierarchical level, which contain the original subsystems as components. These higher level systems may interact until they hit on a closed pattern of interactions, thus defining a system of a yet higher order. This explains why complex systems tend to have a hierarchical, 'boxes within boxes' architecture, where at each level you can distinguish a number of relatively autonomous, closed organisations" (Heylighen, 2001, section 3.6). In a self-organising system, attractors may encourage similar patterns to develop over a range of scales. Mandelbrot (1982[265]) explored fractal models that generated this *self-similarity* (or *self-affinity* where similar patterns stretch or skew as they grow larger). Geologists are familiar with the idea that microfolds can mimic large nappes, and trickles of water on a mud bank can create miniature deltas, presumably because the objects and processes can be scaled up, at least over a limited range.

Over the last few decades new approaches and concepts have been developed for computer backed investigation of complex systems. Complex systems appear to be more familiar to geographers and ecologists than to geologists. Baas (2002[266]), for example, introduces his study of coastal geomorphology with an account of concepts of chaos theory, fractals, attractors, self-organisation and self-organised criticality. Van Wagoner et al (2003[267]) reported a major breakthrough in dynamic modelling of siliciclastic sedimentary bodies. Their investigation, of unprecedented extent and detail, looked at the shape of sedimentary bodies, ranging in length from a few centimetres to 1000 km. "From the shape alone it is impossible to determine the size or depositional environment of these bodies. Thus, shape is independent of scale and place of deposition." Their conclusions are relevant to our present purposes for two reasons. First, if the shape of diverse sedimentary bodies is similar, this bears on methods of interpolation (The geometry of the spatial model 203, Seeking shared

[264] Heylighen, F., 2001. *The science of self-organization and adaptivity*, in: The Encyclopedia of Life Support Systems. EOLSS Publishers Co. Ltd. (in press). http://pespmc1.vub.ac.be/Papers/EOLSS-Self-Organiz.pdf

[265] Mandelbrot, B.B., 1982. *The fractal geometry of nature*. Freeman, San Francisco, 460pp.

[266] Baas, A.C.W., 2002. Chaos, fractals and self-organisation in coastal morphology: simulating dune landscapes in vegetated environments. *Geomorphology*, **48**, 309-328.

[267] Van Wagoner, J.C., Hoyal, D.C.J.D., Adair, N.L., Sun, T., Beaubouef, R.T., Deffenbaugh, M., Dunn, P.A., Huh, C., and Li, D, 2003. Energy dissipation and the fundamental shape of siliciclastic sedimentary bodies. *Search and Discovery* Article #40080. http://www.searchanddiscovery.com/

concepts 247). Second, their arguments are based on the study of complex systems, and have wider applications.

According to Van Wagoner et al. (2003): "empirical and statistical similarities in shapes indicate that these bodies were deposited by a common physics. The physics at the local instantaneous scale are the well-established laws of fluid and sediment dynamics. However, these dynamics do not explain the cause of the global organization of the bodies observed in nature. A deeper, more encompassing explanation is required. We believe that the explanation can be found in nonequilibrium thermodynamics and energy dissipation... All open systems (i.e. systems through which energy and matter are transmitted) evolve toward increasing complexity with time ... as these systems form dissipative structures to minimize gradients..." They conclude: "the sedimentary rock record is built of scale-invariant hierarchies of sedimentary bodies... similar in shape and property distribution... evolv[ing] along a well-defined pathway... scale-invariant and independent of depositional environment." The results obtained by these investigators suggest that the complex systems approach will lead to equally significant findings in other fields of geoscience.

The local systems that a geologist is likely to study are set in the environment of a hierarchy of more general systems, including that represented by The solid Earth systems model (sEsm) 71 (see also The imperfect model 156, Relationships between objects 183). They adapt to that environment by optimising their fitness function. Present-day spatial objects reflect the stage that the adaptation has reached. On a geological time-scale they are still in a dynamic process of change, even although they appear static on the human time-scale of geological survey. Indeed, we may recognise objects frozen in time at different stages along evolutionary paths towards attractors at various levels in the hierarchy, such as temporary lakes and lacustrine sediments in the evolution of a river valley. The distinct environments and separate scales on which major processes operate (such as tectonic, climatological, volcanic and sedimentological processes) complicate the study of geological systems. Nevertheless, the concept that hierarchies of subsystems evolve towards attractors offers a good match to the hierarchical classifications (see The imperfect model 156) by which geologists make their subject manageable.

The study of complex systems seems, reassuringly, to track ideas that have long been part of geological understanding. It does so in a framework that recognises the limitations of reductionist mathematics and the uncertainties of prediction, and that supports an integrated top-down view (see The surveyor's holistic view 60), clarifies concepts, and shares new insights across many applications. It demonstrates how a hierarchy of processes can create recognisable patterns across a range of scales. The details of the overall process may be obscure and the outcome partly unpredictable. Nevertheless, patterns can be mapped, analysed and to some extent explained. Uncertainty has always been recognised in geoscience investigation, and predictions are accurate only in a statistical sense. Reductionist methods, direct, inverse and many other types of model have proved their worth in understanding the Earth and predicting its properties. Complex systems can be seen as another tool that helps us to understand the processes, the uncertainties and the occasional big surprise. The prospect of thinking of the stratigraphical record as snapshots of dissipative structures in a complex system may seem remote. Nevertheless, multi-scalar models that could accommodate such developments already appear to be feasible (see Boundaries: discontinuities and zero-crossings 264).

The aim of this section is to throw light on how we can align current systems developments with future trends and underlying user requirements. One conclusion must be that only a flexible system could cope with the diversity of needs and possible models. It must be possible to accept information of many kinds, levels of detail, sources, and topics, and relate it to one uniform view of the underlying geology that aims to tie observations and predictions to spatial co-ordinates (see At the interface 133) and thence to the real world. This requirement, set against current solutions, points to An object-oriented approach 178 for the spatial models of a geoscience survey, and the development of a comprehensive system (see Reconfiguration 165) in which a range of diverse models can be brought together.

Implementation note: Future developments of The solid Earth systems model (sEsm) 71 are presumably unpredictable, because of its own emergent properties. Nevertheless, when travelling, it can help to have an eventual destination (or attractor) in mind. One approach to finding the unpredictable route is (as here) to *take a view*, tentatively planning for the longer term on a basis of instinctive preferences and prejudices. Another is *contingency planning*, considering a range of possibilities and trying to position a response that secures gains and steers clear of catastrophes. A third is the *second-mouse* approach (named after the one that collects the cheese from the sprung trap) delaying decisions until others have completed the early experiments. A fourth is *sub-optimisation,* building in small improvements as they arise, regardless of longer-term consequences. A fifth is to *wait and see*, ignoring potential risks and benefits.

Reconfiguration

<<Table of contents 1

<<The future geological map 125

<<From map to digital model 145

Reconfiguration 165

Many models, one system 166

Projects and information communities 167

The importance of space and visualisation 169

Reconfiguring the system 172

Representing spatial information and relationships 173

Mark-up and metadata 175

>>An object-oriented approach 178

Abstract: Geological investigation involves widely diverse projects, separated by their objectives but linked by a broad understanding of the subject. Geological survey organisations provide regional geological interpretations and located spatial information on which other projects can build – a view of the anatomy of the Earth and the core of reasoning for many topics in geoscience. They therefore have a particular responsibility for reconfiguring geological information as a component of the emerging comprehensive knowledge base[268] and its supporting advanced cyberinfrastructure[269]. The concepts and rationale behind surveying described in Reasoning, models and reality 127 suggest that the models generated by surveying (described in From map to digital model 145) can and should be brought into a more comprehensive framework. The aim is to provide widely integrated, authoritatively evaluated, computer-based results of geoscience survey, representing a core component of regional geological knowledge. Unlike the traditional map, the representation need not be constrained by the form of visualisation, content, scale, sheet boundaries or number of dimensions. It can provide hypertext links to many sources, with closer integration of text, images, metadata, computer databases, programs and a wide range of applications and models. Mark-up and metadata are a means of linking diverse sources and information environments, leading to An object-oriented approach 178 to geological investigation.

[268] Knowledge base: A dynamic repository for information and methods for accessing and processing it. It is generally machine-readable and online, and may include the means to access expert knowledge.

[269] Cyberinfrastructure: An integrated assemblage of computing, information and communication facilities, deploying the combined capacity of multiple sites to provide a framework to underpin research and discovery, typically with broad access and end-to-end coordination.

Many models, one system

<<Reconfiguration 165

Summary: *Diverse sources of geological information, from many projects, information communities and scientific disciplines can potentially be coordinated by reconfiguring their individual models as components of one, widely shared, more comprehensive system. Spatial information and concepts are important aspects which can be quantified with a methodology that relates to human perception, can be integrated with other types of information, and can be supported by the cyberinfrastructure.*

It is suggested in Abstracting from reality to model 131 that geological mapping reflects the surveyors' holistic[270] view of the geology. The surveyors bring all their relevant background knowledge to bear on the task in hand. Much of their thinking is spatial: recognising patterns and relationships in surface information from landscape features and outcrops, or in subsurface information such as seismic surveys and drill-holes. They may visualise from fragmentary information how rock bodies are positioned and arranged, and how they relate to one another and to the geological history of earlier configurations[271], geological processes and events. This may guide their search for additional observations to clarify and test their evolving view. Such an investigation focuses on aspects relevant to the task in hand, which might concern, say, an academic study of the palaeoecology, or an estimation of the capacity of an aquifer, or a systematic regional survey of the regional geology that could underpin more detailed projects.

Conventionally, the results were communicated as maps, cross-sections and text reports, illustrated by photographs, diagrams and sketches, perhaps accompanied by field records, datasets, samples and specimens. Digital records can be more comprehensive, efficient and flexible, can model the geologist's holistic thinking more exactly, and can link directly to digital tools for data collection, analysis and visualisation. Digital models have been developed for many aspects of geology, not least for regional geological surveys. However, the geologists' holistic view suggests that there is one underlying system[272] in which these models[273] can be considered as subsystems. It should therefore be possible to relate them to a single structure for the geological knowledge system, referred to in this scenario as The solid Earth systems model (sEsm) 71, where they could be serviced by The geological cyberenvironment (gce) 85. A scenario[274] for such an approach is considered in The emerging geoscience knowledge system 42.

The future geological map 125 suggests how the procedures of geological mapping might change in that environment, from the traditional production of maps and reports to an

---

[270] Holistic: A view of a system that emphasises its properties and interrelationships acting as a whole, as opposed to the reductionist approach of studying its components in isolation as distinct entities.
[271] Configuration: The spatial arrangement, pattern, form and shape of objects or their properties.
[272] System: A set of interacting parts that function as a whole. The systems approach involves study of linkages or interfaces between the component activities.
[273] Model: A formalised representation giving a simplified view of aspects of the real (or of an imaginary) world relevant to the purposes in hand.
[274] Scenario: A description of a plausible, though uncertain, outcome.

integrated collection of information within a comprehensive framework. The definition of mapping is seen as evolving to reflect more faithfully the geologists' holistic view. Mapping must continue to respect and enhance the unique abilities of the human mind, particularly in visualising and manipulating spatial information. However, it can also build on quantitative methods made possible by the cyberinfrastructure. The geometrical aspects, involved in recording, simulating and visualising geological processes and their results, have exact algebraic equivalents. The algebra is essential for the computer processing, but also provides the links to geological process models and to rigorous statistical analysis of appropriately sampled data. This implies that there will be major changes in the geologists' approach to surveying, in order to achieve rigorous results that can be more widely shared with those developed in other disciplines (see Seeking shared concepts 247).

It seems likely that most geologists are more adept at handling imagery, visualisation and geometrical concepts than they are at algebraic manipulation and reasoning, and of course they should not rely on techniques that they do not fully understand. The geometry of the spatial model 203 therefore attempts to relate algebraic techniques for processing and analysing geological information to familiar geometrical concepts such as location, slope and curvature. They are fundamental to fulfilling the potential of the systems model to bring together information from many disciplines, from many Projects and information communities 167 , and to achieve a more comprehensive concept of geological mapping.[275]

Projects and information communities

<<Reconfiguration 165

Summary: *Geological Surveys are information communities that offer standard and widely applicable views of the basic geoscience of an extensive area. Many projects can build on this foundation in response to specific business needs, with results that may be published in journals. Bibliographical searches can access the content of many journals electronically, but a geographical and stratigraphical framework is required to structure and access a comprehensive spatial model. Project articles are assessed by editorial and peer review, whereas the Survey applies internal procedures for quality assessment.*

Any geoscience investigation is likely to involve a project. Each project is undertaken within a particular business setting. This might be oil exploration, land use, water extraction, geological research, education, or some other activity. Each project has its own objectives, priorities, operational definitions, sampling schemes, and the like. These are tailored to the business needs, and determine the procedures and products. This context must be explained to give the results meaning beyond the project. Rather than every project starting from

---

[275] Mapping: Conventionally, geological mapping leads to a graphical depiction, usually on a flat surface, of spatial relationships and forms of geological features or properties in a selected area of the Earth's surface or subsurface. As defined in mathematics, mapping relates the elements of one set to those of another. A broader definition of geological mapping could be 'relating elements of geological observation or interpretation of the solid Earth to corresponding elements in appropriate models in the geoscience knowledge system'.

scratch and proceeding independently, it is more efficient to build on a shared knowledge of the basic geology. The task of providing and maintaining this shared resource is so complex that it is generally delegated to established information communities, such as geological survey organisations (of which there are more than 150 worldwide). Geological Surveys aim to provide (in the form of maps and reports) a quality-assured, standardised, view of the basic geoscience of an area.

It is assumed that, for reasons of cost and convenience, the internet, the semantic Grid and their successors will become the primary means of delivering geological information. Users will expect, not only geological maps and reports meeting current global standards, but also a wide range of related information, all accessed through one familiar user interface and compatible with their own information systems. The implications for geoscience require careful consideration by geoscientists. The present scenario is only one of many possibilities for longer-term development of geoscience survey, and of course, assumptions may be premature and views incorrect; we must not squander our unrivalled legacy of conventional maps and reports; and new methods should not be allowed to displace good science.

The internet and World Wide Web give rapid, convenient, low-cost access to the hypermedia knowledge repository called cyberspace. This contains a chaotic, swirling sludge of trivial and often objectionable ephemera. It also offers the serious scientist (at a cost) an up-to-date archive of peer-reviewed, catalogued articles, arranged more conveniently than in any library, instantly available, and linked to references and multimedia extensions at a mouse-click. Taking the scientific literature as its model, it is well suited to recording the outcome of individual projects. But long-standing information communities require a different model to build and maintain a coherent comprehensive view of their chosen topics (see Geological survey documentation 6). In Geological Surveys, spatial models require a well-structured geographical and stratigraphical framework (see The geological framework model 105) to organise the records of a Survey's core activity. Survey organisations are thus well placed to provide the core of regional geological information in the comprehensive knowledge base[276] foreseen by e-scientists as a component of the cyberinfrastructure[277]. It is obviously more likely to flourish within a culture where exchange of information is encouraged and rewarded, than in one where information is stockpiled as a weapon in the internal power struggle.

Like a network of major roads, the framework should provide a means of reaching significant points of interest. From there, it should be possible to transfer to a branch network of minor roads and footpaths for accessing more obscure but related points. Like a map, a spatial model might indicate the stratigraphy of an outcrop and its surroundings, but it could also link, say, to descriptions of fossils from the same locality that had been collected independently in an academic project, to detailed logs, and to alternative interpretations.

---

[276] Knowledge base: A dynamic repository for information and methods for accessing and processing it. It is generally machine-readable and online, and may include the means to access expert knowledge.

[277] Cyberinfrastructure: An integrated assemblage of computing, information and communication facilities, deploying the combined capacity of multiple sites to provide a framework to underpin research and discovery, typically with broad access and end-to-end coordination.

The hyperlinks to other items in the hypermedia knowledge repository, including databases and project-based scientific literature, place the spatial models in a wider context.

An information community responsible for routine surveys evaluates information by a process that differs from the peer review of the scientific literature. It has internal procedures to ensure that standards are met, and to maintain the integrity of its 'brand name'. Its view implies an evaluation of external information and vice versa. If, for example, the fossil descriptions just mentioned suggested a different stratigraphical position from that recorded by the Survey, the Survey might revise its opinion. If it did not change, users can at least identify the conflicting opinions, make up their own minds, and if appropriate record their own view in a scientific paper.

A Survey might gain by offering an archival store for external contributions, as these add value to the maps and models and vice versa. The framework for basic geoscience knowledge must also connect with a larger body of related information, structured in other ways. The British Geological Survey developed their GeoIndex to explore such essential links (Adlam et al, 1988[278]; BGS, 2010a[279]) building on The importance of space and visualisation 169.

Implementation note: Computers can help in the preparation of indexes to a Survey's published maps and reports, including graphical indexes of areas covered. Standards and software are well established. The results can be made available on the internet. Indexes should be complete, accurate, and up-to-date (even if the material to which they refer is not) as otherwise they may mislead. They should be reliably available on a permanent basis, evolving with changing standards. This implies that a credible curatorial group must take responsibility in order to safeguard the users' investment in training and equipment, including a commitment to migrate to new systems when appropriate.

The importance of space and visualisation

<<Reconfiguration 165

***Summary**: The map or spatial model locates the geoscience* objects *(things of interest) and their property variations, establishing patterns and spatial relationships – a key to applications and to interpreting geological events and untangling their history. The* interpretation *integrates diverse sources of information, fills gaps in the observable record, and completes a realistic view of the three-dimensional geology.* Computer visualisation *combines the scientist's abilities to conceive, analyse, interpret and compare images with the computer's strengths in data handling and graphical presentation. The spatial model should support visualisation from a stable model conforming to developing global standards.*

---

278 Adlam, K.A.McL., Clayton, A.R., Kelk, B., 1988. A 'demonstrator' for the National Geosciences Data Index. *International Journal of Geographical Information Systems*, **2**(2), 161-170.

279 BGS, 2010a. GeoIndex. http://www.bgs.ac.uk/geoindex

The user of geoscience information is frequently concerned with spatial information and its visualisation. Where is the mineral deposit? What is its spatial form? Where are the geological hazards? Where might the unconformity be exposed? Where do the faults intersect, and in what sequence? The map or spatial model can give a direct answer to such questions. Spatial considerations also have particular value in integrating aspects of geoscience. It is ultimately through spatial relationships that geological events can be related and their history untangled, as William Smith determined long ago (Winchester, 2009[280]).

Widely differing properties of the Earth's crust may exhibit similar spatial patterns in the same area, for the obvious reason that they reflect the influence of the same geological processes. Consider, however, a comparison of a map from a gravity survey with one from geochemical analyses of stream sediments. Each might map the same hidden granite pluton, but show a rather different pattern. One reflects the density distribution at depth, the other mineralogical changes around the granite, modified by later stream transportation. Comparison of individual points from the two data sets is unlikely to be informative, particularly as the sampling points do not coincide. Mechanical comparison of mapped data might also fail to identify the common cause. The interpretation depends on comparing complete patterns and on knowledge of geological processes and of the surveying procedures. It is only geoscientists who possess this knowledge and can therefore integrate results from the separate projects. With appropriate support, including geographical information systems, they are well placed to visualise the observed patterns and, using their background knowledge, compare them with likely geological scenarios. The spatial model can assist the integration by assembling, correlating and displaying information from diverse sources.

The spatial model is also central to studies of the evolution of an area through geological time. For example, a similar contour pattern might be displayed by each of a sequence of subsurface formation tops and horizons. This could be explained by geological controls of deposition and deformation remaining in place for prolonged periods, and evolving in a comprehensible fashion. The elevations are likely to be measured at the same wells, which makes comparison easier. However, a seismic survey of the area might also show a similar pattern despite referring to different sampling points and different properties. Again, visualisation and interpretation by human geologists are essential. They provide the link from the observations and measurements of the real world (possibly from disparate sources sharing one spatial framework) to an integrated understanding of their causes, relationships and consequences in space and time.

We can thus identify visualisation[281] as essential both to creating and using the spatial model. Computer visualisation is a powerful technique for clarifying spatial patterns and relationships, and supports queries such as "What is?", "Where is?" and "What belongs together?" (MacEachern and Kraak, 1997[282]). It exploits the computer's strengths in data

[280] Winchester, S., 2001. The Map that changed the World. Penguin Books, London.

[281] Visualisation: Transforming quantitative data (including the results of interpolation) into sensory information – images that the eye and brain can interpret and visualise.

[282] MacEachren, A.M., Kraak, M-J., 1997. Exploratory cartographic visualization: advancing the agenda. *Computers & Geosciences*, **23** (4) 335-343.

handling, processing and graphical presentation by integrating them with the abilities of the human eye and brain to conceive, analyse, interpret and compare images. It enables the brain to utilise background knowledge (not available to the computer) of geoscience; the history of the geological states, processes and their consequences; and the objectives and procedures of the study.

Existing maps and sections are fixed visualisations within this framework. Spatial models can create many visualisations, selected and controlled by the user rather than the surveyor, but geologically consistent because they originate from the same model. However, the user interface (the browser on the scientist's desktop or portable computer in the field) must access much more. The overall model is dependent on a wide range of supporting material that should be indexed and preferably available in full to the browser. The records are varied and may be from many business settings and of many kinds: maps, sections, stratigraphical tables, sketches, logs, photographic and other images, text explanations and descriptions, data, databases, metadata, computer programs, models, and so on. The available information varies greatly from one area to another, and will change rapidly as digitisation proceeds. A browser with spatial and stratigraphical indexes (BGS, 2010a[283]) that enables users to see what is currently available, and that provides access to incomplete data and to visualisation procedures, is therefore the essential user interface to a developing model.

This suggests that the spatial model should be designed around visualisation of the three-dimensional geology, outcropping at the Earth's surface and known in the subsurface from boreholes, tunnels, mines and wells. As geophysical and other methods may be crucial to the interpretation, it must be able to incorporate results from all branches of geoscience. It should support a stable model shared by the information and user communities, and capable of adjusting to match global standards as they develop. Standard visualisations (like the published geological map) are essential for efficient comparison, although individual users might also create their own visualisations for specific purposes. The importance of spatial pattern and relationships places them at the core of geoscience, and thinking in terms of a spatial model rather than a map requires Reconfiguring the system 172.

Implementation note: Documents such as reports and maps can be scanned, and the digital images stored and printed on demand, to reduce storage costs. They can be remotely accessed through the internet, with charging if required, for local access by the user. Their availability, including edition, quality and date, can be noted on the index, so that users can find items of interest even within an incomplete set. Scanning can be undertaken in-house, or outsourced to specialists.

[283] BGS, 2010a. GeoIndex. http://www.bgs.ac.uk/geoindex

Reconfiguring the system

<<Reconfiguration 165

Summary: *With older technology, the map format shackles together field survey, archiving, and visualisation. The infrastructure now enables us to survey and store each object on its own terms; interpret, generalise and integrate them; and select scale and content when visualising them. With older technology, maps, reports, data and images, and personal knowledge are communicated with different formats and availability. Newer technology can intimately link multimedia information types.*

Interesting scientific opportunities can arise from technological reconfiguration of systems. Some procedures that had previously been linked may be separated, and some that had been split apart may be integrated. The system of geological surveying offers examples. For instance, with traditional methods there are strong similarities between the field maps prepared by a geologist while surveying, the archived copies on which the resulting information is stored, and the final published maps that reach the user. In contrast, the spatial model can separate the activities of field survey; storage and retrieval; and visualisation. Each of the three activities can be carried out on its own terms without unnecessary constraints from the others (see An object-oriented approach 178). The format, representation, and content of documents can differ markedly in each activity, as well as the techniques and responsibilities. Field survey, for example, can generate maps, diagrams, sketches and photographs at any scale. The results could be evaluated and stored as objects, while visualisation can provide a uniform view with the option to zoom in for greater detail where it is available.

Other system boundaries, on the other hand, may lose their importance. Traditionally, for example, information in geological maps, text reports, data, images like sketches or photographs, and personal knowledge, have each been communicated quite separately, even if they all refer to the same topic in the same area. The spatial model can integrate these different information types[284] more conveniently (see Integrating information types 62). For example, anomalies due to a hidden pluton were mentioned in The importance of space and visualisation 169. In a traditional report describing and explaining the anomalies, diagrams extracted from maps of the area could illustrate the relationships. In the electronic version, readers of the report could click at appropriate points in the text to view the actual maps, with the points under discussion highlighted. They thus move from narrative text into the spatial environment of the map (or visualised spatial model). They might wish to look there for analogous cases by zooming, panning and overlaying with maps of other topics, using tools specific to the spatial environment. The maps could provide links back to the text environment of the report or other reports, and possibly links leading to a database environment, photographic imagery or whatever. The user should thus be able to move freely among the environments of different information types, or view them simultaneously in separate windows side by side on the screen. Hybrid information types are also possible, as on an annotated field map. For example, an electronic map could be viewed either as

[284] Information type: The manner in which information is represented and processed, for example, as spatial images, narratives, data, algorithms, tacit and background knowledge.

graphical information alone or as an annotated map with additional icons on which the user could click for pop-up text notes and images (Voisard, 1998[285], Wikipedia 'Google Maps', 2010[286]). We therefore look at Representing spatial information and relationships 173, and in Mark-up and metadata 175 at how the content can be kept consistent across environments.

Implementation note: Spatial indexes and maps can be overlaid on the screen, relating points on the map to additional documents. Where the infrastructure is adequate, hypertext links can allow the user to move freely between maps, text, databases and other information types.

Representing spatial information and relationships

<<Reconfiguration 165

***Summary:** The map enforces geographical precision and imposes rigid sheet boundaries. It ensures that space is filled and that widespread comparisons can be made at a uniform scale. But the results can be misleading, for observations are limited and interpretations uncertain. Stratigraphical tables, generalised vertical sections, sketches and diagrams, and text descriptions can represent some spatial configurations and spatial relationships more appropriately. They can all be closely linked through mark-up and metadata.*

Geological maps are obviously concerned with spatial aspects of geology, but force their representation into the rigid framework of Euclidean geometry. Representing the third dimension leads to awkward conventions, and uncertainty is hard to depict. Map sheets lock the description (including accompanying reports) into arbitrary rectangular areas. Fixed scales lock representations into possibly inappropriate resolutions. The map defines a global reference point or datum, orientation, and scale or measure of distance. It therefore forces precision even where the information is imprecise or incomplete. If, say, a fault or a formation boundary is known to be present but its position is unknown, it must be shown on the map with geometrical precision, thereby introducing geological connotations other than its mere existence (see Ambiguity and map representation 148).

In an accompanying report, descriptive text can handle spatial statements that are less specific, such as: "There is a dyke swarm in this area." If more was known, additional comments could be added about the spatial characteristics of the dykes (and the area), perhaps mentioning their number, density, breadth, length, shape, regularity, average orientation, spread of orientation, parallelism, whether they join and so on. In text, what is known can be described to any level of detail, and what is unknown need not be mentioned. But draw a map of the dyke swarm, and inevitably the values of many of these properties are implied, whether they are known or not.

---

[285] Voisard, A., 1998. Geologic Hypermaps are more than Clickable Maps! Proceedings of the International ACM GIS Symposium, ACM Press, New York, November 1998. http://citeseer.ist.psu.edu/150047.html
[286] Wikipedia 'Google Maps', 2010. http://en.wikipedia.org/wiki/Google_Maps#cite_note-2

Diagrammatic sketches in the report have properties between the extremes of text and map. The sketch may not have a north arrow, and need not specify a particular orientation. The scale and position on the ground may not be recorded, and the scale and orientation need not be the same throughout the sketch. With diagrams of processes in the geological past, we generally assume that these aspects are uncertain. Other important information with looser geometrical constraints is carried by the marginalia on a geological map. The map key lists the stratigraphical objects shown on the map in time sequence, and a generalised vertical section may show the sequence and thickness range of the objects (but not their position) within the area of the map sheet. This marginal information provides a useful summary of the map content, and is a candidate for regional summaries in a spatial model.

Spatial relationships may be clearer in a text description than on a map. For example, we can state in words that a geological boundary converges upstream with a river. The representation of this on the map, however, can create problems. Map overlays may introduce spurious relationships. On scale change, topographical features may be generalised on different principles to the geological ones, thus distorting the relationship. Spatial relationships between entities, both observed and reconstructed, include: coinciding with; near to; above; inside; containing; bounding; overlapping; parallel to; converging with; crossed or cut by; displaced by; faulted against; unconformable on; oblique to; asymptotic to; continuous with; grading into; interfingering with; adjacent to; touching; branching; accentuating; together with their opposites and approximations.

Spatial relationships may be significant in analysis, interpretation, retrieval and display and should therefore be held in the model. They apply to probability envelopes as well as to interpolated surfaces, on the grounds, for example, that there is zero probability of separate and distinct rock bodies simultaneously occupying the same space. Although no comprehensive account can be given here, it appears that many spatial relationships can be defined in terms of equalities or inequalities, continuity, restricted geometrical transformations, dependencies and set memberships, all of which can be handled in computer analysis. It is desirable, not only to record spatial relationships in a model, but also to link in the inferred consequences. An explicit text statement of spatial relationships could be helpful on occasion, supplementing the map display on clicking on an icon.

The map may be an imperfect tool, but it is fundamental to geological survey. It helps the geologist to visualise spatial patterns and spatial relationships, building an interpretation that reflects the observations. Because it represents all points within the chosen space, it establishes the important relationship of what geology is thought to be present at any geographical location. The interpretation must fill the map space – a useful discipline, as there are neither gaps nor shared space in the real-world geology. The interpretation cannot always be accurate, but at least it should be feasible. Because distance and orientation are measured similarly throughout a map series, the sizes or shapes of spatial objects can readily be compared. Because maps for a range of topics follow widely agreed standards and may be keyed to the same topographical base, the spatial patterns of many properties can be seen and compared within the same spatial framework. Spatial relationships can be inferred that were not even considered when the maps were made. The benefits of the map should be retained and enhanced in the model and, to ensure compatibility with past records, the conventional map should be one feasible product. Reports, sketches and other information types supplement the ability of the map to communicate spatial information. The different information types can be integrated using Mark-up and metadata 175 (see also Integrating information types 62).

Implementation note: Optical character recognition (OCR) of the text in scanned documents gives a more flexible, character-based representation. Map features can be vector-digitised on-screen from a scanned map, and data added for each feature. Detailed searches for words and word combinations or map features are then possible within the documents. New documents, on the other hand, may be created from the keyboard without initial scanning. The gains may include more efficient typing or drawing, editing, on-line checking, proof reading, and printing, with remote access for several authors and editors if need be. All the results can be archived in a digital object store, and their format and availability recorded in the index. The procedures are well established.

Mark-up and metadata

<<Reconfiguration 165

Summary: *Mark-up languages, such as HTML, enable us to tag words, phrases and long sections of text for cross-reference or to indicate the topics, objects or properties to which they refer. The labels can be read by computer software but are normally concealed from the human reader. Geographical Information Systems (GIS) software can similarly tag segments of mapped items. Metadata, defining terms and their relationships, can ensure that the labels carry the same meaning across information types. Tiling schemes impart a spatial hierarchy. Mark-up makes it possible to mouse-click between corresponding map and text objects.*

Spatial information in geoscience typically refers to objects (the things of interest, such as rock bodies, shear zones or stratigraphical formations), their properties, composition, and the processes of their formation, subsequent geological history and investigation. It may deal with their disposition (where they are), their configuration (arrangement, pattern, form and shape), and their spatial relationships (Representing spatial information and relationships 173). Nouns typically refer to objects, adjectives to properties and verbs to processes. Some languages have a locative case for nouns, in which the word ending indicates that the noun is referring to the object's spatial aspects as opposed to its other characteristics. English of course has no locative case, and prepositions indicate the spatial relationships, to the satisfaction of the human reader, but the possible confusion of the computer program. Mark-up[287] languages, such as HTML, SGML or XML, which are now widely used in scientific communication, provide an opportunity for an author or editor to tag words, phrases or lengthy sections of text, to identify their use and relationships in a

[287] Markup: Symbols inserted in a document in a markup language such as SGML, HTML or XML, to tag the beginning and end of character sequences that can be interpreted by machine, and can be omitted for displaying to the user.

spatial context. The tags are visible to the computer software, but are normally concealed from the human reader. Similarly, sections of the text can be tagged according to the topics they are concerned with or the objects they deal with. Software is available to analyse text and assist in assigning keywords (enter 'Semantic text analysis' into a search engine to find available products).

In parallel with text mark-up, similar procedures are possible with graphical information, such as maps, images and sketches. Geographical Information Systems software can segment cartographical objects such as formation boundaries into smaller cartographical items or line segments which can be described individually, and rejoined as required – in effect a mark-up of the cartography. Spatial mark-up languages have been developed such as VRML (Virtual Reality Markup Language), or GML (Geography Markup Language) an XML encoding for the transport and storage of geographical information (see Some related initiatives - GML 352). Not only can a spatial data item be tagged according to whether it is, say, a formation boundary or a fault or both, but also individual objects, such as named formations, faults, outcrops or boreholes can be identified. Furthermore, links can be established between objects on the map and objects in marked-up text or sketches, and vice versa. Thus, a description of spatial relationships observed at an outcrop could be tagged in a report, and accessed by clicking on a linked icon on a map. The user should be able to read the text with graphical material displayed alongside. Mark-up makes it possible to move readily from text descriptions to the corresponding objects highlighted on the map, or to select items on the map and call up the corresponding text. There can also be marked-up links to and from programs, sketches, photographs, databases, people with expert knowledge, and metadata. The attributes of spatial or other marked-up entities may be held in a database for more flexible selection and retrieval.

The integration of maps, reports and other material depends on a shared conceptual model and coding scheme, that is, ensuring that the same objects, processes and relationships are identified, and have the same meaning, in all parts of all the information sources. This is the role of metadata[288], the data about data where terms are defined in feature ontologies and data dictionaries and their relationships are identified in UML entity-relationship or similar diagrams (see Other relevant fields – UML 358, and Other geological initiatives 343). Global standards (such as the ISO 19*** set, see Other relevant fields - ISO standards 357) must be followed where possible, for geological spatial modelling, like the geological map, is part of a world-wide activity. But metadata are required at various hierarchical levels – for the model as a whole and for individual areas where, for instance, some lithostratigraphical terms have only local significance. Local metadata are required for projects, to define the procedures, standards and operational definitions used in the investigation (see Semantic Web and Grid 50). It may also be necessary to record what the Epicentre model calls "activities" for identifying methods used to obtain alternative results, such as measured and visually estimated values of porosity for the same sample. Metadata are essential to understanding

[288] Metadata: Metadata is a description of data (often used in a broad sense of representations of information or knowledge) and its context. It is structured to assist the user or computer to find, manage, control and understand the data.

results. Of course, they need not be repeated for work that exactly follows the standards of some larger, fully documented project.

The metadata hierarchy has a matching spatial hierarchy. Computer archives of spatial information are best stored in such a way that nearby locations are held together. The usual solution is based on *quadtrees* or *octrees* (see Raper, 2000[289] or most GIS texts), where each of a set of uniform and probably rectangular tiles or boxes contains a set of smaller tiles or boxes, and so on in a defined hierarchy (see Making a mesh 207). The separation in a spatial model of field survey, archiving and visualisation means that the tiling scheme need affect only the archive and not data collection or visualisation. It should be possible to accept legacy data from different map sheet boundaries into a quadtree. Nevertheless, the spatial hierarchy may prove most valuable with new data, as an efficient means of relating data and local metadata and accessing information (including generalised vertical sections) at various levels of spatial detail.

Mark-up and metadata provide mechanisms for aligning ideas across information types and projects. Visualising the results on a computer screen as maps and supporting text appears at first to be straightforward. But computer models are not constrained by the older technology of pen, paper and printing press. To benefit from this freedom in longer-term developments, we need to look more closely at An object-oriented approach 178.

Implementation note: New or reworked reports can be marked up for internal indexing and cross-reference using simple HTML commands. This is obviously of value only where the documents are to be accessed electronically (on the internet or intranet). XML offers more complete facilities, but standard document type descriptions are not yet widely agreed in geoscience. Shared metadata should lead to consistent records that can be widely understood. As global standards may not be available, those adopted by the larger players might be followed if appropriate, despite the inevitability of future modification. Local standards should be recorded and any divergence from other standards monitored. But the complex global linkage of stratigraphical tables, maps and reports remains an evolving area.

---

[289] Raper, J., 2000. *Multidimensional Geographic Information Science*. Taylor and Francis, London. 300pp.

An object-oriented approach

<<Table of contents 1
<<The future geological map 125
<<Reconfiguration 165
An object-oriented approach 178
The object-oriented perspective 179
Object instances and classes 181
Relationships between objects 183
Reconciliation 186
Microdocuments and the threads of reasoning 190
Object-oriented survey 195
Benefits of an object-oriented system 199
>>The geometry of the spatial model 203

Abstract: Geological surveying increasingly depends on support from the cyberinfrastructure[290], where the object-oriented[291] approach fits well with geoscience reasoning. Real-world geoscience entities can be located by surveying, represented as object instances and classified within object-class hierarchies. Relationships between object instances[292] or classes[293] (such as: is a part of, is a kind of, follows, is linked to) can be explicitly recorded to structure the objects as hierarchies and sequences, and join up spatial features and chains of events. The object-based reasoning process may reveal inconsistencies between strands of thought, which must be reconciled[294]. The expected behaviour of object classes, records of the reasoning process, and procedures for visualisation[295] can be incorporated within the model as microdocuments[296]. The full benefits of object-orientation call for a unified approach, leading to more flexible, comprehensive, robust and informative systems.

---

[290] Cyberinfrastructure: An integrated assemblage of computing, information and communication facilities, deploying the combined capacity of multiple sites to provide a framework to underpin research and discovery, typically with broad access and end-to-end coordination.

[291] Object-oriented: An approach to analysis, design, and classification, which can support many aspects of thinking about objects and their relationships including linking them with interweaving threads.

[292] Object instances: Representations of specific, identified, real-world or hypothetical objects.

[293] Object class: An abstraction giving a general description of the expected properties and behaviour of the objects belonging to that class. They are a means of categorising object instances within larger groupings.

[294] Reconciliation: Kent (1978, pp. 202-203) points out that people have different views of reality, and that these change with time. But the views overlap and so can be *reconciled* with varying degrees of success to serve different purposes. "By reconciliation, I mean a state in which the parties involved have negligible differences in that portion of their world views which is relevant to the purpose at hand."

[295] Visualisation: Transforming quantitative data (including the results of interpolation) into sensory information – images that the eye and brain can interpret and visualise.

[296] Microdocument: A short note or module of information, typically referenced by a URI and seen here as a system component, such as a minimum revisable unit.

The object-oriented perspective

<<An object-oriented approach 178

Summary: *Geoscience survey gains a new perspective when based, not on notions of geological mapping drawn from traditional cartography, but on the developing cyberinfrastructure. The focus changes from container to content – from map sheets and their explanatory documents to objects that have significance in geoscience.* Object-oriented analysis *specifies the system requirements, and* design *defines a consistent framework that enables software to integrate, manage and analyse the information, and helps users to locate relevant material. Computers manage the complexity; networks and global standards enable widespread information sharing; but geoscientists must define how the system should work and develop.*

The tools we use affect the way we think. The tasks of geoscience survey look different if we view them, not as based on notions of geological mapping drawn from traditional cartography, but in the light of the developing cyberinfrastructure. A geoscience survey map typically refers to a rectangular area of fixed extent projected onto a plane. The content is filtered for presentation at a fixed scale, deals with a limited range of topics and conforms to one particular interpretation. In its preparation, the authors tap into many sources of information and many aspects of geoscience (see The surveyor's holistic view 60). The final products are unlikely to refer directly to all the sources, particularly those that do not bear directly on the interpretation, even where they could be of value to users. The detailed reasoning that led to the marks on the map may remain largely unrecorded.

The spatial model also calls upon a wide diversity of material to support its core view (see Diverse objectives and products 150), but rather than regarding documents (such as map sheets and accompanying reports) as the primary objects, the model[297] frees the content from these containers. It focuses instead on objects that have significance in geoscience, taking each on its own terms. The object-oriented approach can build on and formalise the procedures that are followed, more or less, in producing a conventional map. However, it also provides an opportunity (though not the necessity) of organising the records with direct links to chains of reasoning and supporting material. It offers greater flexibility of presentation (see Objects, ontologies and systems 48).

Objects (see Mark-up and metadata 175) simplify our view of geology by the familiar procedure of lumping together a number of ideas as a single named concept (see The imperfect model 156). The procedures of object-oriented *analysis* specify and evaluate the requirements in terms of objects. *Design* procedures can then define a consistent framework for computer applications that organise the objects, their properties, relationships and interfaces in order to meet these requirements. The objects can be organised to enable

[297] Model: A formalised representation giving a simplified view of aspects of the real (or of an imaginary) world relevant to the purposes in hand.

software to integrate, manage and analyse the information and assist users to locate relevant items as required.

Computers are essential to manage the complexity, but the system[298] depends on the background knowledge of geoscience experts. Just as a map involves collaboration between geoscientists and cartographers, so design of the framework[299] must involve collaboration between geoscientists and technical experts. The scientists should bring a clear view of how the system should work now and develop in the future to meet the needs of the science and its users. E-scientists must relate those needs to the conventions required in designing a system that can interface with appropriate hardware and software.

The object-oriented view is supported by analysis and design procedures (see, for example, Coad and Yourdon, 1991[300], Budd, 2000[301] or OMG, 2010[302]) and up to a point by software. To improve efficiency, the design (not the analysis) may be modified for implementation by techniques other than object-orientated database management. Object orientation relates well to geoscientists' ways of thinking (see The imperfect model 156) and ideas of emergent[303] systems (see Complex and emergent systems 159). An object-oriented computer system breaks the rigid ties between field survey, archiving, and presentation of map information (see Reconfiguring the system 172), allowing each of these three subsystems to be handled according to its own individual characteristics. This may help to overcome some of the inflexibility (see The imperfect map 145) and ambiguity (see Ambiguity and map representation 148) of the geological map. Furthermore, the data analysis assembles data dictionaries and data models, leading to a more rigorous outcome by defining the terms and procedures used in creating the spatial models.

As suitable international standards are agreed and implemented, as attempted by the oil industry (see Other geological initiatives – Hydrocarbons geology 347), the geological community (for example, see Other geological initiatives – GeoSciML 345), and international bodies (for example, Other relevant fields – INSPIRE 356) they will enable us to share detailed geoscience information through computer networks across organisations and countries. Where subsystems and interfaces are well defined, functions such as data collection or information management and dissemination could be handled independently and might even be outsourced to specialist organisations. Compared with the oil industry, there may be fewer commercial pressures and greater information diversity in the academic community and Geological Surveys, but in the long run, global standards offer similar solutions and benefits. Object-oriented methods (see Object instances and classes 181) can change our viewpoint and enable us to trace ideas from field observation through interpretation to presentation.

---

[298] System: A set of interacting parts that function as a whole. The systems approach involves study of linkages or interfaces between the component activities.

[299] Framework: A logical structure and guidelines giving a broad overview for classifying and organizing complex information, within which detail can be added as required.

[300] Coad, P. and Yourdon, E., 1991. *Object-oriented design.* Yourdon Press, Englewood Cliffs, NJ. 197pp.

[301] Budd, T., 2000. *Understanding object-oriented programming with Java.* Addison-Wesley, Reading. 420pp.

[302] OMG, 2010. The Object Management Group (OMG.) http://www.omg.org

[303] Emergence: Complex patterns, properties and systems resulting from relatively simple interactions.

Implementation note: Object-orientation leads to a more flexible view of the information on a digital geoscience map, and allows for greater diversity of input and output. Links to external contributions can readily be added. Repackaging the same content as visualisations[304] to meet a range of customer requirements and levels of understanding expands the market at little extra cost.

Object instances and classes

<<An object-oriented approach 178

Summary: *Object instances are representations, in a map or model, of specific, identified, real-world or hypothetical things of interest. Object classes are abstractions giving a general description of the expected properties and behaviour of the object instances in that class. The class can serve as a template[305] for describing an object instance. Classes typically form a hierarchy, inheriting some properties from higher levels. Spatial location is the link between surveyed instances of objects in the model and their real-world counterparts. The objects in store can be selected by area, resolution, and topics of interest, for example for visualisation.*

Individual instances of objects, such as specific occurrences of a particular sedimentary structure, each have their own unique properties. But they can also be placed within an object class (say the class of ripple marks) that refers to that type of structure in a more abstract sense and deals with its general properties and behaviour. The distinction between instances and classes of objects is important. The *instances* are representations of specific, identified, real-world or hypothetical objects in a map or model. The *classes* are abstractions giving a general description of the expected properties and behaviour of the objects belonging to that class. Like the metadata, they may help to define and explain the properties of an object instance.

Object classes can be structured as hierarchies[306]. The class of ripple marks, for example, might be seen as a subset of sedimentary structures, which in turn might be regarded as a subset of the class of geological spatial objects. Each object class has properties, some of them inherited from classes higher up the hierarchy. The inheritance need not be confined to a single line of descent. Ripple marks, for instance, might also be seen as a class of flow-generated objects, inheriting other properties from that source.

Instances of the objects can be parts of informal hierarchical groupings. The objects described in the field are likely to be interpreted, generalised and integrated to create objects at higher levels in the object instance hierarchy. This is the counterpart of procedures for generalising from field notes to maps at survey scale and thence to a

---

[304] Visualisation: Transforming quantitative data (including the results of interpolation) into sensory information – images that the eye and brain can interpret and visualise.

[305] Template: A file (or paper form) providing pre-prepared elements in repetitive documents and guiding their completion.

[306] Hierarchy: An organised body of things (ranked in classes one below the other) branching downwards as an inverted tree structure.

sequence of smaller scales. For example, an object instance might be a single record of a small item, such as a measurement of bedding orientation. A group of similar objects could be assembled as a spatial object at a higher hierarchical level, such as a description of a complete outcrop – a set of observations referring to an actual occurrence of an outcrop that we can name, survey, visit and examine, and which has properties that conform to the abstract concepts of the object class of outcrops.

Because the object instance starts with the corresponding object class as a template, the description can concentrate on its specific features, taking for granted the general features that it shares with the object class. What is known on general grounds need not be repeated in specific instances. An account of an outcrop, for example, can assume that the reader is aware of the general implications and characteristics of the concept of an outcrop as an object class and need mention only features of specific interest that deviate from the expected norm. The description of the outcrop in turn could be an element in a higher-level object instance, such as a structural map, a facies map, or a model of a formation over its full extent. The higher levels of the object hierarchies, like maps generalised to a smaller scale, simplify the geology by reducing it to fewer, more general, items. The grouping of instances as a higher-level object instance is likely to be handled by specifying relationships (see Relationships between objects 183), rather than formalising it as a classification in the metadata.

Object instances, as held in the computer, are representations, such as any combination of images, maps, diagrams, data, descriptions, explanations, software and models. Where the distinction is important, their equivalents in the real world may be referred to as entities, rather than objects. As stressed in Reasoning, models and reality 127, surveyed objects are linked to real-world entities through their spatial location. Thus the elevations of a formation top might be represented in a model by a precisely located grid of numbers (see Making a mesh 207), or its boundaries at outcrop might be recorded as lines denoted by the co-ordinates of chains of points. The representations could be held in an object store along with records of their spatial relationships, other links, and positions in the various hierarchies. The objects may be directly accessible at a desktop browser if they are held as digital records in the object store. Inevitably, however, many will only be digital identifiers (possibly with additional comments) referencing items that must be retrieved by conventional means. They might refer to conventional published or archived objects, including references to the literature, cores, samples and specimens, or to features observed in the landscape or at outcrop, or even to vague concepts like "a pattern of river valleys".

The core of the spatial model thus emerges as a set of surveyed object instances, which can act as a bridge between model and reality (see At the interface 133). On the one hand, they interface with the abstract world of object classes and relevant aspects (including local geological history) of the general, largely non-digital, geoscience model (see The imperfect model 156). On the other hand, they interface with the real world through geographical co-ordinates, either directly, or indirectly through geographical features shown on an overprinted topographical map. A local map or graphical model might be regarded as the visualised result of applying a series of filters[307] to the model, restricting it to the specific area, projection, resolution, topics, objects and properties of interest. In the model, as opposed to the conventional map, filters are applied at the time of visualisation[308] rather than during the initial survey. Relationships between objects 183 are an important aspect of positioning them in the wider system.

Implementation note: The specialised and complex task of data analysis should lead to the design of a framework for assembling and retrieving the information, usable by all who share the same (preferably global) standards. An object-oriented design seems appropriate. It must be capable of growing to accommodate new methods as they emerge. Geoscientists must fully understand the conceptual models and share in the analysis, working in collaboration with e-scientists to ensure that the design can lead to a workable implementation. Management's long-term backing and understanding of the organisational implications are essential to its success.

Relationships between objects

<<An object-oriented approach 178

Summary: *Object relationships play a vital role in the implementation. They define hierarchies of classification ('is a kind of') and composition ('is a part of'). They link stratigraphical units into sequences and hierarchies – a key to generalisation. They help to define the geometry of the model through spatial relationships, linking spatial elements to establish boundaries, stratigraphical correlation, interpolation, fault blocks, continuity, the outcome of processes, and so on. Relationships implied on a map can be made explicit in the model. They may apply to object instances, classes or both.*

The procedures of geoscience survey were seen in 'Abstracting from reality to model' 131 as a process of abstraction,[309] starting from field observation and leading to classification and explanation, working back and forth between observation and interpretation at many levels of detail. Object classes can support this process of abstraction, enabling us to bring together a set of analogous ideas and handle them as a single, more general, concept. The hierarchical organisation of object classes, where lower-level objects are seen as more specific cases of higher-level objects, establishes one type of *relationship* between the classes. This relationship between the lower and higher level object might be expressed as 'is a kind of' or 'is a'. For example, Monograptus is a kind of graptolite, orthoclase is a feldspar, feldspar is a mineral. Hierarchies of classes already exist in geoscience, as in the familiar classifications of stratigraphy, rock types or palaeontology. Stratigraphical object classes, in particular, are likely to appear in a map key, cross-referenced by colour or

---

[307] Filtering: A process that selectively enhances or reduces specified components of the information stream.

[308] Visualisation: Transforming quantitative data (including the results of interpolation) into sensory information – images that the eye and brain can interpret and visualise.

[309] Abstraction: reducing the information content of a concept or an observable phenomenon, typically in order to retain only salient information, relevant for a particular purpose

ornament to the geographical distribution of instances shown on the map face. Likewise, a visualisation selected from a spatial model should include a key to the object classes selected for display.

Another important relationship can be expressed as 'is a part of' or conversely 'has'. For example, a quartz grain is a part of a sandstone bed, or, taking the converse view, the sandstone has grains of quartz. Similar examples of subdividing objects into parts can be seen in geographical subdivisions and co-ordinates of space – the Matterhorn is a part of the Alps, the grid square referred to as 735426 is a part of the larger square 7342. These relationships are concerned with composition rather than classification. Maps of geochemistry, mineral content, ore quality, porosity, sand-shale ratio, fossil content are concerned with the composition of the material. The corresponding visualisations from a spatial model are concerned with 'part of' relationships.

Object instances are linked not only within hierarchies but also through time or stratigraphical relationships, analogies and chains of reasoning. A general method of recording such links was proposed by Berners-Lee et al. (2001[310]) in his work on the Semantic Web and has been carried forward to the Semantic Grid (see Semantic Web and Grid 50). Of particular relevance to spatial models are the *spatial relationships* (see Representing spatial information and relationships 173, Seeking shared concepts 247) between the objects. For example, they make it possible to express stratigraphical correlation by relating individual observations to other instances with analogous properties that are, or were thought to have been, contiguous parts of a continuous unit. This generalisation by extension (see Stratigraphical units in space and time 141) relates observations, bringing them together as instances of higher-level objects such as specific beds, formations, seismic maps or gravity fields. In turn, these may be related and regarded as components of larger or more general objects. A different spatial relationship, 'is discontinuous with', defines positions in space where abrupt changes of properties occur. It might mark a fault or a facies or formation boundary, and is fundamental to mapping and to stratigraphical correlation. Some spatial relationships, such as 'lies unconformably on', have complex connotations. They might be simplified by regarding 'the unconformity' as an object, referring to the missing stratigraphical units. The relationship of 'lies on' refers to two objects. In this case, one might be a formation, the other an unconformity.

Spatial relationships (see Representing spatial information and relationships 173) help to define the geometry of the spatial model. For example, the model might record the position of several adjacent formations that had been folded and faulted as a group, perhaps restricted to, say, those above a major unconformity within a particular fault block. Each formation within it might be regarded as a component spatial object, instances of classes in a stratigraphical framework. From the point of view of structural geology, however, the group might be combined as a single structural object within the defined area, where it might be related to other structural features like faults and folds. Fault segments bound the area and define the extent of the fault block. Spatial relationships, such as 'is continuous with' and 'bounds', link the segments to one another and to the area of the fault block they

[310] Berners-Lee, T., 2007. The Giant Global Graph (blog post) http://dig.csail.mit.edu/breadcrumbs/node/215

enclose. The geometry of the model is formed around such spatial relationships (see The geometry of the spatial model 203).

The framework must take account of processes[311]. Instead of objects and the relationships between them being the primary concern, the process models emphasise *processes* (in which the objects take part) and *interactions* between processes (constrained by spatial relationships). The behaviour of objects taking part in a process may be predictable from their properties and relationships. The account of self-organising, emergent[312] systems (see Complex and emergent systems 159) suggests that their process models might be placed within a hierarchy of component subsystems. An object seen in the field might be regarded as the outcome of such a process model, linked to it by the relationship 'creates' or 'is the product of'. An object may contain processes, processes may involve objects, and the two can be handled together for present purposes.

The properties of a single object may be the result of various different processes, classified in separate hierarchies. For example, examination of hand specimens might involve identifying and separating the effects of stratigraphical, metamorphic and structural processes. On a conventional map, the consequences might be implied by the mapped stratigraphical units, by the mapped metamorphic zones, and by the structural situation implied by orientation of bedding, position of faults, and the geometry shown on cross-sections. In an object-oriented framework, the specimen might be seen as an object in its own right, related to object classes in stratigraphical, structural and metamorphic hierarchies, to larger object instances that showed their geographical distribution, and to models of the processes that created them. An appropriate framework with computer support enables us to manage the objects within their complex structure of diverse, cross-cutting, hierarchical classifications.

These examples refer to spatial relationships between instances rather than between classes. However, the two must work together, and more general relationships apply between object classes. For example, the principle of superposition of strata leads to general spatial relationships (such as 'overlies') linking depositional and stratigraphical sequences; spatial cross-cutting relationships can be placed in a time sequence; the symmetry of folds may reflect their positional relationship to larger structures and to stress fields in the folding process. The object-oriented model provides an opportunity to distinguish between what is specific to a spatial-object instance, what the corresponding object class and its relationships imply, and what is inherited from higher levels of the class hierarchies. This is comparable to the processes of abstraction to build explanations and interpretations that fill gaps between observations and can be tested against new data. It provides the means to clarify (and possibly defer) decisions about the degree of interpretation involved in visualisation. A range of images could be available along the spectrum from observation to interpretation – from a map limited to observations, to a visualisation based solely on a theoretical geological interpretation.

---

[311] Process: A particular course of action intended to achieve a result, or a series of natural occurrences that bring about change.
[312] Emergence: Complex patterns, properties and systems resulting from relatively simple interactions.

The design of an object-oriented implementation focuses on interfaces (and therefore relationships) between objects. *Encapsulation* within the object hides content that is not required beyond its immediate context. The information of wider value is tagged and made available through the interface. Software can send messages to the object interface, eliciting information relevant to other objects. Such technicalities are unlikely to concern many geoscientists. However, an analogy can be drawn with objects, such as field notes, that the scientist creates in the field. Much of that material is of limited interest outside the context of the field observation, but it might have a bearing on, say, the lithology of the formation or its structure at a particular resolution. The surveyor might flag the field notes to indicate these relationships. On a conventional map, a symbol or stratigraphical code representing an item might imply links to similar items on the map face and to relationships implied by the map key. In the spatial model, relationships between objects can be recorded explicitly and more flexibly. Relationships that reconcile objects and establish their place in the interpretation are of fundamental importance, considered in the section on Reconciliation 186.

Implementation note: It is a huge task to create a comprehensive framework for geoscience objects. Even where the framework attempts to formalise an existing pattern of thought, the unfamiliar procedures may prove unusable in the field. Cautious development of local and interim frameworks, to support experiment and development by sympathetic users, are suitable starting points (see Overview of the geological framework model 35), preferably conforming to an existing framework, such as GeoSciML 345 (CGI, 2009[313]). Step-by-step extensions can follow if justified, when familiarity has been gained and snags ironed out.

Reconciliation

<<An object-oriented approach 178

Summary: *A Geological Survey builds a coherent picture of the geology of an area from evidence from many sources, integrating models, observations and other spatial objects, in effect building a sequence of composite objects from a diverse set of component objects. They reflect different objectives and views of reality that may change through time. Conflicting views can be identified as a 'mismatch' relationship between the specific objects. If possible, they should be reconciled and the reasons recorded in the model.*

A clear framework for a spatial model does not ensure that its contents make scientific sense. A well-structured object repository could contain a jumble of contradictory items. When viewed as part of an interpretation, objects may conflict. This is an important relationship ('conflicts with') between objects that participate in the same reasoning process but lead to different conclusions.

[313] CGI (Commission for the Management and Application of Geoscience Information), 2009. GeoSciML. http://www.cgi-iugs.org/tech_collaboration/geosciml.html

To overcome conflicts in making a map, geoscientists must assess and balance many aspects and sources of information (see The surveyor's holistic view 60). Similarly, to arrive at a spatial model of assured quality, they must reconcile a diversity of possibly inconsistent objects. In his account of data and reality, Kent (1978[314], pp. 202-203) points out that people have different views of reality, and that these change with time. But the views overlap and so can be *reconciled* with varying degrees of success to serve different purposes. "By reconciliation, I mean a state in which the parties involved have negligible differences in that portion of their world views which is relevant to the purpose at hand." He points out that reconciliation is growing in importance as technology increases the interaction between people, and integrates processes to serve more and more purposes.

As an example to clarify these issues, consider the task of contouring a formation from well data. The elevation of the formation top might be picked at each of a number of wells, with each individual observation and identification recorded as a spatial object. The picks could be combined (by linking the objects through their spatial relationship, rather than by repeating their content) as a composite object – the dataset. The dataset could then be related to a hypothetical surface, with geometrical properties based on ideas of its origin and history. For example, geologists might expect from evidence of adjacent surfaces that the dip would increase gradually towards the west. By analogy with situations nearby, and arguments from dynamic stratigraphy, they might expect to find, say, a series of northeast trending sandbars.

The hypothetical surface could then be related to the dataset by adjusting the surface to tie in with the data while retaining its expected properties. An obvious approach is to sketch a surface with appropriate properties on a map on which the data points have been plotted, and adjust it by trial and error to arrive at an acceptable match. The two objects, the dataset and the hypothetical surface, come from different starting points: one from interpretation of individual observations, the other from an interpretation of the local stratigraphy and structure. They are combined to form a new object – the interpreted surface – by a process of reconciling one with the other. The new object should be in harmony with the spatial properties of both parent objects.

The interpreted surface, itself a spatial object, might then be reconciled with an adjacent contoured seismic reflector horizon. This represents yet another viewpoint, stemming from a different facet of knowledge from specialists in another field. As subtly different aspects of the Earth's crust influence the two surfaces, this reconciliation would require a geological appraisal. It might result in modifications to one or both surfaces, or simply a statement of why they differ. It might call for revisiting the original reconciliation, revising the well data, or possibly weighting the compromises differently to take account of the new evidence.

The result might not agree with the conclusions of an article, published earlier in a scientific journal, which was based on less evidence. A note explaining the disagreement and its resolution might be agreed with the author and referenced from the spatial model. If the journal was published electronically, the editor might retrospectively insert a forward reference from the article to the note.

---

[314] Kent, W., 1978. *Data and reality*. North-Holland Publishing Company, Amsterdam. 211pp.

Inconsistencies between and among objects take many forms, and can be reconciled in various ways. At present, they are usually resolved informally by procedures that are taken for granted. Considering them explicitly may help with subsequent decisions about which should be formalised and made an integral part of a computer-based knowledge system. Some examples may make this clearer.

1. The business setting (see Abstracting from reality to model 131) of the original study influences the procedures and products. The objectives guide the interpretation (see The imperfect map, Diverse objectives and products 150) and constrain the results. Users must be able to understand this background, take it into account, and recognise why differing interpretations were preferred for particular purposes. The provenance and derivation of the information should therefore be recorded in the metadata or in an explanatory note.
2. Properties of the Earth's crust (see The stratigraphical framework 139), such as gravity anomalies, structural features, fossil occurrences, or geochemical analyses, may throw light on the same geological model, but reflect different aspects of the complex results of interacting processes. Some objects representing the properties may not conflict and so cause no problem, but others will certainly require reconciliation. Some differences may reflect arbitrary choices in an uncertain situation, and might be resolved by discussion. Some reflect different opinions and could be expanded and negotiated. Some arise through different approaches or methods (see DSIs, FEMs and their geometrical significance 235) and could not be directly resolved, but might be linked through a more general model (as in resolving gravity and geochemical models through a geological model in 'The importance of space and visualisation' 169). Some reflect conflicting evidence and might be resolved through further field investigation. Some reflect separate objects that are only loosely spatially related (such as diachronous objects resulting from processes that operated on separate time scales), and compromise solutions might be found that are in accord with all the evidence. Some objects might be represented (or reflect processes) at different scales or levels of resolution, and knowledge of the generalisation procedures might clarify whether comparison is appropriate. Reconciliation depends on knowing how the objects are related and the sequence of steps that led to the final conclusion.
3. Like the edge-matching of conventional map sheets, discrepancies may have to be corrected where adjacent surveying projects meet, and more generally to maintain consistency within a larger area. For instance, stratigraphical correlation starting from different points with subtly different criteria may fail at the boundary to join what is ostensibly the same bed. Arbitrary decisions may be appropriate, or procedures may have to be retraced and other evidence considered in order to pinpoint the reasons for the divergence. Adjustments, possibly requiring clarification of standards, might then be agreed.
4. Alternative hypotheses may need to be considered. They may start from different premises or reach different decisions about, say, analogies for processes or models, thus

leading to the need to reconcile and integrate them as part of the dialectic[315] (see The dialectic model 129). Threads of hyperlinks could record the chains of reasoning (see Microdocuments and the threads of reasoning 190) by which conclusions were reached. They should help to clarify the reasons for favouring the chosen overall interpretation, or for reaching an open verdict on multiple hypotheses.

5. The information comes in various types (see Reconfiguring the system 172, Representing spatial information and relationships 173, Integrating information types 62) for example as text, diagrams, maps, contours, elevation grids, map keys, stratigraphical tables and generalised vertical sections. Digital records should be held in standard formats, and tagged to enable the markup language or other software to recognise the information type and process the information correctly (like the .html and .gif suffixes in Web references). However, there may only be loose links between geometrical, graphical, mathematical, statistical and text models. Some, such as a map and cartoon model, may refer to different geometries (see Ambiguity and map representation 148) and only some spatial relationships can be expected to conform. Such differences should be clarified by the metadata. It might be possible to reconcile one-dimensional projections such as vertical sections and stratigraphical or time sequences with two-dimensional projections, such as maps and cross-sections, by considering their relationships in a conceptual three-dimensional model. They might also help to reconcile evidence from different geometrical properties, such as location, orientation and arrangement. A sketch of the model in an appropriate explanatory note might adequately resolve or explain their relationships. Mathematical considerations, considered in 'The geometry of the spatial model 203' and 'Transforming space' 224, could support a more formal account.

6. Spatial relationships (see Representing spatial information and relationships 173) and their geometrical implications may be hard to disentangle and reconcile. To reduce ambiguity, a simplified, standard set could be identified and defined in the metadata. GIS and other software that conform to the standard should not give unexpected effects on spatial representation and display. Non-standard relationships could be flagged for the software to present as pop-up notes for manual checking during visualisation.

7. The expected behaviour of the model and the mode of processing must be consistent. The visualisation of spatial data should be consistent with its expected behaviour and, more generally, data should be analysed with appropriate software. It is therefore desirable to record the behaviour of objects as metadata in a format that software can access, thus ensuring that objects and models are consistent, when objects are reused in various contexts.

A Geological Survey must build a coherent overall picture of the geology of an area within the shared framework of space and geological time. It must assemble evidence from many sources that offer views of the same underlying spatial entity (the real world) from different standpoints and perspectives, in diverse presentations. It must organise, relate and reconcile

---

[315] Dialectic: The theory and practice of weighing and reconciling juxtaposed or contradictory arguments for the purpose of arriving at the truth.

this plethora of models, observations, and other spatial objects, in effect combining them into hierarchies of composite object instances. It must bring together and test the relevant information, and present its findings as one consistent view that end-users can relate to, or reconcile with, their own data and requirements.

A Survey assures the quality of its regional maps and models through internal procedures (see Projects and information communities 167): the scientific literature assures the quality of project-based contributions through editorial control and peer review. Hyperlinks extend the scope and detail with which each impinges upon (and, by implication, must assess and be reconciled with) the work of the other. Quality assessment and assurance could be extended to resolve the inevitable conflicts by documenting the process of reconciliation. The reconciliation takes place as part of the reasoning process. Conventionally, the reasoning process shaped the map but was recorded only in the map explanation. The section on Microdocuments and the threads of reasoning 190 considers the possibility of incorporating the reasoning process and reconciliation as intrinsic parts of the spatial model.

Implementation note: Collaborative processing, with separate remote sites accessing the same interactive screen display, can enable experts to discuss and document opposing views. It may also enable users and experts at different sites to work together to select and visualise appropriate material (see Workflows, collaborative networks and Linked Data 53). This should efficiently reconcile and combine the users' knowledge of their own requirements and the experts' knowledge of the content. Geoscientists must first get accustomed to the collaborative techniques in-house, for example in multidisciplinary map production.

Microdocuments and the threads of reasoning

<<An object-oriented approach 178

Summary: *The properties and relationships of objects determine their behaviour within models of the processes in which they participate. Interactive processing may help scientists to identify conflicts, reconcile them, and record the reasoning as short explanatory notes attached to the mismatch relationships. Notes should more generally explain why particular views have been reached. They are equivalent to the traditional map explanation, but split into short, more self-contained sections, tied closely to surveyed objects. They should be linked to, or embedded in, a more general explanatory object – the marked-up map explanation document. Hypertext links can record the threads of reasoning. Scripts of computer instructions can generate a range of documents, maps and other visualisations. Objects can be held once in the object store and reused (through indexes, links and relationships) in many contexts.*

Comparison with similar problems addressed in other fields may help to clarify ideas in geoscience. For example, the Physiome Project, 2010[316] (see also Clapworthy et al., 2008[317])

---

[316] Physiome Project, 2009. http://www.physiome.org

is a medical equivalent of a solid Earth systems model. Each bone in the human body is described as an object in its own right. Establishing the spatial relationships between them makes it possible to reconstruct a complete skeleton. Establishing their relationships to musculature, nervous system, blood supply and so on, can make it possible to establish a model that helps to understand the mechanics and dynamics of the human organism. Object classes can be defined, as can their hierarchies (a leg has a foot, a foot has toes), their spatial relationships (...the knee-bone's connected to the thigh-bone...) and the processes in which they take part (walking, standing, kicking). The attributes and properties of the object classes clarify their range of behaviour within these processes (the knee bends one way only). Comparison of individual instances with the class properties may indicate anomalies (say, a malformed leg) that could be related to other anomalies and disease, injury, diet, genetics or whatever. The Project is seen as having relevance to databases, models, network access, and science, education, exploration and dissemination of knowledge. Their models "include everything from diagrammatic schema, suggesting relationships among elements composing a system, to fully quantitative, computational models describing the behaviour of physiological systems and an organism's response to environmental change." Their object-oriented approach seems to be scientifically appropriate, as it formalises an existing viewpoint and takes a comprehensive view, as well as being convenient for computer implementation. These features are equally necessary in geoscience.

In surveying and describing geoscience objects, we establish object properties and relationships that should help to clarify their behaviour (as in the Physiome Project). *Models* (constructs referring to some aspects of reality) represent the processes where the objects manifest this behaviour. The terminology is loose and ambiguous, and the distinction between object, process and model is not at all clear. However, the underlying concepts also overlap and are equally blurred (see Kent, 1978[318]). The flexibility makes it easier to extend our thoughts and adjust to change, and ambiguity and analogy help to extend ideas. Greater terminological rigour would make computer implementation more straightforward. That may follow in due course, but would be inappropriate and inhibiting at this early stage of exploring ideas.

Individual models might describe processes within the general model (see The solid Earth systems model (sEsm) 71, The imperfect model 156), such as depositional, structural, metamorphic or geochemical processes. Others might describe procedures within a Geological Survey model. Examples are: a field survey model (describing surveying procedures for each object class), a filtering model (to select information of interest), an interpolation model (to fill gaps between observations), a visualisation model (for looking at the results), a generalisation model (to adjust the amount of detail), or an explanatory model (to describe the reasoning process). A few of these models are computer-based and implemented in software that predicts the response of specified processes acting on particular objects. Most are text instructions, explanations, or manuals with diagrams and other information types, and may or may not be referenced in a computer index or available

---

[317] Clapworthy, G,. Viceconti, V., Coveney, P.V., Kohl, P., 2008. Editorial. Phil. Trans. R. Soc. A 366, 2975-2978. doi: 10.1098/rsta.2008.0103 http://rsta.royalsocietypublishing.org/content/366/1878/2975.full.pdf+html

[318] Kent, W., 1978. *Data and reality*. North-Holland Publishing Company, Amsterdam. 211pp.

in digital form. The procedures might be brought together as aspects (subsystems) of The geological cyberenvironment (gce) 85 (*Figure 34*).

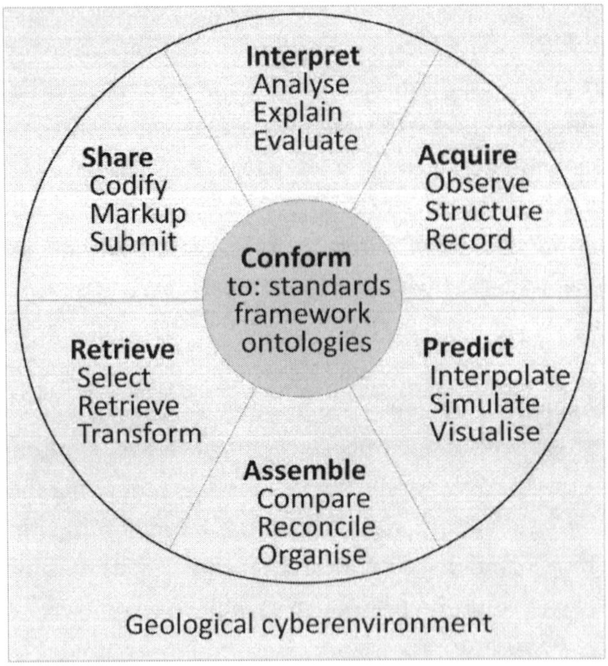

Figure 34: Stages of geological investigation (duplicate of **Figure 17***)*

In order to reconcile spatial models of different aspects of the real world, we need to know the premises on which the models were based, and what might account for the differences. It would be easier if they all referred to one consistent set of objects. But there is no unique hierarchy of object instances or matching object classes. Classification depends on which aspects are observed and the relative importance assigned to each. Even stratigraphical classification (see The stratigraphical framework 139) takes many forms, such as biostratigraphy, chronostratigraphy, lithostratigraphy, or seismic stratigraphy, possibly with diachronous relationships between them. A great strength of the object-oriented approach is the ability to handle this diversity of models through semantic links (see Semantic Web and Grid 50) and to help scientists with Reconciliation 186 of the results. Automation of the actual reconciliation decisions is not an immediate prospect, but it should be feasible to assist scientists' decisions with interactive processing, and record the results through links to explanatory notes and microdocuments (see A framework for the reasoning 135).

The notes should explain to the user why a particular view has been reached, its limitations, and perhaps how it might be amended for particular purposes. The notes serve the same function as the traditional map explanation, but are split into shorter, more self-contained sections, generally as a text explanation but tied closely to surveyed spatial objects. They are likely to be embedded in, or at least have links to and from, a more general explanatory object, thus establishing a hierarchy of explanations describing the reasoning process. In their conventional form, these general objects are documents, leading us to think of the smaller component objects as *microdocuments* (also referred to as granules or minimum revisable units).

Microdocuments can define the paths (workflow) that the surveyors followed in their thinking (Ludascher et al., 2006[319]). The dialectic of field survey (see The dialectic model 129) could be recorded as the surveyor links, relates and reconciles observations in the field. Reconciliation can result in objects being amended to conform to new evidence and therefore replaced. The relationship 'is superseded by' could bulk large in the object store. In general, reasoning processes can be recorded as *threads of reasoning* – sequences of ideas, represented by a chain of hypertext links (see Mechanisms 45). The sequence winds through the repository, relating objects to their place in flows of ideas or lines of argument (Loudon, 2000, part J[320]). They are not part of a database schema, but resemble the links between and within Web documents. The links could extend to the general geoscience model, metadata, the wider hypermedia knowledge repository and references to non-digital material. Additional notes may describe and clarify the reconciliation and reasoning process. The reasoning could then be tracked and reinforced or refuted by others.

Microdocuments can also define a set of actions, or sequence of instructions to a computer system, as '*scripts*' connecting pre-existing components to perform a composite task or application. Innumerable paths can be established for different purposes and expressing different views. A script could generate the equivalent of a map explanation (largely text with some graphical illustrations) or could provide paths that create standard maps and their accompanying explanations by selecting the required spatial objects and visualisation procedures. Alternative Survey scripts might select objects appropriate for groups of users interested in specific topics or with different levels of expertise. A wide range of individual scripts might also be prepared for more specific purposes. If they are archived, procedures for cataloguing and assessing their relevance may be needed to avoid losing users in the information fog.

A mark-up[321] language, such as XML and its derivatives, provides the means of creating scripts and threads of reasoning by linking objects in a branching sequence, possibly embedded within explanatory text. The spatial index database and the mark-up language should both be able to access the same objects. They represent different information environments, the first addressing human short-term and spatial memories, the second addressing episodic memory (see *Figure 12*). In the first, users select a set of objects that conform to the profile of their requirements, and process them together. In the second, users thread their way from object to object, making decisions about which of the branching paths to follow on the basis of what they have learned so far (see Mechanisms 45). The desktop browser must make it easy to move from one information environment to the other, or to work with both together on a split screen. Each object in the object store must

---

[319] Ludascher, B., Lin, K., Bowers,S., Jaeger-Frank, E., Brodaric, B., and Baru, C., 2006. Managing scientific data: From data integration to scientific workflows. Geological Society of America Special paper 397, p.109-129, doi: 10.1130/2006.2397(08). http://users.sdsc.edu/~ludaesch/Paper/gsa-sms.pdf

[320] Loudon, T.V., 2000. Geoscience after IT: a view of the present and future impact of information technology on geoscience. Elsevier, Oxford. 142 pp. Also available as *Computers & Geosciences*, Special Issue, **26** (3A), A1-A142 http://nora.nerc.ac.uk/2405/

[321] Mark-up: Symbols inserted in a document in a mark-up language such as SGML, HTML or XML, to tag the beginning and end of character sequences that can be interpreted by machine, and can be omitted for displaying to the user.

therefore carry a uniform resource identifier (URI) that enables it to be found, similar to the familiar URL of Web documents. It must have a label, like the URL suffix, which indicates how it can be displayed.

The microdocuments point to some important consequences of the object-oriented approach. One is the blurring of the traditional distinction between map and map explanation, and between graphical and textual representation (see Integrating information types 62). A second is that a hierarchy of composite object instances can be created, where we can define precise limits within which the relationships apply. For example, objects are surveyed at different resolutions, and a set of orientation measurements, say, might have to be generalised as a new object in order to achieve a resolution appropriate for reconciliation with a structural map based on seismic data. A third is that objects are reusable in different contexts, so that many products can come from the consistent basis of one object store. For example, well picks might be combined as a dataset by establishing the spatial relationship linking them as component objects, rather than by repeating their values; and the generalised orientations just mentioned might not be stored, but generated as required from the updated measurements (see Multiresolution survey 259).

Relationships between objects may determine where their reuse is valid. For example, a sample might be collected, described and curated once, but analysed in both a geochemical study and a petrological study, forming a link between the studies. Higher in the reasoning chain, the only link between maps of gravity and geochemistry (see 'The importance of space and visualisation' 169) might be through completed models reconciled to the same geological model. Some general relationships may be determined during data analysis. Most relationships, however, refer to specific instances and are recorded by the surveyor, or in a subsequent editing procedure when experts in the local geoscience build their interpretation. The relationships and their provenance (who established this relationship, when and why) can be recorded and tied to the objects within the object store. Maintaining the relationships of spatial objects requires reconciliation, inserting explanatory notes and establishing links from object to object within and across many levels of the hierarchy. Spatial objects, with links and metadata, can be identified and archived while digitising legacy data such as maps or borehole records. Serious problems will inevitably arise in maintaining a constantly changing store of linked, reusable objects. And the full benefits will not be achieved without a unified system that extends from field surveyors to end-users (see The geological investigation model 98, Object-oriented survey 195).

Implementation note: Likely future developments of suitable equipment and software are largely outside the control of geologists. They must nevertheless be kept in mind during analysis and design.

Object-oriented survey

<<An object-oriented approach 178

Summary: *An object-oriented approach corresponds closely to existing ideas in geoscience, but gaining the full benefits may require a unified system from fieldwork to end-use. This will involve changing the boundaries between long-standing subsystems to provide more flexible procedures for handling multimedia during surveying, explanation, archiving and visualising. Selection by area, topic and level of detail can be deferred until visualised results are required. Recording the dialectic between model and reality can throw light on the significance of the products and the influence of feedback. The field or desktop browser must support a comprehensive, top-down view. The approach should converge with general developments in computer-based knowledge systems, should optimally combine the capabilities of man and machine, should accept high-level legacy objects as well as new detail, but above all should be appropriate for geoscience.*

The object-oriented approach leads to a view of the Geological Survey model as a system of related spatial objects. A complete object-oriented approach, from survey to visualisation, might overcome many shortcomings of the map. The notion of object-oriented geological field survey may seem far-fetched, but in reality, it more or less corresponds to present practice. The range of objects implied in Harrison's account of field mapping (see The surveyor's holistic view 60), for example, can be carried through to the spatial model. The map key identifies object classes, and observations recorded in field maps and notebooks may be identified as instances of the objects. The computer metadata merely formalise earlier procedures. Objects are described in academic studies and in map explanations, free of map sheet boundaries and at a resolution related to geological significance rather than to fixed map scales. In contrast with the Geological Survey map, a computer spatial model can retain this greater flexibility in integrating and reconciling the spatial properties of objects, regardless of scale.

The object-oriented framework is appropriate for computer-aided field mapping (see Internal BGS – SIGMA 343), but can also accommodate information collected using conventional methods. It can potentially provide individual slots for recording information to any level of detail, including each field measurement or observation. However, a low-technology solution, such as a field map, may give an adequate summary at lower cost. The field map has already aggregated much of the detailed information, and thus slots into a higher level of the framework. Its metadata might identify the detailed information content, but it makes sense only in the context of the map and not as individual items. Retrospective untangling of legacy maps can be attempted only by geologists familiar with the area, and only to a limited degree. Much legacy information is so tight-knit that its components cannot be disentangled. For example, a strike and dip symbol might be misleading outside the context of the map in which it is recorded, and so should be seen as a part of the map and not as an independent object in a wider context. Within each project, the level of detail to which objects are separately identified must rely on common sense and perhaps an appraisal of cost-effectiveness. New field studies might formalise the separation of objects to a deeper level, but the geoscientist requires freedom to specify object classes only to a level of detail that proves helpful.

With the object-oriented framework in mind, it may now be useful to reconsider in turn the following ideas from other sections:

1. system reconfiguration (changing boundaries between subsystems) (see Reconfiguring the system 172)
2. abstraction (reducing the volume of information by selective observation, interpretation and explanation) (see Abstracting from reality to model 131)
3. dialectic analysis (juxtaposing and reconciling arguments from different sources) (see The dialectic model 129, Reconciliation 186)
4. feedback (reviewing earlier work as more is learned) (see Geological surveying as reinforcement learning 79, Complex and emergent systems 159)
5. the holistic view (keeping the big picture in mind) (see The surveyor's holistic view 60)

1. The conventional system in which the map acts in various incarnations as a field survey record, a published visualisation, and an archived record can be reconfigured. Each activity (surveying, archiving and visualising) can be viewed separately, with its own objects, procedures and products, linked through defined interfaces. Each of the many sources of information can be described on its own terms, as an object in its own right. This reflects the immense complexity of the real world and the variety of techniques and viewpoints available for its investigation. Interpretation and integration build composite objects to express broader ideas, while retaining and reusing the detailed objects. Relationships among the diversity of objects can be recognised, identifying the level of detail at which they apply (see The importance of space and visualisation 169, Microdocuments and the threads of reasoning 190). The object repository can archive all material deemed to be of potential future interest. Users can browse and explore the contents of the archive, selecting and visualising what seems appropriate to their current interests and objectives. The objects are not confined to one information type, and support relationships across information environments, such as database, spatial or episodic. The information conventionally separated into data, graphical and text documents can thus be reconfigured into one integrated system.

2. Abstraction[322] begins with selective observation. The surveying process that creates the spatial model describes object instances, ties them to geographical co-ordinates and establishes the relationships between them. Each object is surveyed on its own terms at an appropriate resolution and level of detail. Objects can be compared and reconciled, and analogies with similar instances and processes elsewhere lead to explanations of the observed phenomena. Generalisation and explanation of object instances add information to the general geoscience model (see The imperfect model 156). This deals with past states, the hypothetical processes operating on objects and the responses, and the resulting sequences of historical events. It is the context in which the general properties of the object instances can be organised as object classes, and their roles, relationships and classifications defined. The Geological Survey model should record the thrust of this process of

[322] Abstraction: reducing the information content of a concept or an observable phenomenon, typically in order to retain only salient information, relevant for a particular purpose

investigation and reasoning, in which reconciliation has a major role. The spatial model can be thought of as a collection of spatial and descriptive objects and relationships organised within a defined framework, and regarded loosely as a kind of object database, or at least an object store with an index that can be accessed by database management systems (see A framework for the reasoning 135). This provides for further abstracting by filtering or selecting from the model for purposes of visualisation or analysis, on the basis of, say, area, resolution, topic, stratigraphy, and certainty.

3. The dialectic process of juxtaposing model and reality is followed (and taken for granted) during conventional field mapping (see The surveyor's holistic view 60), aiming to provide a consistent overall product. It can start from many viewpoints, based on different interpretations, emphasising different properties, and reflecting the battery of available techniques that narrow down the valid interpretations (see Forward and inverse models 154). Rules may be specified in the metadata, or assumed as a result of training and experience, and not necessarily recorded. If the rules are followed, the surveyor creates an object of a particular type (see At the interface 133) – a mark on the map resulting from what was observed in the field. There is a defined hierarchy of objects, which should be mutually consistent. Readers of the map, knowing the rules, should be able to envisage a situation that caused the surveyor to create the objects they see on the map. The value of the map, after all, lies not in the image but in what it tells us about the real-world geology (as seen through the surveyors' eyes). There is ambiguity here, however, for various combinations of circumstances could have caused the surveyor to draw the map in a particular way. The reasoning process may or may not be clarified in the map explanation. The map gives only the final result of the dialectic process. An object-oriented model, on the other hand, could clarify and record important steps linking objects at any stage of the reasoning. In the model, the threads of dialectic reasoning can be explicit, juxtaposing arguments from various viewpoints and seeking their resolution. The object-oriented framework could provide for reconciliation and assessment of individual objects at any level of detail.

4. There is a cycle involving abstraction during observation, recording, interpretation and explanation. The procedures of abstraction lead to the formulation of ideas about the properties and relationships of classes of object (see Relationships between objects 183) that occur in the spatial model, seen as a component of The solid Earth systems model (sEsm) 71. Metadata formalise and clarify the procedures and vocabulary. This establishes the object classes as the currency of geoscience thought and communication, and determines their significance within geoscience survey. Feedback completes the cycle, the evolving interpretation and explanation determining what is of interest and thus influencing future observation and the surveying of object instances.

5. The browser[323], on the desktop or in the field, must link through object management and retrieval, database management and geographical information system (GIS) facilities to the objects in the spatial model. They should make it possible to select and filter the objects for

[323] Browser: A software application that assists the user in searching for, retrieving and presenting relevant information, typically from the internet or World Wide Web.

the required level of generalisation according to criteria such as those mentioned earlier. It must also be possible to manipulate the spatial objects with GIS functions, such as selecting a specific area and finding overlapping objects, and more generally handling spatial relationships. The browser should eventually have links also to facilities for visualisation, computer graphics, image analysis, statistical analysis, word processing, email, telephone, and in the field to GPS, electronic surveying and possibly augmented reality. Fortunately, the computer industry is moving in this direction (see Representing wider knowledge 283), for it is an area where geoscientists have only limited influence. The browser must be able to reach beyond the individual geologist's model, to relate to global metadata and projects developing the general geoscience model. It should enable the surveyor and other users to adopt and extend a holistic view. It must therefore remain wherever possible within the mainstream of cyberinfrastructure developments.

This leads to what may be the crux of this application of object-orientation, and maybe the crucial issue for acceptability or otherwise of a Geological Survey model. The techniques are designed for computer implementation, arranging the information within a framework where the software can operate. Ideally, it should converge with the developing general procedures for knowledge systems in a computer environment (see At the interface 133), working towards a hybrid system that combines the best features and capabilities of man and machine. Primarily, however, the underlying philosophy and the detailed structure must make sense for the scientists' own purposes. Objects, their relationships, classifications, and explanations should not diverge between a computer implementation and the scientist's thought processes. Geologists should not try to think like computers, but they should surely expand their thought patterns to make full and appropriate use of computer support where, and only where, it may prove helpful. A satisfactory new framework can be achieved only through the full involvement of the geoscience community.

A geologist carrying out a one-off investigation for a specific purpose might find traditional methods effective, efficient and convenient and would not wish to be side-tracked by thoughts of objects or models. A large surveying organisation, on the other hand, is likely to have extensive records. They might include: field notes; project descriptions; map explanations; well logs; borehole descriptions; vertical sections; samples and specimens; thin sections; fossil records; petrographical analyses; and records and maps of topography, airborne and satellite imagery, geophysics, geochemistry, soil science, and engineering geology. Such records fit neatly into an object-oriented framework. Survey mapping in the field (see The surveyor's holistic view 60) suggests a natural fit with the more formalised framework. When the system eventually takes shape, even a one-off academic or commercial project might benefit from taking Survey objects as a starting point and possibly making its own results available in the same standard form.

The Survey model should provide a comprehensive view and so must build on our existing legacy while not restricting the development of valuable new approaches. Despite the inconvenience of change, the potential benefits, such as those listed in 'Benefits of an object-oriented system' 199, suggest that an object-oriented spatial model is the way ahead.

Implementation note: Object-orientation can provide a flexible means of defining objects of interest and arranging and relating ideas. The object-oriented approach provides an opportunity for more rigorous and complete documentation, and may instigate major changes in the organisation, which are worthwhile only if they can be justified by long-term benefits. They involve untested concepts and unreliable software that must be approached with small, tentative steps and with an exit strategy for backing out safely from failed attempts.

Implementation note: Flexibility brings diversity, which must prove unsettling to those accustomed to a single, authorised map accompanied by one coherent text explanation and shared by the geoscience community as a whole. Easy user access to traditional forms of presentation must be retained as a default option. The surveying community and many users have a stake in the status quo, but they also see the internet as a growing source of information where they expect a new perspective. Separation of the surveying function from the dissemination (publication) function could help to ensure that a wide range of information is secure, evaluated, and readily accessible.

Benefits of an object-oriented system

<<An object-oriented approach 178

Summary: *An object-oriented system should be more flexible, comprehensive and robust than conventional methods. More diverse and relevant data can be collected with IT support in the field, and reconciled with input from a growing battery of sources. It can provide a comprehensive, consistent and coherent account, organised (like the geological map) around lithostratigraphical and structural interpretations of the geology at local, regional and national scales.*

Although implementation is a challenging task, an object-oriented system has many potential advantages over conventional methods of geoscience survey, including the following.

1. It is more *flexible*

- each spatial object has its own location and extent, free of map sheet boundaries, and can be investigated with a design and resolution appropriate to its characteristics, significance and the surveyors' objectives
- surveying, archiving and presentation can each be handled on its own terms, for example, the initial procedures of collecting, storing, analysing and interpreting the information are not constrained by a predetermined form of presentation, and may therefore be designed to meet a wider range of objectives
- new data and changing interpretations can be accommodated with minimal delay and disruption, by continual piecewise revision
- the structure, vocabulary, procedures and interfaces can be carefully defined as shared metadata, allowing subsystems to be handled independently and integrated as required
- alternative scenarios can be explored by temporary modification of the data or procedures, reflecting multiple hypotheses, different objectives (see Diverse

objectives and products 150), revised assessment of probabilities or changing concepts
- detail is hidden within an object until it is required and so does not obscure more general views of which the object is a part
- database techniques make it possible to select and visualise only what is appropriate to the specific requirement
- thematic maps for many different applications can be created by reuse of objects from a single object store
- in contrast to the printed map, filtering by area, resolution, projection, certainty, topics, objectives or relevance to a chain of reasoning can be deferred until the user invokes a visualisation model to meet specific objectives

2. It is more *comprehensive*

- information from different sources can be included along with its metadata, bringing the power of a larger battery of techniques (see Forward and inverse models 154)
- explicit links can maintain appropriate connections at each level of detail in the process of abstraction from observation to explanation (see Abstracting from reality to model 131), enabling the user to view the model at an appropriate resolution and drill down to relevant detail or link to related information
- many views can be handled, a wide variety of spatial objects can be considered, and the overall model can be reconciled with or refuted by them
- different forms of information, such as verbal descriptions, chains of reasoning, databases, images, analogies and process models can be combined within an object or linked to it
- estimates of the most likely disposition, probability envelopes and typical configurations could be made available, perhaps clarifying the process of combining spatial information of differing certainty or resolution
- in contrast to the map, the models can extend to three dimensions wherever appropriate
- diverse ideas and alternative interpretations can be accommodated
- discrepant sources of information can be reconciled, with records explaining how any conflicts were resolved

3. It is more *robust*

- the hierarchical framework, metadata, object classes and object instances can each be ring-fenced to avoid unauthorised alteration
- the reasoning process is clarified, and lines of reasoning can be traced through chains of objects
- interpreted objects can be related back to detailed observations in order to correct errors and identify their knock-on effects
- well-defined spatial objects and relationships at an appropriate hierarchical level can be unambiguously identified and their dependencies recorded in the chains of reasoning
- individual spatial objects can be identified uniquely, and their spatial relationships and other links can be stated explicitly

- authorisation procedures for modifications can readily be implemented
- ambiguities can be resolved through metadata that offer fuller documentation including the provenance of each item and the procedures that created it
- geometrical representation and manipulation of spatial objects (as opposed to mental images) introduce more rigorous and reproducible mathematical reasoning
- spatial objects that require different geometrical approaches can be identified and handled appropriately, perhaps with links to suitable methods of interpolation and generalisation
- bounds can be set for the filters and transformations (see Seeking shared concepts 247) that can be applied to a specific spatial object

The technology for recording field observations (complete with GIS, GPS and electronic surveying) will facilitate routine geometrical analysis of spatial objects as the survey proceeds (Brodaric, 2004[324], de Kemp, 2000[325], Kessler et al., 2008[326]). Automation will enable the object-oriented approach to handle greater diversity of field data collection than is practicable with traditional methods. With computer support in the field, numerous documents and models could be managed together, with rapid communication to and from shared archives. Rather than recording all the information on a single map, individual spatial models might include composite objects together with their spatial relationships and other links. Examples are present-day topography, structural geology, individual formation boundaries, Drift deposits, metamorphism, igneous intrusions, petrology, and biostratigraphy. Each of these might be an interpretation of records of many observations, themselves regarded as objects at a lower hierarchical level. At a higher level, the overall model of the local geology might be fleshed out in the field, reconciling it with these composite objects, and with external sources of information such as boreholes, geophysical and geochemical maps and cartoon models of the dynamic stratigraphy.

None of this suggests that a Geological Survey should lose its focus on consistent, countrywide, lithostratigraphical and structural interpretations that can be visualised at local, regional and national scales. However, object orientation might help to clarify the links from these primary records to other sources of information such as geophysical maps or detailed accounts of outcrops. The approach seems tailor-made to accommodate the growing variety of input to geoscience survey and of its thematic visualisation (see The imperfect map 145). It should overcome the ambiguities and barriers to integration in the conventional geological map that arise from the diversity of sources and diversity of forms for expressing and visualising that knowledge.

---

[324] Brodaric, B., 2004. The design of GSC Field*Log*: ontology-based software for computer aided geological field mapping. *Computers & Geosciences*, **30**, 1, pp. 5-20

[325] de Kemp, E.A., 2000. 3-D visualization of structural field data: examples from the Archean Caopatina Formation, Abitibi greenstone belt, Québec, Canada. *Computers & Geosciences*, 26, 5, pp. 509–530.

[326] Kessler, H., Campbell, D., Ford, J., Giles, J., Hughes, A., Jackson, I., Peach, D., Price, S., Sobisch, H-G., Terrington, R., Wood, B., 2009a. Building on geological models : the vision of an environmental modelling platform. In: *Geological Society of America Annual Meeting 2009, Illinois, USA, 18-21 Oct 2009*. Illinois, USA, Geological Society of America, pp. 24-30. http://nora.nerc.ac.uk/8423/

The view that is emerging of the geoscience survey knowledge system is of aspects of knowledge being expressed in different information types (see Integrating information types 62), notably in a network of spatial and narrative models. The processes of surveying include abstraction, to reduce the volume of information, and interpolation, to fill gaps in the observational record, both informed by feedback from knowledge gained from geoscience survey and also from other aspects of geoscience and scientific knowledge in general. If acceptable, this view should refer equally to conventional methods of survey and to a computer-based knowledge system. The system of spatial objects and relationships provides a context for geometrical analysis. The spatial model framework calls for an integrated view of the geometrical representation, visualisation, reconciliation and spatial relationships of many diverse objects. We therefore need to look at the geometry in more depth (see The geometry of the spatial model 203).

Implementation note: The same framework (see The geological framework model 105) would ideally include, or could be extended to include, many topics, such as topography, ecology, soil science, stratigraphy, structural geology, geophysics, petrology or geochemistry. The framework relates to similar surveying procedures in all these fields, where conventional maps are frequently compared at present. In future, objects may be shared and reconciled. A larger market would encourage the development of more effective software and facilities, enabling broader information exchange. Widespread consultation in developing a standard framework now may reduce the need to backtrack later. In due course, it may be necessary to reconsider which organisations and information communities take responsibility for design and maintenance of computer implementations of the core Geological Survey model, the general geoscience dynamic model, and their frameworks, standards and metadata (see Stratigraphical units in space and time 141 and Transforming space 224).

The geometry of the spatial model

<<Table of contents 1
<<The future geological map 125
<<An object-oriented approach 178
The geometry of the spatial model 203
The need to harmonise the geometry203
Making a mesh 207
Drawing the line209
Estimation by interpolation 212
Continuity, fractals, octrees and wavelets 217
A wish list for integrated geometry 221
>>Transforming space 224

Abstract: A unified system of spatial models requires compatible computer representations of the geometry. Some current approaches include interpolation to fit surfaces to nodes on a square grid, using a weighted moving average or by fitting mathematical functions to the data. Variation within a volume can be represented by voxels. Fractals offer a model of processes distributed in space at different resolutions. Wavelet analysis breaks down an overall pattern into superimposed local patterns at different scales. An integrated view calls for a coherent overall system with a shared mathematical framework for all formal procedures from initial observation to final presentation, selected by users to match their specific needs.

The need to harmonise the geometry

<<The geometry of the spatial model 203

Summary: *Digital cartography is more than just automation in the drawing office. It can provide a route to the object-oriented model described in* 'An object-oriented approach 178', *and to mathematical representation and analysis. Interpolation (filling gaps) can be separated from visualisation (creating images for human interpretation and analysis). The geometry of the surveyed object instances must be reconciled with the expected behaviour of the object classes in geological process models, all potentially implemented on the computer. The three-dimensional model can include both surface and subsurface data that refer to the same objects, and so requires a consistent geometrical representation.*

Digital cartography may be introduced for more convenient and efficient storage, editing, updating and presentation of maps, thus automating drawing office functions. Compatible functions can be built into computer-aided recording systems for use in the field. The results can be entered directly into a Geographical Information System (GIS)[327] to provide an interactive interface appropriate to geological needs (see Some related initiatives – GSI3D, SIGMA 343). The GIS enables users to select and visualise sets of relevant objects, for example, to superimpose a map of geological hazards over a regional development plan. But a longer-term benefit may be the support for a radically new framework for geoscience survey.

When a map is digitised, the computer represents the map contents numerically, and the geometry and its transformations algebraically. This brings opportunities to introduce more powerful and rigorous methods to analyse and represent geological findings. Geologists are more likely to be familiar with the geometry of observing and understanding the intricate shapes and scales of geological material than they are with the algebraic representation. They are in the best position to select relevant procedures for processing their information, and it is ultimately the responsibility of the geological community to determine appropriate methods for their work. A geometrical approach is therefore followed here in describing a range of possible techniques, in the hope that geologists can evaluate potential geological implications (remembering that the Scenario is concerned with where we want to go, not how to get there). The algebraic equivalents are essential for implementation, and may help in understanding the constraints and possibilities.

Within the computer model, spatial objects must be broken down into geometrical elements (such as points, lines and areas) for representation and analysis. Mathematical functions (or their surrogates, such as grids of values) may represent the spatial characteristics of objects for simulation, interpolation, analysis, reconciliation, generalisation, filtering and visualisation. Mathematical methods are thus inherent in digital cartography, and, as discussed later, should lead to representations that are appropriate across a wide range of modelling procedures. In comparison with the conventional manipulation of images in the mind, they offer opportunities for explicit mathematical and statistical analysis and more rigorous and effective science.

Digital cartography and GIS are therefore routes into mathematical methods and the object-oriented approach to spatial modelling outlined earlier (An object-oriented approach 178). Content can be freed from its containers. Map sheet boundaries lose their significance. The various information types can be more closely integrated. General properties of object instances can be abstracted and reused within hierarchies of object classes. Each spatial object can be treated on its own terms. The results of a battery of investigative procedures can be archived, and reconciled to record one or more coherent interpretations. Printed maps and memoirs for packaged delivery can be supplemented by an object store from which products can be assembled and visualised as required to meet a wide range of user-specific needs.

Thus, a spatial model is not the equivalent of a single map, but a record of many different objects and properties, linked through spatial position. At its core is a spatial view,

---

[327] Geographic Information System (GIS): An integrated system for the capture, storage, management, retrieval, analysis, manipulation and display of geographically referenced spatial data and its attributes.

dependent on visualisation and geometry. Unlike the map, it is not itself a visualisation, but is structured information that can be visualised in many ways. Whereas interpolation and visualisation are inextricably combined in a map, they can be separated in a computer model. *Interpolation* is concerned with estimating values between known points, essential for filling space and predicting unknown values. *Computer visualisation* is concerned with transforming quantitative data (including the results of interpolation) into sensory information – images that the eye and brain can interpret and analyse. Contours, or more generally lines of equal value, are just one of many methods of portraying a surface for visualisation. Alternatives include hill-shading, continuous colour variation, cross-sections, block diagrams, perspective views, and augmented and immersive virtual reality (see Raper, 2000[328]).

The spatial model aims to provide coherent interpretations by juxtaposing and reconciling the surveyed instances of objects and their spatial relationships with the more abstract object classes and dynamic explanatory models. This calls for a unified approach to describing and analysing the geometry. We need to decide which mathematical concepts and methods are appropriate (on the basis of our knowledge of the physical phenomena), and how deeply we embed them in the spatial modelling process rather than just in its static outcome (the map). Where representations of spatial objects, models and relationships are not directly compatible, they should at least be reconcilable, even if they come from different sources or are expressed in different ways.

The spatial model must accommodate two-dimensional views. Much legacy information is in this form, as there is limited scope for representing three dimensions on a paper map. Geological Survey maps tend to concentrate on the land surface and a few cross-sections, even where subsurface data are available from shallow boreholes and geophysical studies. Computer models, on the other hand, extend naturally to three dimensions. The arguments for filling space on a map (see Representing spatial information and relationships 173) apply equally to a three-dimensional model where there are sufficient subsurface data. The rock units that fill the space are usually represented by the surfaces that bound them. Mathematical interpolation between subsurface observations, such as seismic lines, wells or boreholes, has been studied more extensively than interpolation of boundaries in field mapping (Tearpock and Bischke, 2003[329]). But the underground surfaces and their lines of intersection with the land surface are geologically and geometrically related, and each can throw light on the other.

Ideally, known points on a line, like those on a surface, might be recorded and joined up by methods of mathematical interpolation justified by explicit geological reasoning. The lines depicting intersections with the land surface on a geological map might be converted to three dimensions by fitting them to a digital terrain model, thus freeing them from the topographical base map for display, generalisation and analysis. We might take into account inferences from secondary observations of, say, soil types and landscape features, just as a subsurface map based on downhole logs may be influenced by secondary sources, such as

---

[328] Raper, J., 2000. *Multidimensional Geographic Information Science*. Taylor and Francis, London. 300pp.

[329] Tearpock, D.J. and Bischke, R.E., 2003. *Applied Subsurface Geological Mapping*. 2nd ed. Prentice-Hall, Upper Saddle River, NJ. 822pp.

seismic data. In practice, legacy maps cannot be interpolated in this way because of the volume of information and the difficulty of retrospectively tracing the intricate reasoning process. Digitisation must therefore be based on the existing map representation until the area is resurveyed.

The extension to the third dimension implies that the digital spatial model will also include subsurface information. Geoscientists prepare contour maps as a routine method of interpolating, depicting and studying the variation in space of properties of spatial objects, such as elevations of subsurface horizons. Computer contouring is widely used in subsurface geology, geophysics, and other topics. It can be quick and convenient, particularly if the data are already available as a computer file. It encourages systematic recording and storage of the data, and can produce an attractive product without recourse to a drawing office. The methods are particularly helpful for large volumes of data. For example, an oil company might maintain a regional study of a sequence of subsurface formation tops and horizons. The interpretation might be updated regularly in response to new data or new ideas – a more efficient process with computer support. Many aspects of the geology in the same area share similar spatial patterns, and each surface can be seen as an integral part of a broader spatial model, where each entity throws light on the others.

A basic overview follows, from a geological viewpoint, of some approaches to computer representation of geoscience surfaces in current use. The problems lie not so much in understanding the mathematical models, which are explained in many textbooks, as in harmonising the diverse representations of the geometry and matching them to the underlying geological reasoning.

Implementation note: Digitising existing maps is a useful starting point, but does not offer the full flexibility of a spatial model. Opportunities will be missed if there is no analysis of the longer-term geoscience objectives and design of an appropriate system. The spatial model has the potential to include digital map data and diverse other sources of information in a full three-dimensional framework, but tracing the reasoning and separating observation from interpretation calls for input of expert knowledge, and may involve resurvey.

Implementation note: Implementations for use by geologists must correspond to the pattern of their datasets and the diverse geological models that guide their observations. Methods of interpolation, including kriging, are in wide use, but the geological applications of some other approaches discussed here have not been fully explored. One promising long-term approach is to build applications around a framework familiar to the geologist (see Some related initiatives – GSI3D 342) that can act also as a test-bed for experimental approaches.

Making a mesh

<<The geometry of the spatial model 203

Summary: *The Digital Elevation Model (DEM) is one useful computer representation of a three-dimensional surface, covering the surface with a grid of squares and recording the elevation of each. It mimics aspects of human vision, matches square size to resolution or to scale, and lends itself to efficient computation. Geological data are generally scattered and incomplete, but may be interpolated to a grid as a preliminary to analysis, comparison with other similarly gridded surfaces, and visualisation. The separation into two stages is theoretically unsatisfactory but practically convenient. A hierarchical grid (quadtree) can cope with varying density of information.*

One basic model for contouring, which may influence many geologists' thought processes, can be traced back to procedures for topographical surveying. As the surveyors can see the land surface, they can include information about all relevant features that are large enough to show at the map scale. For a 1:10 000 scale map, these might be features of some 10 metres across (1mm on the map), or 50m for a 1:50 000 scale map. Smaller features judged to be significant might also be included, and even exaggerated to give them appropriate prominence. Contours are probably surveyed more carefully at critical points to ensure that they correspond exactly to features such as cliffs, summits, rivers or ridges. A similar approach could apply to mapping two-dimensional features such as the line taken by a road.

The contour map depicts a smooth surface. The real landscape is rough. When we view a landscape from a distance, however, our eyes cannot detect the small jagged features and the surface may appear smooth. At a greater distance, medium-sized features blur and we see only their broad outline. It is as though there were a small circle on the landscape, or more exactly, a narrow cone pointing from the eye intersecting the scene, within which the eye cannot discriminate individual points. Move further away and the cone encompasses a larger area on the ground. It thus determines the *resolution* (the least distance apart of two points on the ground that are individually detected by eye or shown on an image). Similarly, detail is lost from a contour map, or from computer visualisation of a surface model, when it is generalised for display at a smaller scale. Reducing the scale is like viewing it from a greater distance. Salient features, such as roads and rivers, can be exaggerated to retain continuity or significance that does not depend on size. This model is appropriate for visualisation and interpolation, for it mimics the way we see the real world and enables us to use our lifetime learning of visually interpreting our surroundings.

A hand-drawn map showing the relief of the land surface might start with contour lines. The computer equivalent generally starts with a digital elevation model (DEM), where the land surface is tiled on a square grid and the average elevation of each tile (perhaps 10 metres across) is recorded. The DEM is more flexible than contours, being sampled at a uniform geographical resolution. The DEM can be visualised in many forms, including contours, and is appropriate for analysis and calculation of spatial characteristics at any resolution equal to or coarser than the tile size. It can readily be generalised to a coarser resolution for presentation at a smaller scale.

In the ideal case of the topographical surveying model just described, each tile was visible to the surveyor and its relevant characteristics could be recorded at the resolution required for the model. Satellite imagery, digital terrain models and three-dimensional seismic surveys come near to this ideal data pattern, but most geoscience data are unevenly distributed. Geological exposures are small, scattered features on the land surface. The geological map consequently contains more interpretation than direct observation. Maps of subsurface horizons are typically based on data from points where wells or boreholes were drilled for reasons other than map making. Some contour maps refer, not just to a scattered sample, but also to a discontinuous population. For example, although the contours of trace element concentrations in stream sediments may cover the map area, such sediments exist only in the streams. As always, the model differs from reality.

Modelling methods might therefore start by interpolating from the available data to points on a uniform grid. The result would be analogous to a DEM, and could be seen as a first step to subsequent processing, including more precise interpolation. This would be particularly valuable in a Geological Survey model, which must handle many surfaces, derived from diverse data within many projects, possibly analysed by different specialists with their own software (Wikipedia[330], and references therein). If all these processes generated similar grids, the surfaces and their relationships could more readily be visualised and analysed together.

Some parts of a geological surface may be known in more detail than others, and can therefore be visualised at a finer resolution. This does not imply that their geology is more intricate, merely that there are more data. It can give a false emphasis but need not necessarily obscure the underlying pattern. We are accustomed to seeing most detail near the centre of our field of view, with a fuzzier image of the surrounding context, which is why you are moving your eyes as you read. Artists build on this by, for example, painting the foliage of a few trees in detail. The detail catches the viewer's eye, and the rest of the forest can be shown as simplified tree shapes, leaving the viewer's imagination to extrapolate the detail. A hierarchical grid, known as a quadtree, or octree in three dimensions (see Mark-up and metadata 175), can store more detail as required, and is thus not confined to one level of resolution.

On theoretical grounds, the two-stage process of analysing a grid of interpolated points is unsatisfactory. The grid is an artefact created on one set of assumptions. It is then subjected to subsequent processing, for example for visualisation, possibly based on conflicting assumptions that could invalidate the result. Care is needed to avoid attributing patterns to geological causes when they merely reflect data distributions or interpolation procedures. Hand-drawn contours do not normally involve a grid, but similar, if less explicit, assumptions are made that may not be apparent to a user who subsequently draws conclusions from the completed map.

On practical grounds, the two-stage process may be necessary. A consistent base is needed for visualising a shared model derived from many sources. The initial gridding process can provide a broadly relevant interpretation that is sufficiently general to be useful in a wide range of specific and possibly unforeseen applications. The results can be represented as a grid of data, like a DEM, which can stand in as a surrogate for the functions that originally

[330] Wikipedia 'Mesh generation', 2011. http://en.wikipedia.org/wiki/Mesh_generation

generated it, and can be processed efficiently by computer. The grid spacing can be selected to give a reasonable estimate of the value of the function at any point by simple interpolation of the grid, rather than rerunning complicated calculations on the original data. The grid can be analysed further to generalise the information or to look for statistical relationships or pattern within it.

For visualisation, the surface should smooth over the irrelevant but eye-catching irregularities of the data distribution, offering a consistent view based on approved grid values. However, the need for consistency may conflict with the variety of objectives. Other types of surface model may be required for estimation or study of patterns or alternative explanations. Recalculation from the raw data is desirable for these special purpose models. The results of such models may add sufficient value to justify storing them also in grid form. They are likely to be confined to a local area, and if they are tentative scenarios, should not affect the model as a whole. If they are needed only occasionally, they could be regenerated as required.

The metadata should store information about the methods that generated each DEM and might recommend links to procedures for its detailed interpolation, generalisation and visualisation. These procedures should be selected to avoid inconsistency between creating and expanding the grid, between successive stages of generalisation, or between DEMs created for different topics that reflect the same geological circumstances. If we think of the spatial model as recording the dialectic between model and reality (see The dialectic model 129), the process of interpolation juxtaposes and reconciles the spatial model with observations of the real world, thence providing testable predictions of unknown values of its properties. Some simplified examples of how geological assumptions can influence interpolation are considered in Drawing the line 209.

Implementation note: A structure of hierarchical three-dimensional grids can bring together information interpolated from, say, surface mapping, shallow boreholes, deep wells and geophysical and other spatial models. Standard grids simplify comparisons within a sequence of surfaces. Different levels of detail can be handled by quadtrees. The result is a comprehensive record of a succession of three-dimensional surfaces, with flexible visualisation. The original data should be stored to handle updating, uncertainty and special purpose modelling, and may prove essential for later revision.

Drawing the line

<<The geometry of the spatial model 203

Summary: *Even simple interpolation between points on a marker horizon along a line of cross-section raises basic questions. Do nearby points have more similar elevations than distant ones? Does an adjacent elevation overrule more distant data? Should we consider the same questions about slope or curvature as well as elevation? Can we take a known pattern into account? It is easy to demonstrate that different answers lead to different cross-*

sections. Analogous issues arise when interpolating a triangulated surface, but the geological conclusions may not be so obvious.

In computing grid values for a surface, we might take into account: the objectives (see Diverse objectives and products 150); observations of the surface in the real world; the model representing relevant aspects of our opinions about the properties of the surface; the spatial distribution of data; and the standard grid patterns already in use. The effects of some basic decisions can be illustrated by the simple example in *Figure 35*. This shows five vertical cross-sections each drawn along a straight line. Along each cross-section, eight wells already penetrate the surface, and the elevation of a marker horizon is shown (by a circle) for each. Estimates (shown by stars) are required of the elevations of the marker at proposed drilling locations shown by vertical lines. The geologist must make a number of decisions to arrive at a valid prediction. Cases A to E in *Figure 35* illustrate the effects of various decisions.

A starting assumption could be that the target point is likely to resemble nearby points more closely than more distant ones, otherwise any form of prediction would be difficult. Assuming resemblance between nearby values in *Figure 35*, the estimate of the unknown point would be closest in value to the two nearest wells. The other wells might resemble the target to a lesser extent, being more distant. All eight could be taken into account, placing most weight on the nearest values (case A). For many geological properties, such as the concentration of a trace element, one can think of arguments to support this approach. For the elevation of the marker horizon, an alternative view might be taken. The evidence of nearer points might be taken to override or block out information from more distant points, and only the closest values would be considered (case B). Possibly the slopes of the surface might also be taken into account. The regular slope in case C suggests that the predicted point should be placed to lie on the slope.

Another alternative would be to consider that the surface followed a known pattern, such as concentric folding, and estimate the value by fitting the assumed pattern through the data points (case D). Local clusters of wells as in case E might provide more information about the pattern and the degree of uncertainty introduced by local variation.

Analogous assumptions might be made if an entire surface were taken into account, rather than just a short cross-section. The surface can be divided into a set of triangular facets with a data point at each apex. The triangles can be chosen to be as close to equilateral as the data distribution allows (Delauney triangles). If the triangle sides are bisected, and a polygon drawn around each apex through these midpoints, the result is the so-called Thiessen polygon. The polygon might be regarded as the “area of influence” around the apex or data point, as all points within the polygon are closer to that data point than to any other data point (see, for example, Bonham-Carter, 1994[331]). Assumptions could then be made about Thiessen polygons on a surface, generalising from the simpler case of points along a line. The concepts of elevation, rate of change of elevation (slope), and rate of change of slope

[331] Bonham-Carter, G.F., 1994. Geographic Information Systems for Geoscientists: modelling with GIS. Elsevier, Oxford. 398 pp.

(curvature), the patterns they impose on a surface, and the relationship between their predictive value and their distance from the predicted point are fundamental considerations. They are used subconsciously in manual contouring and in interpolating boundaries across unexposed areas on a geological map. They carry through to mathematical interpolation, where their consequences are explicit.

A

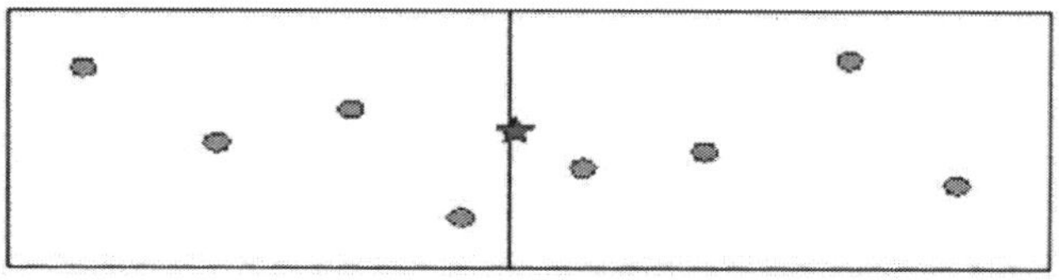

Case A: In vertical cross-sections, the dots show known elevations of a surface. The star is its estimated elevation at a proposed well location, based on the data average.

B

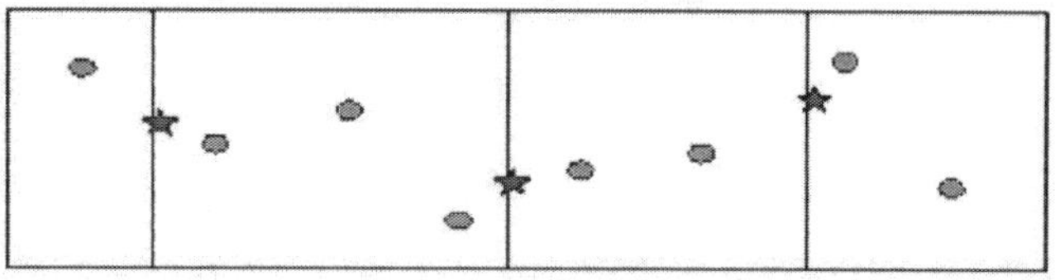

Case B: Here the stars are the estimated elevations of the same surface at three potential well sites, giving most weight to the nearest points.

C

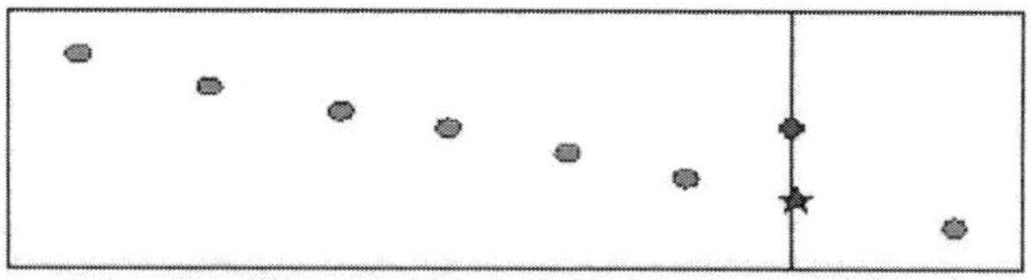

Case C: Here, a surface shows a regular slope. If this is taken into account, the estimated value might be placed at the star, rather than near the average, shown by the cross.

D

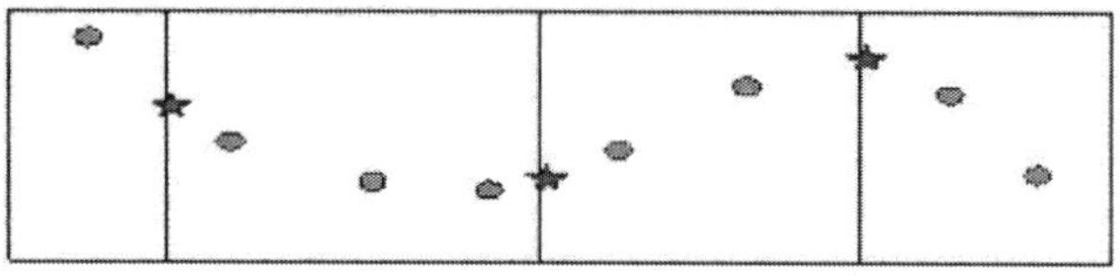

Case D: Here, the stars are estimates that take the curvature of a surface into account. The star on the right lies above the data points.

E

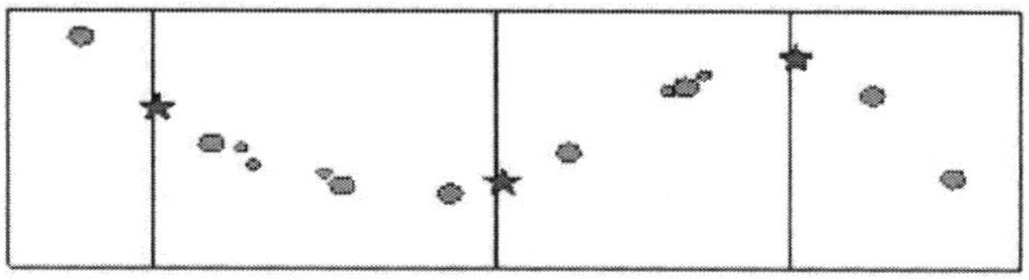

Case E: In this case, knowledge of small-scale variation, as shown by the small circles, would not greatly affect the estimates, although it would suggest greater uncertainty.

Figure 35: Interpolating a line or surface implies that any point on it can be estimated.

Assumptions could be based on visualisation of the results, and deciding what looks right and what does not. Indeed, this may be necessary to detect the unwitting introduction of geological features, such as the small central anticline implied in case A of *Figure 35*. Where applicable, scientific reasoning is usually more powerful than rules of thumb, but the geological arguments are unclear even for the simple examples in the figure. Mathematical expression of the geological assumptions can give a more rigorous approach, but lacks geological insight. An immediate task for geoscience survey, however, is more mundane, namely to provide a reasonably neutral view of spatial aspects, on which specific interpretations could be based. This requires a simple, transparent process of smoothing the data values for clearer visualisation. Where there are valid reasons and general agreement, other procedures could be introduced and their justification recorded. Some issues concerning computer techniques for interpolation are considered in Estimation by interpolation 212 at a basic level, pointing to some existing possibilities and setting the scene for the future developments considered in Transforming space 224 and Seeking shared concepts 247.

Implementation note: Suitable software for interpolation and visualisation is complex and expensive, as it must meet the exacting requirements of geoscience. Mainstream developments are therefore limited by the software options. The cold light of feasibility suggests that if computer procedures work reliably and give acceptable results, even if they do little to advance our geoscience reasoning, they can still be helpful and are at least a means of gaining experience with realistically large data sets.

Estimation by interpolation

<<The geometry of the spatial model 203

Summary: *Mathematical interpolation calculates the elevation (or other quantitative property) of a point on a surface. The values of nearby data points can be weighted in various ways, including geostatistical kriging with a variogram, to calculate the required points. Or an appropriate mathematical function can be fitted to the data. The function optimises criteria, such as discrepancies between data and interpolated values, or bending energy of the surface (splining). The data points can be interpolated exactly or approximated by a smoother surface. The area of interest can be fitted as a whole, or local patches can be blended together as a smooth surface.*

Creating a grid of elevations requires a procedure to obtain the elevation at a point with given grid co-ordinates such as easting and northing or latitude and longitude. Mathematically, this might be written as $z=f(x,y)$, where z is the elevation at the grid co-ordinates x and y, and f() indicates a *function*, in this case a set of calculations to be done on x and y, giving a result equal to the value of z at the required location. In the computer implementation, the set of rules for calculating the function is part of the algorithm formalised in the software. Two possible approaches are to calculate the grid values (or indeed any other required value) as an average of nearby data values, or to read the values from a mathematical surface that has been adjusted to fit the data points.

The first approach leads to a *moving average* where each successive grid point is calculated as the average of the nearby data points. In a *weighted* moving average, nearby data are given greater influence than those further away, ensuring that the extent of a data point's contribution to the average gradually diminishes with distance (case B of *Figure 35*). A *weighting function* relates weight to distance. It can be selected to give 100% importance to the data point where it coincides with the estimation point, if it is thought that the surface should pass exactly through the data points. Or a weighting function can be selected that smoothes and generalises the surface, rather than honouring data points precisely. It can be selected to ensure that what is felt to be a reasonable number of data points are generally included in each calculation. It can be chosen with a form that avoids sudden breaks in the surface as points move beyond the weighting zone, by gradually reducing the weight to zero. It can balance the weights for data points within each octant around the grid point, to reduce the directional bias from clusters of data points. These subterfuges help to prevent the data distribution dominating the model and its visualisation, but are somewhat arbitrary.

The weighting of data in moving average techniques assumes that nearby points are more similar than distant ones. The study of *geostatistics* quantifies this approach (Isaaks and Srivastava, 1989[332]), and leads to some advanced mathematical analysis (Mallet, 2002[333]). A so-called *variogram* plots the dissimilarity of the elevations of points against the distance between them. The relatively small number of data points for a single surface is likely to give an irregular variogram. However, it may be enough to indicate the type of surface, and a smoother variogram may then be available based on experience of similar surfaces elsewhere (thus using information about the object class to interpolate the object instance).

A process known as *kriging* estimates grid values, using the variogram to weight data points by distance. Furthermore, it can reduce the weighting of points in a cluster, to take into account the fact that closely spaced values each provide similar information. Geostatistics can thus avoid some arbitrary decisions for calculating a moving average. It can include all the points that are likely to have a significant effect on the estimate, and the weighting is gradually reduced to zero, thus avoiding sudden breaks in the surface caused by the data distribution. '*Cokriging*' can take into account information from a better-known, correlated surface. Geostatistics is not restricted to point data. Reflecting its origin in mining estimation, it can for example predict the values of blocks within an ore deposit from smaller volumes, such as samples. For many purposes, however, the more arbitrary decisions of the simpler moving average techniques might give an adequate approximation to the geostatistics solution.

The second approach to estimating grid values is based on surfaces that can be described mathematically, that have properties that are known and appropriate, and that can be adjusted to fit the data. Given a set of (x,y,z) values (such as easting, northing and elevation) for data points on a surface, it is possible to calculate the coefficients of a mathematical function to fit these points. One mathematically tractable function (*Figure 36*) is the *polynomial surface*: $z = a + bx + cy + dx^2 + exy + \ldots + nx^iy^j$. The basis functions for a cubic polynomial, and the combined curve from adding them all together, are shown in A. In B, the coefficients are altered to give a different curve, which is still smooth, has the same number of inflection points, and heads for plus or minus infinity at each end.

---

332 Isaaks, E.H. and Srivastava, R.M., 1989. *Applied Geostatistics*. Oxford University Press, Oxford. 592pp.
333 Mallet, J.-L., 2002. *Geomodeling.* Oxford University Press, Oxford. 599pp.

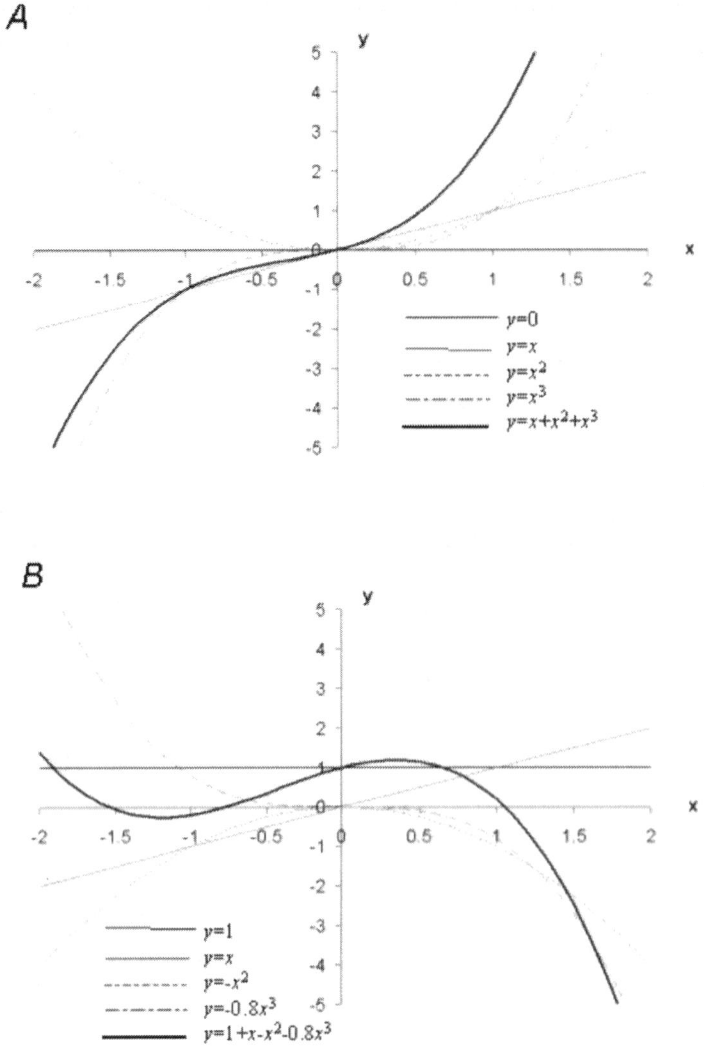

Figure 36: Generation of polynomial curves.

Another is the *Fourier series*: z = a series of sine and cosine terms, such as $c\sin y$ or $d\sin 2x$. The forms of these mathematical surfaces can easily be demonstrated in two dimensions by simple experimentation with a spreadsheet. *Figure 37* shows how sine waves can be combined to approximate even awkward shapes, such as a square wave. The individual sine waves in the lower diagram are added together to give the blue wave in the upper diagram – an approximation to the target square wave shown in red. A single wave offers a first approximation, which can be improved by combining it with appropriately weighted harmonics, shown individually in the lower diagram.

The elevation, slope and curvature of the surfaces change gradually, with no sudden breaks (as in case D, *Figure 35*). As more terms are included in the function, the surfaces become more complex. However, surfaces that can be vertical or have more than one z value at a location (x,y), as in an overturned fold, cannot readily be represented on a horizontal grid. They require parametric functions, relating x, y and z to different functions (say, f,g and h) of two independent parameters, say s and t: x=f(s,t); y=g(s,t); z=h(s,t).

The function can fit all the data points exactly if the number of terms equals the number of data points, or more strictly if the number of degrees of freedom in the function equals that in the data. If the equations are *underdetermined* (more terms than data points), then additional information is needed to give a unique solution. This can be provided by additional constraints, such as slope data, pseudo-data points, or by an objective function using a method known as linear programming. If there are more data points than terms, the function is *overdetermined*, and will not give a perfect fit to the data. However, a best-fit surface can still be calculated to optimise a criterion based on its deviations from the data values. One widely used optimisation procedure is to minimise the sum of squares of the deviations of the data points from the fitted surface. Where the data values have significant random errors, the best-fit surface may smooth over them and clarify the underlying pattern.

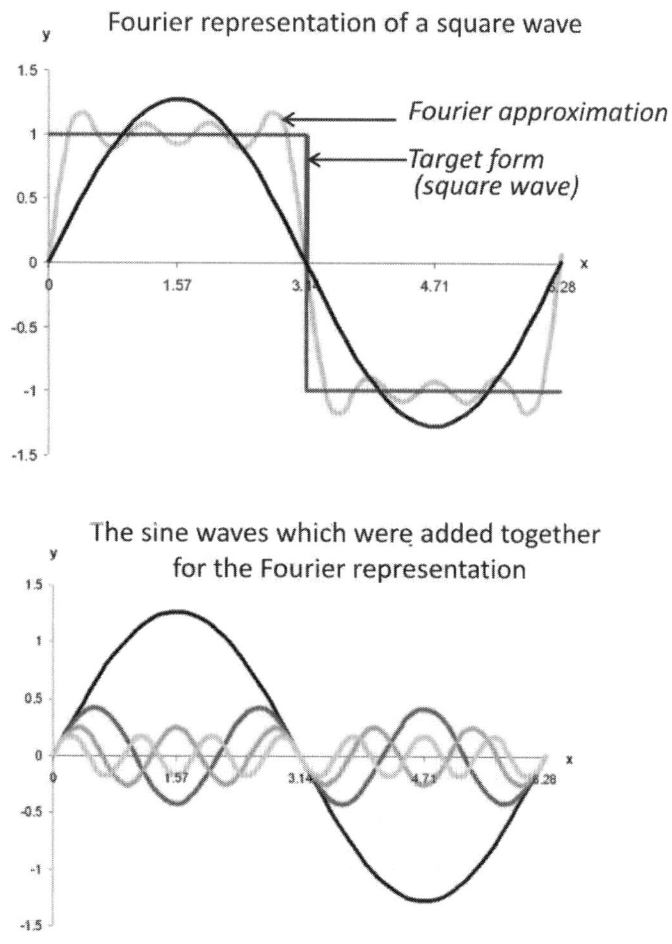

Figure 37: A complex periodic curve

Another method, solved algebraically by a simple matrix inversion, is the *thin-plate spline* model. This calculates the form taken by a metal plate that is deformed in three dimensions, causing it to pass through the data points while adopting the form that minimises bending energy. In two dimensions this is similar to the use of a draughtsman's spline (a flexible ruler for drawing smooth curves) and is referred to as *splining*. It subdues the variability of the surface elevation between data points, and might conform to geological expectations of, say, a gently folded surface. The variability can be subdued further by recalculating the surface with selectively increased tension, thus pulling the surface taut as though tightening the guy ropes or using a less flexible spline.

Adding more terms to the equations improves the optimised fit. If the terms are *orthogonal*, each can be added independently, as in the Fourier series or orthogonal polynomials, but practical computational problems restrict these to gridded data. If the terms are non-orthogonal, as in general polynomials, the entire equation has to be recalculated when a new term is added. A high-order surface (one with many terms) requires heavy computation, and may introduce fluctuations that are geologically misleading. A low-order surface, typically including polynomial terms up to third power (cubic), can be fitted to a large area with many data points. It will not fit the data precisely, but is a best-fit surface that may reflect large-scale variation and is known optimistically as a *trend surface*. Deviations from the trend may point to local anomalies. The trend surface may thus separate the results of superimposed geological processes operating at different scales, such as regional folding and local patterns of erosion or deposition. The *deviations* from the trend surface can be mapped separately to show small-scale variation that would otherwise be masked by the trend. However, caution is needed, for their literal interpretation is simply that they show the difference between the geometry of the sampled points on a geological object and a rather arbitrarily selected mathematical function.

An alternative approach is to fit functions *locally*, that is, to small numbers of adjacent points, representing a small area or *patch* of the surface. Each patch might contain as few as three data points. The surfaces of adjacent patches can be smoothed across the boundaries by splining or by applying a so-called blending function. Local fitting can be applied to gridded values or to patches based on data points, such as the Delauney triangles mentioned earlier. The resulting surface may or may not be selected to pass through each data point precisely. A stack of nearly parallel surfaces might yield simpler and better-fitting surfaces by treating the intervals between them as isopach maps, subsequently converting them (by addition to the better known surface) to elevations for contour mapping. If one surface, real or artificial (perhaps obtained by principal component analysis), is more representative of the stack or is known in more detail than the others, it might be taken as the base for calculating the isopach values, at the risk of a rather opaque geological justification.

The various approaches of moving averages and fitting functions produce different results, but have several features in common. They generate a smooth surface from a set of scattered data points. Either they can be fitted precisely to the data points, or they can pass smoothly nearby. There is no obvious geological significance to the form of the mathematical surface, but the absence of any sudden break or discontinuity subdues the visual effects of the data distribution, thus making it easier to visualise the underlying geology. They are both candidates for a neutral view for standard presentation. Even here, however, the choice between them can affect the view of the geology (compare case A and D of *Figure 35*). In fact, the concept of a neutral view is somewhat dubious. At best, it is an attempt to avoid commitment to a specific view based on inadequate evidence. The 'neutral' view may help viewers to consider a range of possibilities by offering a clearly understood mechanical representation, defined within the metadata, as a base from which they can imagine more elaborate scenarios.

The algebraic procedures just described are technically convenient, geometrically expressing and controlling the values of elevation, slope and curvature. But geologically they do not take us much beyond the rules of thumb that might be used for contouring by hand. They do not even throw a clear light on the questions asked in 'Drawing the line 209'. They are, however, basic techniques that are likely to underlie any extension to a more powerful system. In this account so far, the emphasis has been on using established computing techniques to organise the representation of spatial aspects of the results of geoscience survey. This is itself a major task of long duration, and its final outcome is unknown. The section on Continuity, fractals, octrees and wavelets 217 mentions other geometrical concepts that explore the boundaries and limitations of this approach, and point to future possibilities. Additional information on interpolation and available software can be found in Wikipedia.

Implementation note: Undertaking to move geoscience cartography to mainstream computer standards as they develop, and maintaining conformity as they evolve, is a major long-term commitment. The timing of involvement is crucial. Many developments will eventually be discarded in favour of new approaches, but inaction now may merely delay familiarisation with the problems and movement towards longer-term goals.

Continuity, fractals, octrees and wavelets

<<The geometry of the spatial model 203

Summary: *In contrast to smooth interpolation, fractals offer spatial models of geological phenomena that lack continuity but show similarity of pattern over a range of scales. Voxels represent patterns that vary in three-dimensional space, and octrees cater for their variable resolution. Wavelet analysis handles local patterns superimposed in different locations at different resolutions. They remind us of the disparities between reality and the diversity of incomplete models – a piecemeal approach that does not fully satisfy the requirements.*

Images of the real world and its interpretation are greatly simplified in geologists' minds and in map representations. They are visualised, with or without computer help, as simple geometrical shapes representing lines and surfaces (mostly the boundaries of objects) rather than as real geoscience entities in their full complexity. It might seem that the geologist is thinking within a mathematical framework, working with geometrical ideas that Euclid might have recognised – readily manipulated and analysed by computer. This could be a dangerous assumption, for geoscience visualisations carry baggage and imply perceptions that would baffle a mathematician. Mismatches between representation and reality are generally accepted and taken for granted by geologists. A mathematician, introducing the fractal model, called attention to some of them.

The surfaces that bound geoscience objects represent discontinuities where the surveyor has detected abrupt changes in the properties of the rock continuum. Detection of sudden breaks and changes in the sequence is crucial to classification and thus to interpretation. *Continuity*, in mathematics, refers to a characteristic of a mathematical function. It implies the possibility of creating a very small zone about a point, within which the value of the function does not significantly change. *Discontinuities* arise where this is not possible. If we view bounding surfaces, such as formation tops, as mathematical discontinuities, this would imply that they mark significant abrupt change in probably undefined functions representing properties of the rock mass above and below the discontinuity. Discontinuities in the elevation or slope within a bounding surface, such as faults and hinge-lines, also tend to be seen as geologically significant. However, Mandelbrot (1982[334]) suggested that in natural phenomena, including most of those represented by lines and surfaces on a geological map, we are unlikely to find mathematical continuity at any scale.

Mandelbrot described mathematical forms that he termed *fractals*. The process generating the fractal starts with a simple geometrical object (the initiator). Its form is replicated by a process (the generator), which incorporates smaller versions of itself at appropriate points in the initiator, these in turn incorporating yet smaller versions, and so on, repeated at ever-diminishing size until stopped for practical reasons. The forms of the resulting objects depend on the initiator and the generator and are very varied, perhaps resembling a cloud, a sponge, a fern, or one of the computer graphics patterns familiar on the Web. They share some properties with natural objects. Because of the way they are generated, they can show *self-similarity* in which a small part, when enlarged, resembles the whole. There is no continuity at any scale, no tangent, no measurable slope, and nothing special about a discontinuity.

Fractals may be an appropriate model for some physical processes, such as ore emplacement (Turcotte, 2002[335]) or rock fracturing (Paredes and Elorza, 1999[336]). They relate well to ideas of emergent systems (see Complex and emergent systems 159) and can generate lifelike visualisations of mountain ranges and landforms. They give rise to measures, like fractal dimension, of value in map analysis. However, they are mentioned here to stress that, while spatial models may invoke various mathematical analogues that mimic aspects of real physical processes, and although their analogies may help our understanding, we must always bear in mind the limited aspects of the real world to which each one refers. For practical reasons, the geoscience community may accept a single interpretation, such as a geological map, conforming to the current paradigm. But regardless of the number of signatures on a map, it has no fundamentalist status as an embodiment of absolute truth. Its validity stems from tests of its predictions against the real world. It illustrates a view, which was seen as appropriate for a particular purpose, but may no longer be appropriate for many applications of a spatial model. Fractals are a salutary reminder of the artificiality of lines on a map.

Mapped lines may illustrate the intersection of conceptual geological surfaces with the ground surface. Smooth surfaces are presumably important in the conceptual model

---

[334] Mandelbrot, B.B., 1982. *The fractal geometry of nature*. Freeman, San Francisco, 460pp.

[335] Turcotte, D.L., 2002. Fractals in petrology. Lithos, 65(3), 261-271.

[336] Paredes, C. and Elorza, F.J., 1999. Fractal and multifractal analysis of fractured geological media: surface-subsurface correlation. *Computers & Geosciences*, **25**(9), 1081-1096.

(fractals notwithstanding) because they approximate to surfaces where rapid changes in properties are seen as reflecting the influence of historical events. For example, depositional or erosional surfaces are formed with only fluids and loose grains above them: the density contrast and gravity tend to even out such surfaces with results that could be approximated by a smooth surface. Also, shear zones, faults and maybe igneous intrusions may approximate to gently curved surfaces, required geometrically for relative movement of the two sides. In some statistical sense, the surfaces and lines on the map represent interpretative features that geologists would recognise although, like the equator, there is no line on the ground. Through interpretation by the author of the map and subsequently by its users, the surfaces and lines are linked to observable phenomena. But each observation is itself a geological interpretation and it is the geological, rather than mathematical, properties of conceptual surfaces that determine how they behave in the model. For example, slopes on visible objects that the geologist thinks are related to the conceptual surface may be estimated by strike and dip measurements and recorded on the map. The measurements may refer to an area as small as a field notebook, which would not normally be visible even on a large-scale field map. They therefore do not correspond to the resolution of the mapped lines and surfaces. The geologist takes this into account in reading the map, but a mathematical analysis might be able to deal with them only as a separate sub-model with its own characteristics.

Spatial entities need not be bounded by surfaces. Seismic data, for example, can be viewed in terms of properties varying throughout a volume of three-dimensional space. The variation can be represented in the computer by voxels, each *voxel* being a cubic unit of volume centred on a point in a three-dimensional grid. It is the counterpart of the pixel in a two-dimensional raster. Each voxel has values associated with it that represent properties such as density, grain size, chemical composition or whatever. A hierarchical grid or *octree* is the counterpart of the quadtree structure for pixels (see Making a mesh 207). The development of non-invasive medical imaging techniques, such as tomography, has generated extensive research in this field. Geoscience spatial models, particularly in geophysics, are likely to require some volume-based models, perhaps referring to the internal structure of a separately defined object. However, they must generally be combined with surface-based models to locate them in the geoscience knowledge base where the ability to analyse and interpret relies on imagining, recognising, defining and relating surface-bounded objects. Just as points and lines in vector format can be superimposed on a raster, so points, lines and surfaces in three dimensions can be superimposed on a volume represented by voxels.

Self-similarity is a feature of fractals, and of many processes of geological interest. Most geological processes operate over a range of scales, and create similar patterns of many sizes from microfolds to nappes or tiny ripples to giant dunes. As in simple fractals, the position of small-scale geological features within a larger-scale feature is likely to affect their form, depending on how the large and small-scale processes interact. For example, microfolds might alter shape across a contemporaneous larger fold or in other circumstances be refolded by a later one, probably of a different size.

Wavelet analysis (see Graps, 2004[337]) is one method of detecting and separating pattern at different scales. It searches for defined patterns by fitting a local function, akin to a bounded segment of a Fourier series (see Estimation by interpolation 212), representing the pattern. It compares the pattern at different scales to each area of an image (or, potentially, of a surface) by a process known as convolution. It thus provides a mechanism to locate the patterns at any scale in any region within the grid, and measure their contribution to the overall surface variation in a so-called scalogram. Wavelets are widely used for *image compression*, retaining meaningful pattern while reducing the volume of information, a process not unlike map generalisation. During compression, 'thresholding' can avoid smoothing out sharp structures that may carry significant geological information. Wavelets have been used more in geophysics (see Holden et al, 2001[338]) than for detecting three-dimensional patterns of geological significance on a surface.

To speculate a little, a surface might be created as a combination of appropriate patterns and adjusted by moving the patterns about and weighting the different scale components to fit the data points. The wavelets could then carry geometrical representations of features across from dynamic to survey models and vice versa. But the functions, scales and positions vary independently. There is no limit to the number of possible solutions, and no obvious criterion for selection. Additional information is needed to find a unique solution. Superficially, this may resemble geologists manipulating and comparing patterns in their minds. The conventional geological map shows pattern that is interpreted from background knowledge and field observations. It is a composite pattern reflecting influences from many events and processes that operated over a range of scales. It may not be possible to disentangle their effects by analysing the map, nor indeed from the field evidence. It is small comfort that conventional methods have the same problems as wavelet analysis.

Overall, the methods just described for handling the geometry fall short of satisfying the requirements. One limitation may be the piecemeal approach. A wish list for integrated geometry 221 may help to identify some features of geometrical representations that could help to integrate information from a wider range of sources and thus narrow down the possible solutions (see Forward and inverse models 154).

Implementation note: The spatial model highlights some issues in conventional mapping where object-oriented methods offer fresh insights. Computer contouring and geographical information systems provide a means of handling geometrical aspects, but prove inadequate for some requirements. We can use only what works, but meantime can extend the framework so that potentially it can include a more coherent set of geometrical methods applicable to a wider range of situations.

---

[337] Graps, A., 2004. An introduction to wavelets. http://www.amara.com/IEEEwave/IEEEwavelet.html

[338] Holden, D.J., Archibald, N.J., Boschetti, F., Jessell, M.W., 2000. Inferring Geological Structures Using Wavelet-Based Multiscale Edge Analysis and Forward Models. Exploration Geophysics, **31**, 4, 67-71

A wish list for integrated geometry

<<The geometry of the spatial model 203

Summary: *A more integrated approach is needed to pull together the plethora of approaches to the geometry in dynamic and static models. Because they refer to the same model, activities such as observation, measurement, analysis, summary, interpolation, prediction, reconciliation, visualisation and generalisation would benefit from a shared mathematical framework, handled consistently from initial observation to explanation to final presentation. Solutions must fit within the overall system view.*

A geoscience spatial model involves reconstructing and describing the geometry of objects and their properties by interpolating from very limited data. Current surveying procedures lead to a diversity of representations of spatial information (DSIs, FEMs and their geometrical significance 235) at separate resolutions. Spatial models currently implemented on the computer also take various incompatible approaches. It can be difficult to picture the geological significance of their algebraic manipulations, even at the level of the simple geometrical issues raised in 'Drawing the line 209'. The main link to geological thinking is the rather unsatisfactory one that after some ad hoc manipulation the results 'look right'. Scientific validation, on the other hand, requires a clear record of the procedures in moving from initial observations through the various steps in their interpretation (but see also Unexpressed knowledge 63). The record is also valuable for other applications (see Diverse objectives and products 150) which may call for reworking the information with different sampling and analytical methods. To improve on this situation for generic geoscience survey, we might aim to integrate the geometrical knowledge within a single knowledge base where field observations, descriptions and summaries can be more rigorously linked with geological interpretations. With that in mind, we can list features seen as desirable in the methods of interpolating, generalising, manipulating and representing the geometrical data.

- Knowledge of the geological behaviour and characteristics of the objects should be reflected in their interpolation. Our views of the spatial characteristics of the objects depend on ideas of their origin and evolution, and therefore dynamic process models should share a compatible framework.
- Techniques for observation, measurement, analysis, summary, interpolation, simulation, prediction, reconciliation, visualisation, filtering and generalisation must all be applied to the same model, and should therefore be seen as aspects of the same mathematical framework. The geometry should be handled consistently from initial observation to final presentation.
- Interpolation methods for points, lines, areas and volumes should be compatible, because they may be aspects of the same object, such as stratigraphical surfaces and their lines of outcrop.
- The spatial information must be represented algebraically for computer processing, while retaining geometrical significance to enable users to picture the processes and their results. The computer representation must support visualisation to communicate with the human user.
- The perception of a visualised model should have much in common with the perception of the real world. The key properties of the model should be identifiable in the real

world and should be measurable by visual estimation, by instrumentation or by statistical summary.

- Different information types, such as text, maps, images, sketches and numerical data, should use spatial representations that can potentially be integrated.
- Knowledge of the existence, topology, configuration and disposition of objects (see Ambiguity and map representation 148) affects our views differently. We should be able to distinguish them in the representation but bring together their implications where appropriate.
- Standard views at defined resolutions must be available by default, but where practicable users should have flexibility in selecting alternative methods of visualisation and interpolation. To make full use of human background knowledge, the process must be interactive, enabling the surveyor and possibly the user to explore modifications (*what if?*) of the model. This suggests that key properties of the geometry of the model should be accessible to users for controlling the visualisation.
- Incomplete information implies uncertainty and requires statistical methods to identify and elucidate it. Statistical analysis may include description and recognition of spatial pattern, and its use in analysis, explanation and prediction. It is thus an integral part of the interpretation.
- Geological knowledge derived from statistical studies of process models and field observations should be assembled as metadata describing the behaviour of the object class, available to guide interpolation and generalisation.
- We should be able to make local changes in response to new or revised information without unnecessary change to the surrounding region or to the model as a whole.
- We should be able to handle dynamic models that lead to composite or superimposed patterns, such as a model of deposition of a stratigraphical sequence, followed by a different model to reflect its subsequent deformation. Where feasible, we should be able to separate their effects on observed patterns.
- Generalisation and abstraction are central to creating and using the model, and it is clearly undesirable that a model's properties should alter unexpectedly on changing scale, although generalisation should reflect the range of scales over which individual processes operate.

The concept of integrated geometry is inseparable from the aim of a more comprehensive system of geoscience survey. Despite earlier reservations (see At the interface 133), therefore, the way ahead may be to reconsider the object-oriented spatial models discussed in 'An object-oriented approach 178', and see how they might be taken forward into a computer-based knowledge system. This leads into a research area beyond current routine implementations. It points to future integration of a greater range of observations and interpretations. It impinges now on our view of the spatial modelling framework, and gives a basis for a more holistic approach. We therefore turn next to future systems, suggesting more radical approaches to the geometry, including greater use of geometrical transformations (see Transforming space 224). Even aspects that are some way from implementation, or for various reasons may never be implemented, may nevertheless influence the direction of our current thinking.

Implementation note: The results of spatial modelling must be stored efficiently on the computer, with flexible links to further processing modules including those for visualisation. The storage scheme must adjust readily to match evolving standards.

Transforming space

<<Table of contents 1
<<The future geological map 125
<<The geometry of the spatial model 203
Transforming space 224
Rationale 224
Geometrical transformations 228
Invariant properties and classification 232
DSIs, FEMs and their geometrical significance 235
Unevenly spaced data 238
Spatial variation and uncertainty 241
The geometry of interpolation 243
>> Seeking shared concepts 247

Abstract: The quest for a comprehensive view of spatial characteristics in geology leads to composite geometrical transformations, built from components like translation, rotation, scaling and projection. The invariance of properties of objects under specific transformations can throw light on their significance and behaviour during abstraction and feedback. Spatially invariant properties such as slope and curvature have a bearing on interpolation, the consequences of unevenly spaced data, and uncertainty envelopes. Considering interpolation as a geometrical operation, linked to algebra for computation, could tie it more closely to geological interpretation, linking the strengths of computer methods and human insights. A systems approach to the knowledge base helps to clarify the complexities of their interactions and the links to dynamic stratigraphy.

Rationale

<<Transforming space 224

Summary: *Integration of spatial information of many kinds from many sources requires a comprehensive approach to the geometry. Spatial information requires special handling, for spatial correlations constrain statistical analysis. However, the homogeneity of three-dimensional space permits transformations of spatial objects within a flexible reference frame, while retaining links to non-spatial properties. The hierarchies of object classes and models of the geometry must match the top-down methods of field survey. Conceptual visualisation of geological transformations of the form and configurations of the objects finds its counterpart in spatial transformation, a key to integration.*

As this is a scenario, we look at what might be, not at what will be, and consider where further research might be fruitful, rather than where it has been put into practice. Earlier arguments called for a changing emphasis from map to digital spatial model in the procedures of geoscience survey: loosen the ties to cartographical concepts, with maps and their explanations as the standard product; introduce information technology concepts (such as webs, systems, objects, models, processes and events) with an object store as the standard product; support the object store with a spatial index and flexible visualisation; introduce the spatial model at its core as a step towards a knowledge-based system.

In prospect (see The system framework 310) are powerful, robust, portable computers for use in the field, integrated with a global network through broad-bandwidth wireless communication, and with electronic tools for locating and surveying. They potentially provide comprehensive knowledge system support from initial survey to end use, and therefore call for further development of the form and content of the spatial models that represent the results. Limitations of conventional approaches were identified (see From map to digital model 145), while the computer methods mentioned (see The geometry of the spatial model 203) rely on numerical and statistical calculations that jar somewhat with the field geologist's concepts and insights.

Successive hierarchies of linked objects are utilised by Mallet (2002[339]) as a flexible structure for spatial modelling, and by Berners-Lee et al. (2001[340]) as a means of relating data and concepts in the semantic web. The links therefore potentially offer a powerful structure to support procedures of spatial abstraction and feedback (see Abstracting from reality to model 131), from individual observations to stratigraphical objects to comprehensive interpretations.

The emphasis of Mallet (2002) is on applications to subsurface geology, but his approach could be extended to field geology. Compared with the battery of instrumental techniques supporting subsurface geology, field geologists are more concerned with direct observation of exposed rock, and in particular with the spatial disposition and configuration of its observable properties. Few field geologists will fully understand Mallet's rigorous algebraic formulations. However, geometry can provide a bridge from geological thinking (where field observations, reasoning and visualisation depend on the ability of the human brain to handle spatial information) to the mathematics of the spatial model. The aims of the spatial modelling can be represented and visualised in the geometrical terms and concepts that underpin field geology, although the corresponding software must rely on their algebraic counterparts.

Object-oriented analysis (see An object-oriented approach 178) points to the need for integrating spatial information of many types from many sources. Although the computer representation, interpolation and contouring of spatial information (see The geometry of the spatial model 203) may efficiently handle a variety of special cases on their own terms, they do not satisfy the wish list (see A wish list for integrated geometry 221), which reinforced the

---

[339] Mallet, J.-L., 2002. *Geomodeling.* Oxford University Press, Oxford. 599pp.

[340] Berners-Lee, T., Hendler, J. and Lassila, O., 2001 (May). The Semantic Web. Scientific American, 284 (3). http://www.scientificamerican.com/article.cfm?id=the-semantic-web

objective of a unified representation of the geometry of spatial objects and processes, ultimately as part of the standardisation procedure. Furthermore, the knowledge-based approach (see Mapping geology into the knowledge system 282) points towards embedding three-dimensional geoscience spatial models in more general systems.

Geoscience reasoning ties together our views of the tiny fragments that are all we can observe of the outcome of the general model (see The general geoscience spatial model 293), and is therefore fundamental to the surveying process (see The geological investigation model 98) from observation to explanation and generalisation. In order to accommodate this reasoning, the spatial models must extend to the processes and dynamic changes of object configurations throughout geological time. We therefore need to consider how spatial information from these diverse sources can interact and contribute to the holistic view (see The surveyor's holistic view 60). This calls for another look at methods for integrating the geometry, starting from well-established methods that do not involve computers, and aiming at a uniform mathematical approach.

In field mapping, geologists assess and take into account the genesis and historical development of the rocks that they depict on the map. For example, siliceous grit directly overlying the granite from which it had been eroded would not be mapped as part of the granite, for despite their contiguity and similarity of appearance and physical and chemical properties, they are the results of quite different processes in contrasting environments. The interpretations, central to geoscience surveying, thus associate dynamic models with the static outcome (see Representing wider knowledge 283).

As an example of the conventional approach (see The surveyor's holistic view 60), consider the procedures of mapping folded sediments, and the lines of thought that could lead from observations in the field, alongside consideration of the originating processes, to marks on a map. Perhaps scattered exposures of various extents might give indications of the main structure, based on evidence from small folds, intersections of cleavage and bedding, and so on. Variation in the thickness, grain size, sedimentary structures and type of beds within the sediment (as plotted on field maps) might suggest the form of, say, an elongate sand bar with siltstone on either side. From stratigraphical correlation across the exposures, it might be possible to put together a generalised vertical section and estimate the stratigraphical position of at least some outcrops. The variation in size and shape of the folds might relate to variation in the mechanical properties of the sediment, conforming to the pattern expected from the folding mechanism, or even, conceivably, from experiment.

The fragments of knowledge are organised and extended to fill the gaps between observations. In this example, the resulting pattern might be constructed to be consistent with: the observations; nearby observations and descriptions of type sections; analogies between small and large-scale features; deductions about underlying processes and their consequences; the evidence of symmetry and pattern of deformation; an interpreted view of sediment thickness variation within the likely depositional environment; the expected results of tectonic processes interacting with heterogeneous material.

In the field geologist's mind, perhaps helped by informal sketches, images are created that encapsulate a vast amount of knowledge relevant to a realistic representation of the geology on a map. The images reflect the range of likely patterns imposed on the objects as a result of the processes, and the compromises that resolve conflicting or competing evidence. Their representations on the completed map are designed to communicate to the user as much as possible of the relevant conclusions and implications. Not surprisingly, in view of its evolutionary background, the human mind is adept at grasping and manipulating the subtle interplay of vague patterns and, not surprisingly, the results can be ambiguous (see At the interface 133, The imperfect map 145, The geometry of interpolation 243).

As we move towards system-supported survey (see Object-oriented survey 195, The solid Earth systems model (sEsm) 71) various points about this view of the geometry may be noted. The geometry refers to hierarchical classes of objects and processes. These concepts act as frames into which increasingly specific objects are fitted, corresponding to the context in which ideas are embedded in the human mind (Minsky, 1981[341]). As a starting point, the sediment, more or less by definition, is seen as having a source, transport mechanism and depositional site. As the survey proceeds, specific object classes at a more detailed level of the object hierarchy might tentatively be assigned. Perhaps the sand was seen as brought from distant mountains by a river complex and deposited eventually as an offshore sand bar by long-shore currents. Further survey might lead to ideas about the extent and relative position of these features, and eventually their absolute location and direction. The tentative ideas are subject, not just to refinement, but also to complete reassessment as new evidence is uncovered, or the regional setting is reviewed.

Hypotheses about the nature of observed or interpreted objects and about their geometry might be altered separately. The 'sand bar' might be reinterpreted as a beach deposit of the same size and shape, or its identification might be retained while the shape was reassessed as a set of smaller overlapping units. We therefore need the ability to deal with the geometry (spatial characteristics, spatial properties and spatial relationships) of the objects and processes separately from their other properties. Although we may concentrate on the locative case (see Mark-up and metadata 175) in computer spatial models, we must retain the vital connections to other aspects, by links through metadata, database, GIS and hypermedia (see, for example, McCaffery et al., 2008[342]). Conventional methods create a similar division. They physically separate the geometrical information on maps and sections from related properties of the same objects discussed in text accounts and datasets. But they retain connections through links, such as keys in the map marginalia, which can be related (by a trained geologist) to keywords in the field notes and headings in the map explanation.

The methods of spatial statistics are constrained by the probability of a property's nearby values being more alike than distant values (see Estimation by interpolation 212). This violates the null hypothesis in classical statistics that observations are independent of one another. On the other hand, the three dimensions of space are of the same kind, enabling us to position and manipulate our reference frame more freely than we could with other

---

[341] Minsky, M., 1981. A framework for representing knowledge. Reprinted, pages 95-128 in Haugeland, J. (editor), *Mind Design*. MIT Press, Cambridge. 368pp.

[342] McCaffrey, K.J.W., Feely, M., Hennessy, R., Thompson, J., 2008. Visualisation of folding in marble outcrops, Connemara, westen Ireland: an application of virtual outcrop technology. Geosphere, 4, 588-599. doi:10.1130/GES00147.1

variables that we might choose to regard as dimensions. This in turn gives us access to other methods of representation and analysis, and can exploit our visualisation skills honed by a lifetime's intensive training.

As in conventional survey, the spatial model should support geologists in their top-down analysis, moving, for example, from broad notions of deposit and source to increasingly specific classification of rock and source-type, and should be able to handle the geometry at various levels of detail, from vague notions of existence to more specific ideas of direction and distance. Appropriate characteristics should automatically be inherited by the object classes in moving to more detailed levels of the object hierarchy. Where the evidence is reassessed, and the current view of either the classification or the geometry is changed, the system should respond accurately to the consequences.

When geologists visualise in their mind's eye the processes by which rocks formed and reached their present state, they see continual transformation. Processes of dynamic stratigraphy transform ancient landscapes by the erosion of mountains and movement of sediment, building up deposits with beds stacked one on another, buried, squeezed, folded and faulted. Metamorphic, igneous and sedimentary processes transform material to create new geological objects, and along with tectonic processes continually transform their location and configuration throughout geological time. In building spatial models of a system pervaded by transformation, we naturally turn for mathematical support to the methods of transformational geometry and hope to find there a key to spatial integration.

Implementation note: Progress towards an integrated computer-based knowledge system for geoscience survey will inevitably take time, with many years of unsatisfactory and incomplete experimental systems. Unifying the representation of the geometry might be achieved more quickly, bringing immediate benefits as well as speeding future development.

Geometrical transformations

<<Transforming space 224

Summary: *Geometrical transformations, such as translation, rotation, magnification, scaling and projection, can clarify the geometry of the dynamic and static models on which much geoscience reasoning is based. Complex transformations can be built from simple elements and computed using matrix algebra. They underpin computer graphics and visualisation and help to define the behaviour of spatial objects and relationships.*

Thompson (1942[343]) showed how spatial transformations of diagrams of fossils could shed light on similarities of form among various species. He superimposed a rectangular grid on a diagram of the skeleton of one species, and then manipulated it as though it were drawn on a rubber sheet; moving, stretching, shrinking and turning various parts of the diagram to align it with corresponding points in the anatomical structure of a different species. He retained the spatial relationships between anatomical elements and introduced no gaps,

[343] Thompson D'A. W., 1942. *On growth and form* (revision of 1917 edition). Cambridge University Press. 1116pp.

tears or overlaps. Thus he did not contravene what he knew of the processes by which the skeletons had once operated. The deformed grids illustrated the differences and similarities of shape among the species, and clarified the changes of spatial form by illustrating them as geometrical transformations.

Griffith (page iv, in Griffith and McKinnon, 1981[344]) points out that: a "powerful notion – that of a transformation – has served as the backbone of twentieth century mathematics. It has altered disciplinary focus from the study of individual mathematical systems to the study of relations between mathematical systems." He refers to d'Arcy Thompson's work on spatial transformations and cites a wide range of later applications to topics including cartography and measurement of biological shape. More recent references can be found in morphometric websites, such as SUNY (2010[345]). In geology, despite various attempts, the methods have hardly lived up to their promise. Maybe it is time to try again, hoping to find a route that clarifies the developing configuration of objects in a spatial model and leads us around the dead end of ad hoc static models (see The geometry of the spatial model 203).

During surveying, the spatial properties of the objects that are recorded usually include location, and may refer to slope or orientation, texture and pattern. Descriptions might carry information about symmetry (such as, fold axes trend east-west), curvature (the crest of the fold was broadly rounded), and overall form (conical or cylindrical folds). Spatial relationships between objects or parts of an object (see Representing spatial information and relationships 173) may also be noted. The procedures of knowledge representation for observations of static properties such as these should be included in the spatial model (Brodaric and Gahegan, 2006[346]). But on their own they are incomplete.

The objects that geologists create and manipulate are abstractions from reality. For example, stratigraphical surfaces correspond to discontinuities in the model (see Continuity, fractals, octrees and wavelets 217), interpreted from observations of the real world by the exercise of a geologist's imagination and reasoning powers, rather than by any mechanical inference. When we visualise, say, the underlying dynamic stratigraphy and tectonics (see Representing wider knowledge 283) we might build a mental image of sediment being deposited in layers, and undergoing compaction, folding and faulting: unlike the observations just mentioned and the static model shown on a map, this dynamic model describes change and movement. Knowledge representation in the spatial model should therefore extend beyond static observations to the dynamic reasoning on which they are based, linking spatial properties of the static and dynamic models. For example, Paton et al. (2007[347]) illustrate the application of computer-aided design and 3D visualisation in a complex geological setting.

In a static spatial model, such as a digitised map, the information is broken down for computation into geometrical objects, such as: points; lines, represented by points to be

---

[344] Griffith, D.A. and McKinnon, R.D. (eds), 1981. *Dynamic spatial models*. Plenum Press, New York. 443 pp.

[345] SUNY, 2010. Morphometrics at SUNY Stony Brook. http://life.bio.sunysb.edu/morph/

[346] Brodaric, B., and Gahegan, M., 2006. *Representing geoscientific knowledge in cyberinfrastructure: some challenges, approaches and implementations*. GSA Special Paper 397, pp1-20.

[347] Paton, D, Carr, M., Trudgill, B., Ortner, H., Medwedeff, D.A., 2007. Alpine-scale 3D geospatial modeling: Applying new techniques to old problems. *Geosphere*, **3**, 527-549.

joined in sequence; areas, represented by the enclosing lines; surfaces, represented by a function or a grid of points; and volumes, represented by the enclosing areas and surfaces. In a dynamic spatial model, the spatial changes to these assemblages of objects can be broken down for computation into their constituent spatial transformations, sometimes referred to in mathematics as 'mappings'.

Some basic transformations that change the position or form of an object relative to the chosen origin, scale and axes, are:

- translation – bodily movement of defined distance and direction
- rotation – turning of the object about an axis through the origin
- magnifying or dilating – multiplying all distances by the same factor
- stretching – different magnification along different axes
- projection – reduction of the number of dimensions
- perspective projection – diminution of size (of objects in a 2D projection) with distance from the viewpoint, mimicking the effect of perspective

Complex transformations can be defined by a sequence of basic operations of this kind. For example, rotating an object about its long axis might involve: translation to centre the axis on the origin; rotation to align the long axis with a coordinate axis; rotation about the long axis; and rotation and translation back to its original position. The order of the operations affects the result, as you can demonstrate by thinking through a few geometrical examples. Corresponding operations in matrix algebra bring about the same result, establish a link between geometry and algebra, and provide the means of computation (Foley and van Damm, 1995[348]). The transformations are basic elements of computer graphics, where they may be represented by quaternions rather than matrices (Schneider and Eberly, 2002[349]). The non-mathematical geologist can readily appreciate the geometrical significance, and leave the algebra to the computer. The availability of morphing software makes it easy to explore the visual effects of spatial transformations, but caution is needed as the user interface may obscure their three-dimensional geometrical implications.

The transformations can be applied to the points, lines, areas, and volumes of digital cartography and to slopes, surfaces, curvature and functions (see DSIs, FEMs and their geometrical significance 235). Statistical measures of probability (see Spatial variation and uncertainty 241) and shape can also be chosen to respond appropriately to transformations. In three dimensions, each point would be represented by three co-ordinates or, for greater generality allowing for translation and perspective change, by four homogeneous co-ordinates (Foley and van Damm, 1995[350]). Slopes can be represented by direction cosines and analysed statistically (Loudon, 1964[351], Watson, 1966[352], Koch and Link, 2002[353]).

---

[348] Foley, J.D., van Dam, A., 1995. *Computer graphics: principles and practice in C*, 2nd ed. Addison-Wesley, Reading. 1200pp. ISBN-10: 0201848406

[349] Schneider, P.J. and Eberly, D.H., 2002. *Geometric tools for computer graphics.* Morgan Kaufmann Publishers, Amsterdam. ISBN-13: 9781558605947

[350] Foley, J.D., van Dam, A., 1995. *Computer graphics: principles and practice in C*, 2nd ed. Addison-Wesley, Reading. 1200pp. ISBN-10: 0201848406

[351] Loudon, T.V., 1964. *Computer analysis of orientation data in structural geology.* Technical Report No. 13 of ONR Task No. 389-135, Northwestern University, Illinois. 138 pp.

Multiplying the four homogeneous coordinates of a point by a 4×4 matrix, representing a geometrical transformation, alters the coordinates to give the new location of the point. A single 4×4 matrix can also represent an entire sequence of basic transformations. This composite matrix is obtained by multiplying together the matrices representing the component transformations in the correct order. A complex transformation is thus built from a sequence of basic constituents. It can then be applied point by point to every recorded point on the object.

One application is to transform the reference frame, allowing us to deal with each object separately and assemble a joined-up version later. The geographical reference frame consists of a scale of measurement, and three axes at right angles, referring to a standard datum as origin, and to conventional directions, such as east, north and up (the spheroid is not considered here). However, each object that makes up a spatial model can be investigated on its own terms at an appropriate resolution, and related to its own coordinate system. The spatial objects might then be integrated by transforming their local reference frames to create a composite object with a shared reference frame, which might refer to, say, inferred directions of principal stress. Eventually, composite objects might be transformed again to bring them together in a single, global, geographical framework.

Spatial relationships must be resolved as the objects are brought together (see Seeking shared concepts 247). For example, the relative position of measurements within an outcrop may be known to a millimetre, and could retain that accuracy internally within the composite object representing the outcrop. However, the location of the set of measurements relative to, say, a satellite image may be known only to some tens of metres. The spatial relationship between imagery and outcrop would therefore refer to the entire composite object at an appropriate resolution, rather than to individual component measurements.

Geometrical transformations underpin computer graphics, where they support the computer processes for visualising spatial models. They might equally come to underpin the mathematical framework for a computer model designed to provide an integrated view of the spatial concepts, observations, abstractions, models, reconciliations and visualisations that are part of geoscience survey. From the geologist's viewpoint, transformations throw a new light on interpolation methods. Instead of trying to determine the geological significance of an algebraic function that represents, say, a surface, the geologist could reason instead from the geometrical representation of the function (see DSIs, FEMs and their geometrical significance 235). Focusing on the geometrical context of a spatial model makes it easier to relate field observation, computer interpolation, narrative description and visualisation. It may also clarify links to dynamic geoscience models and procedures for reasoning, abstraction and generalisation, all of which involve geometrical transformation.

When spatial objects are transformed, their spatial relationships, such as those linking sequences of points to form lines, areas, surfaces and volumes, must of course be retained

---

[352] Watson, G.S., 1966. The statistics of orientation data. Journal of Geology, 74, 786-797.

[353] Koch, G.S., Link, R.F., 2002. *Statistical analysis of geological data.* Dover Publications, 832 pp. ISBN: 0486495124

to identify and preserve the composite objects and their relationships. Some spatial relationships act as constraints on valid transformations and thus on the *behaviour* of the object, in this context its response to the operation of a computer process. This introduces the concept of Invariant properties and classification 232, which may help to clarify the significance of various components of geological maps and their role in the spatial model.

Implementation note: The results of geoscience survey have long been represented as static maps. The reasons for this may soon be out-dated, but old habits die hard, and the legacy is irreplaceable. The initial priority for a unified spatial representation must therefore be to cope efficiently with static models. But greater generality and future flexibility can be gained, and backtracking reduced, by designing the framework to include dynamic models.

Invariant properties and classification

<<Transforming space 224

Summary: *Some properties are invariant (unaltered) under spatial transformations. The existence of an object is not affected by any such transformations. Some objects and relationships are invariant under topological (rubber-sheet) transformations, others under affine transformations (rotation, enlargement and stretching) or rigid-body transformations (translation and rotation). This can clarify constraints on the significance and behaviour of spatial objects as they are processed within a computer-based knowledge system.*

Geometry is the branch of mathematics that deals with the properties of space and of objects in space. Spatial transformations help mathematicians to generalise the ideas of geometry by creating internally consistent mathematical systems where defined geometrical properties of objects are *invariant* under certain classes of transformation (that is, they remain unaltered after the transformation). Such systems, for example projective geometry, do not necessarily follow the postulates of Euclid.

At a less exalted level, similar ideas can throw light on ambiguities within a geological map (see The imperfect map 145, Ambiguity and map representation 148) where features with different geometrical behaviour are not separately identified. One aim, therefore, is to enable geologists to specify and record invariant properties in an object class, in its specific instances, and in spatial relationships. This can clarify the significance of items shown on a map. In the model, the invariance can constrain behaviour during the procedures of interpolation, generalisation and visualisation, ensuring that the results mimic more closely the behaviour of their real-world counterparts. The same concepts are also relevant to quantifying hitherto Unexpressed knowledge 63.

In the most general case we may be aware of the *existence* of an object somewhere within the area being surveyed, without knowing where it is located. This could arise, for example, if a major fault occurred on either side of the area and must be assumed to pass through it. Or, formations A and C might occur within the area, and external evidence could indicate that formation B was always associated with them, and must therefore also occur within the area. Or evidence from other sources, such as finding a loose fossil or fragment that must have come from B, might indicate its existence without determining its location. The occurrence of B must be somewhere in the limited area of immediate interest but could be placed anywhere within it. Existence is unaffected by (invariant under) any spatial transformation within the area.

A slightly less general case is invariance under *topological transformations*. If we imagine a geological map drawn on a totally elastic rubber sheet, the objects it depicts could be translated, rotated, squeezed and stretched by deforming the sheet, as described by d'Arcy Thompson (see Geometrical transformations 228). Most of their spatial properties and relationships would be altered, but some would not be affected. These are known as topological relationships, and topology is the branch of mathematics in which they are studied. A line that bounds an area, or two points that coincide, would retain these topological relationships after deformation. Adjacent beds remain adjacent and fossil localities remain within the correct bed during rubber-sheet transformations. On the other hand, two points that were not coincident, no matter how close together they were, could be moved any distance apart by sufficient stretching. Topological relationships are important in digital cartography. When a map is digitised, the order of points on a line, the coincidence of two lines (say, a fault and a formation boundary) or the point of intersection of two lines, are examples of relationships that must be retained during transformations for generalisation or visualisation, such as the basic transformations listed in Geometrical transformations 228.

When, in the eighteenth century, James Hutton postulated that the natural agents at work in the Earth's past showed a general uniformity with those in operation now, he implied the existence of immanent properties and principles (see Abstracting from reality to model 131, Invariance and processes 56), which by definition are invariant under translation in geological time and space. Knowledge of the processes that created an object may give us information about its invariant properties, as opposed to the historical configuration relative to other objects and processes that affects the outcome. This is the basis of much geological reasoning.

The so-called *affine transformations* include linear transformations like rotation, stretching and enlargement, but not perspective change. Geological processes, such as development of sand bars, deposition of turbidites, or folding of sequences of layers of different viscosity, typically operate over a range of scales (invariant under dilation within limits). The vertical axis has a special status in many geological processes on which gravity has an influence. Such processes would not be invariant under rotation about a horizontal axis.

In general affine transformations, an object might elongate disproportionately as the size of the object increases. Thin turbidite layers might, for example, be more extensive relative to their thickness than thick ones because thickness and extent were controlled by different factors, involving interactions of mass, density, viscosity or slope. They might nevertheless retain a broad resemblance that is similar apart from the change of scale and stretching. That is, like some fractals, they are *self-affine* (see Continuity, fractals, octrees and wavelets

217). Lewis et al. (1999[354]) describe the mathematics of this anisotropic scale invariance. They are the result of processes that are invariant under affine transformations.

In other cases, the process, within its range of operation, may be invariant under enlargement but not under stretching: enlarge one aspect of the process and the rest scale up in unison. The results of such processes (including many fractal processes), which are invariant under translation, rotation and enlargement, are known as *self-similar*.

Processes that are invariant under *rigid-body transformations*, that is, transformations that do not alter the size or shape of an object, namely translation and rotation, create self-similar objects of the same size. Most geological processes operate over a range of scales, but man-made objects may be created by procedures that fix their size. A map may be printed to a specific scale, or a diagram or visualisation fitted into a predetermined space on a page.

The geometrical transformations under which a geological object is invariant reflect the nature of the processes that formed it, and thus the reasoning about its origin. They influence its likely behaviour on interpolation, generalisation or reconciliation, and could be recorded as metadata in the spatial model. The metadata could then constrain the object's representation in visualisations, by resolving the ambiguities mentioned in Ambiguity and map representation 148. In a visualisation, they might indicate, for example, whether a pattern (such as a band of tight folding) is positioned where it has been observed and therefore should not be moved; or is included to illustrate the pattern known to occur in the area and its position might be adjusted; or that an object has been shown merely because it exists and must appear somewhere, although its location, size, shape and orientation are unknown and might be adjusted by rubber-sheet transformations.

Geometrical transformations may also be helpful in placing objects within an existing classification, or in specifying the range of an imprecise term. Classifications and descriptions, including those of surveyed objects and patterns, may refer to size and shape. This can give rise to questions of how to deal with approximate and ambiguous cases (see Representing spatial information and relationships 173). The natural world studied in geology tends not to follow strict mathematical rules. A pattern of folding might be identified as, say, isoclinal, but it might change across an area to a similar pattern with somewhat gentler folds of different size. This raises the question of when we must identify it as a different pattern. More generally, how do we handle natural variation and relate spatial observations of object instances with variable properties to the overall properties of the object class?

Rather than classifying objects on the basis that each of a set of quantitative properties should fall within predetermined limits, a more natural solution might be to illustrate a typical example, and, in the spirit of d'Arcy Thompson, specify the range of acceptable transformations that delimit the object class, probably including changing the size (magnification), possibly as part of a composite transformation that altered other spatial properties in step. This approach has the advantage that it creates a link between spatial

[354] Lewis, G.M., Lovejoy, S., Schertzer, D., and Peeknold, S., 1999. The scale invariant generator technique for quantifying anisotropic scale invariance. *Computers & Geosciences*, 25 (9) 963-978.

descriptions in words and geometrical descriptions that can be adjusted and visualised. Furthermore, it enables distortions that arise, for example from oblique views or tectonic deformation, to be taken into account (see Jain, 2004[355]).

Cartographers have long been concerned to select map projections to preserve chosen geometrical properties of a nearly spherical Earth surface when depicting it as a flat map. Various map projections alter some of the properties, like distances, angles, areas, parallelism, or straightness of lines, while others remain invariant under the projection. The appropriate projection is selected on the basis of which properties should be invariant for the purposes in hand. The spatial model delays cartographical issues by storing three-dimensional information and deferring projection until the information is visualised. Users, who know more about their specific requirements than the surveyor, can then select the cartographical projection and procedures best suited to their needs.

Geometrical transformations thus have a role in classifying surveyed objects, identifying important differences that representation on a static map conceals. Some constraints on the behaviour of spatial objects and relationships (during surveying procedures and computer processing) are related to their invariance under specific transformations. The constraints might therefore be recorded as metadata that can be recognised by relevant software, including that for spatial abstraction and feedback (see Abstracting from reality to model 131, The geometry of interpolation 243), and for interpolation to fill gaps between observations, considered in DSIs, FEMs and their geometrical significance 235.

DSIs, FEMs and their geometrical significance

<<Transforming space 224

Summary: *Mathematically fitted lines and surface can be difficult to relate to geological concepts and processes. But flexibility is gained in Mallet's Discrete Smooth Interpolation method by separating point data of elevations from the links or spatial relationships between them. Finite Element Methods show how patches can be fitted to visualised or measured geometrical properties (such as elevation, slope and curvature) giving compatible representations of lines, surfaces, grids and solids, and a better match to geological thinking.*

The fitted functions for interpolation (see The geometry of the spatial model 203) have useful properties, such as avoiding visual distractions from sharp discontinuities in position, slope or curvature. The slope and curvature can be calculated by differentiation at any point, and slope maps can readily be drawn. The volume below the surface can be calculated by integration. The functions may give an unrealistic representation, in that discontinuity is ubiquitous in the real world, but for many purposes give an adequate match to the geologist's visual model of the surface.

---

[355] Jain, A.K., Ross, A., Prabhakar, S., 2004. An introduction to biometric recognition. *IEEE Transactions on circuits and systems for video technology. Special issue on Image- and Video-Based Biometrics.* **14**, 1, 4-20. http://www.csee.wvu.edu/~ross/pubs/RossBioIntro_CSVT2004.pdf

One form of function mentioned in Estimation by interpolation 212 was local fitting of small patches of the surface, perhaps Delauney triangles, and blending them smoothly together. Mallet extended this powerful approach to geoscience modelling with the method he called Discrete Smooth Interpolation. "In this discrete approach, the geometry of any object is defined by a finite set of nodes (points) in the 3D space, while its topology is modelled by links bridging these nodes... For example, if the object to be modelled is composed of surfaces, then the links can be arranged in such a way that the mesh so defined generates triangular facets. These facets can be interpolated locally by flat triangles or, if need be, by curvilinear triangles. It is not difficult to imagine how this strategy can be extended to the modelling of curves and volumes" Mallet (2002[356], page v).

Separating the nodes from the links that join the nodes adds considerably to the flexibility of the representation. For example, the introduction of a new node, say with results from a newly drilled well, affects only a few links, not the existing nodal data. Geological faults can be introduced by breaking the links along the line of intersection with the fault. The concepts of emergent systems (see Complex and emergent systems 159) correspond well with those of local fitting. Extending the idea of links to general relationships, as in the Semantic Grid (see Semantic Web and Grid 50), would provide a means of representing a wide range of spatial relationships (see Representing spatial information and relationships 173) in many applications, and could allow explanatory reasoning to be incorporated in the spatial model (see Mapping geology into the knowledge system 282).

Spatial transformations make it possible to move readily from one reference frame to another (see Geometrical transformations 228). With local fitting, the computations for each patch can be, and frequently are, carried out in terms of local coordinates, relevant only within the patch. The transformation from local to global coordinates can then be applied to fit the results to a global coordinate system. More extensive local systems, referring to a set of patches, can make good sense in some contexts. The natural coordinate system for most geoscience investigations is unlikely to be the present-day east, north and up. For a structural study, for example, the principal stress directions might provide a more appropriate local reference frame.

A related approach to the representation of lines and surfaces is that of Finite Element Methods (FEM). Although FEM are strangely neglected in many areas of geoscience, search engines reveal Web sites and extensive literature in other applications. For example, they are widely used in engineering including engineering geology, and medicine including the Physiome Project (2009[357]) – the medical equivalent of The solid Earth systems model (sEsm) 71. They are mentioned at this point, not because they provide a cost-effective or even practicable approach in geoscience survey (more work would be required to establish that), but because, as part of this scenario, they lead to interesting possibilities that are relevant to any mathematical approach to geoscience interpolation, and its correspondence with geological thought processes.

[356] Mallet, J.-L., 2002. *Geomodeling.* Oxford University Press, Oxford. 599pp.

[357] Physiome Project, 2009. http://www.physiome.org

The Finite Element Method represents a surface or the boundaries of a volume as a combination of adjoining triangular and quadrilateral patches separated by lines. One method of particular interest in this context is the use of quintic polynomials to fit local patches on a triangular mesh. These are polynomial functions $z=f(x,y)$, which include x, y and xy terms up to the fifth power, such as x^5 or x^2y^3. As the function has more terms than the points to which it is fitted, it is underdetermined when fitted only to the elevations of the three nodes of a triangular facet (see Estimation by interpolation 212), and additional information is required to give a unique solution. That additional information can be the slope and rate of change of slope of the function, reflecting the curvature. The curvature may change along a fold axis, for example, as the fold becomes tighter. But the orientation of the fold axis and axial plane may also change along the length of the fold. This tendency (twist) can also be measured at the nodal points and included in the FEM.

The algebraic representation for a triangular facet, in terms of a quintic function, can be computed from a geometrical representation in terms of location, slope, twist and curvature at its nodes, or vice versa (Strang and Fix, 1973[358]). Furthermore, similar procedures can be applied to quadrilaterals, such as the rectangles of a gridded surface. They can be represented by bicubic polynomials, the edges of which are compatible with those of the quintic triangular facets, with the elevation, slope and curvature of the surface changing continuously across the boundary. The lines that mark the edges of the triangles, quadrilaterals, and the junctions between them, are cubic polynomials. Because the functions are selected to be compatible and are calculated from data at the same nodes, line, surface and volume data merge seamlessly. Piecewise blending of cubic line segments could represent the line of outcrop of a formation top. If the corresponding surface is also represented, it could be partitioned into elements by Delauney triangles, quadrilaterals, or a combination of both. Thus triangulated and gridded areas of a surface can be merged smoothly with one another and with lines of outcrop.

Continuity of elevation, slope and rate of change of slope can be maintained across the patch and line segment boundaries, by specifying the shared values at the nodes where the patches meet. At each data point, the value of the slope and curvature can be computed from the surrounding patches over a suitable area (see Unevenly spaced data 238). Interpolation of adjacent patches is then based on shared values of the elevation, slope and curvature at the nodes. This ensures that patches merge smoothly and their shared edges are the same (cubic) curve. The local fitting of patches provides flexibility, while the blending of adjacent patches provides a consistent overall solution. The same forms can also be extended to represent solid objects. For example, the base of a formation could be represented by the same method as its top, with the actual formation represented by solid elements bounded by patches on the top and base closed by cross-patches and patches depicting faults and the edge of the area of interest.

For many spatial models, the geoscientist can work more intuitively with geometry than with the algebra of the computation. Unlike algebraic functions, the geometrical properties are

---

[358] Strang, W.G. and Fix, G.J., 1973. *An analysis of the finite element method*. Prentice-Hall, Englewood Cliffs. 306pp.

familiar to the geologist, can be estimated by eye or measured in the field and represented in a readily recognisable form in a computer visualisation. Interpolation can be based on elevation, slope and curvature at data points, separating properties that behave differently under geometrical transformations. Slope and curvature (unlike location) are invariant under translation, and thus have a role (see Unevenly spaced data 238) in abstraction and feedback.

This approach might encourage more rigorous field estimations and measurements of shape characteristics, including slope variation and curvature over defined areas, as these could contribute directly to the interpretation (see Pearce et al., 2006[359]). Potentially, it might cope with integration of field observations, subsurface models and dynamic models, and of measured and visually estimated information. Thinking geometrically allows visualisation to proceed in a smooth transition from vague hand waving, through rough sketches, to a precise geometrical formulation. This could help to clarify the geological reasoning and procedures of analysis that underlie interpolation, and lead to clearer ideas about what is known, what can be estimated, and what must be surmised.

Whether or not it proves convenient or desirable to implement such methods, they provide a mathematical framework that more closely resembles patterns of geological thinking, with an immediate role in clarifying some of the underlying issues. The next sections therefore continue to explore these concepts, starting with the important issues of selecting areas to measure geometrical properties from unevenly spaced data.

Implementation note: For generality, spatial relationships between nodes could be identified by URI's similar to links in the semantic Web. The nodes and links could still be structured as tables for efficient database management without necessarily compromising their flexibility for hypermedia reference.

Unevenly spaced data

<<Transforming space 224

Summary: *Slope and curvature are properties of the smooth model, representing average values over defined areas of the rough natural surface. Their value depends on the area over which they are measured. It could be matched to the resolution of the interpolation, which itself may vary. Field observation and background knowledge throw light on the appropriate interpolation model.*

The concepts of slope and curvature may sometimes be taken for granted by the geologist, but they call for careful thought in any attempt at surface fitting. To a mathematician, slope refers to the tangent of a curve at a particular point, in other words, the first derivative of a function representing, say, the elevation of a formation top along a line of cross-section. But the geologist in the field has the task of measuring the orientation of a rough, natural

[359] Pearce, M.A., Jones, R.R., Smith, S.A.F., McCaffery, K.J.W., Clegg, P., 2006. Numerical analysis of fold curvature using data acquired by high-precision GPS. *Journal of Structural Geology*, **28**, pp. 1640-1646.

surface, where tangents are undefined (see Continuity, fractals, octrees and wavelets 217). The usual recourse is to measure the slope of an artificial flattish surface, like a field notebook, placed to approximate an average slope over a notebook-sized area. The measurement represents not the real surface, but the smooth surface of a spatial model. And the decision about the size of the selected area is critical.

The average slope on an irregular surface depends on the size of area to which it refers. On a subsurface horizon, Delauney triangles (see Drawing the line 209) are facets of the surface, each with a slope that can be determined from the measured elevations at the corner points. The facets are triangular, of widely differing size and shape. The nodes of the triangles are points, such as boreholes, where the elevation is known. However, the slope and curvature at each node, from which a quintic polynomial (see DSIs, FEMs and their geometrical significance 235) could be calculated and a smoothly curving facet fitted, have to be estimated from the surrounding facets. The slope might refer to an approximately circular patch around the node, if all directions are thought to be of equal significance.

In the ideal case of evenly spaced data points linked by a mesh of equilateral triangles, the slope of each triangular facet can be measured over the same area. Since the aim is to merge each facet smoothly with the surrounding ones, the area around the node to which the slope and curvature refer might reach out halfway to the adjacent nodes. This defines the so-called Thiessen polygon (see Drawing the line 209) or area of influence of the data point. As the slopes and areas of the adjacent flat facets are known, an average area-weighted slope can be computed for the Thiessen polygon around each node. The curvature can be calculated for a comparable area (this broad-brush scenario dodges the distinction between curvature and its proxy, the rate of change of slope). Having calculated these values for each node, quintic functions (see DSIs, FEMs and their geometrical significance 235) could be fitted to each facet, based on the nodal values. These mathematical procedures generate a smooth surface.

In reality, the data points are unlikely to be evenly distributed and we must be able to deal with, say, widely spaced data from exploration wells combined with closely spaced points from field wells, or a line of closely spaced points, for example from a seismic line. Similarly, in field investigations, observations tend to be concentrated on small areas of outcrop and along bands of exposures such as sea cliffs or river valleys. The three-dimensional reconstruction is therefore built on awkward distributions in space. The elevation at each data point is readily ascertained, but the slope and curvature depend on the size and shape of the area selected.

The slope and curvature at a node must therefore be estimated across a suitable area. The area can be taken as half-way across the facet to be interpolated. However, the facets that meet at a node may be of significantly different sizes, and it is not satisfactory to use different slopes for each, as this would create eye-catching, but geologically meaningless, discontinuities (see Continuity, fractals, octrees and wavelets 217). One possibility would be to measure the slope over an approximately circular area around the node, with a radius that reaches half-way across the largest facet. The average slope of this circular area is the average of the slopes of its component facets, weighted by their areas. But this procedure would obscure the significance of what is known of the more detailed variation.

In field mapping, data at different resolutions (that is, measured across different areas) should develop together, determining where additional observations (data) are likely to throw most light on the overall picture. As we learn more from additional observations, more points are added to the facetted network. Our growing knowledge of the geological setting may help to clarify the likely form of the surface at various resolutions. The form of each helps to determine the most appropriate method for its interpolation. What is known of the surface form comes from observations in the field and from background knowledge, including ideas about the results of the geological processes involved and analogous situations seen elsewhere. These ideas may help to establish how far the influence of observation points extends across the surface geometry. If we wish to interpolate a value at a specific point, we might look only at the nearest neighbours, or decide that more distant neighbours could shed additional light (see Drawing the line 209).

Variations in elevation might be evenly spread or concentrated in narrow zones. The geological mechanisms that result in distant objects having a similar form, such as large-scale eddies in currents or thick adjacent beds forcing a large fold structure, may or may not be known. With data for structural geology, slope and maybe curvature are obviously relevant, and likely to be measured in the field, raising the question of how they can be taken into account for interpolation. If the point to be estimated lies close to data points to the north and south, and farther away from points to the east and west, they might all be considered in interpolation, but then we might be interpolating shapes from processes on two different scales. Unevenly spaced data may well refer to a surface with superimposed forms at various resolutions and orientations.

From a geometrical point of view, there are a number of possibilities. The slope of the smaller facets, and their variation in slope, may or may not be dependent on the slope of the next larger facet in the hierarchy. The variation of slope at a small scale may or may not be related to curvature at a larger scale. At any scale, the spatial properties of elevation, slope, curvature, and their variation, might be influenced only by nearest neighbours or, to varying degrees, by more distant parts of the surface. There may be patterns that extend over many facets, and possibly a range of scales. These possibilities can be tested and analysed directly from the data supplied by the network of nodes and facets. However, the geologist's observations and understanding may give clearer indications in the field of the appropriate model, which can then be tested by further observation.

The more general question is how parts of a surface or surfaces sampled at different resolutions can be jointly taken into account to give a coherent geological depiction. For the moment, however, the additional complexity of superimposed mulitresolution variation in geological surfaces is set aside (see Scale-space 255, Multiresolution survey 259). Another aspect of the unevenly spaced observations is considered in Spatial variation and uncertainty 241.

Spatial variation and uncertainty

<<Transforming space 224

Summary: *The slopes and areas of the flat Delauney triangles of a faceted surface can be calculated and slopes plotted against area. A plot of slope variance against area is the 3D counterpart of the variogram of geostatistics. It is an abstraction from the data, giving feedback on the form of uncertainty envelopes within which the shape, but not the location, of smaller features may be simulated.*

The flat facets of the Delauney triangles can be looked at from another viewpoint. The facetted surface is the product of observations and measurements on the surface. For each facet, the area and average slope can be calculated directly from the elevations of their corner points, and can provide information about the frequency distribution of slopes for areas of different sizes. Abstracting this information, which is invariant under translation, can provide feedback (see Abstracting from reality to model 131) for subsequent parts of the investigation.

Field measurements fill a size gap in subsurface well data, where dipmeter logs measure orientation across a borehole, and the next available area for measuring slope is between wells that could be many kilometres apart. Field measurements of slope or orientation generally refer to an area of the same order of size as the borehole. However, electronic support for surveying in the field simplifies the task of measuring slopes or orientations at a range of sizes. The availability of ground-based laser scanning of outcrops by light detection and ranging (lidar) can provide detailed three-dimensional information (McCaffrey et al., 2005[360]). Enge et al. (2007[361]) showed how its application in a well exposed area can lead to interpretations of value in understanding reservoir characteristics. Seismic data also record variation at many scales. Information can thus be obtained from a range of sources giving the distribution of slopes over areas of different extents.

One aspect of the form or shape characteristics of surfaces can be investigated by plotting the slope of each facet against its area. Graphs of variation in slope versus area (slope variograms) are analogous to the variograms of geostatistics as a means of representing the overall characteristics of spatial variation. The elevations of points that are close together on a geological surface tend to be more similar than those of points further apart, and slopes measured over a small area more variable than those measured over a large area.

In geostatistics, a variogram (Isaaks and Strivastava, 1989[362]) indicates on a graph how the likely difference between the value of a property at one point and its value at another point is related to the distance between them. In geostatistical methods (see Estimation by interpolation 212), the measure of variation in elevation between points at a given distance apart involves the calculation of the mean square difference of their elevations to give the

---

[360] McCaffrey, K.J.W., Jones, R.R., Holdsworth, R.E., Wilson, R.W., Clegg, P., Imber, J., Holliman,N., Trinks, I., 2005. Unlocking the spatial dimension: Digital technologies and the future of geoscience fieldwork. Geological Society (London) Journal, **162**, 927-938. doi:10.1144/0016-764905-017

[361] Enge, H.D., Buckley, S.J., Rotevatn, A., Howell, J.A., 2007. From outcrop to reservoir simulation model: Workflows and procedures. *Geosphere*, **3**, 469-490. doi:10.1130/GES00099.1

[362] Isaaks, E.H. and Srivastava, R.M., 1989. *Applied Geostatistics*. Oxford University Press, Oxford. 592pp.

measure of dissimilarity known as the semi-variance. The semi-variance is plotted against the distance between points (*Figure 38*).

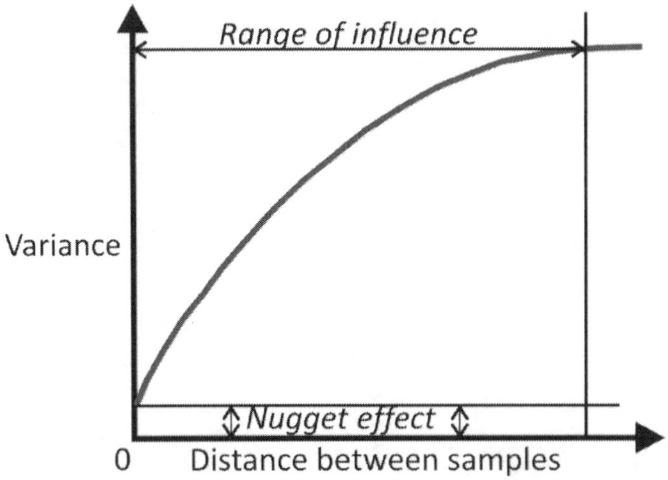

Figure 38: Semi-variance versus separation distance (from Loudon, 2000)

The geometrical expression of the elevation difference at a particular horizontal distance is the slope, in a specific direction, between the two points. Viewing it geometrically can offer greater flexibility. Slopes can readily be visualised and can be estimated, described or measured in the field. The concept of slope extends naturally into the third dimension. Slopes can be geometrically transformed, and thereby related to other reference frames of geological significance. For example, they could be measured with respect to tectonic stress axes; or slopes of small facets could be measured, not from the horizontal plane, but from the orientation of a larger facet of which they are a part.

A slope scattergram could show all the values of slope over different distances or areas, and might be informative. A three-dimensional display might be plotted on a stereonet, where the pole normal to the slope is represented by a dot, and the area of the facet indicated by the dot size or colour. Alternatively, on a computer visualisation, the viewer might watch points being added to the stereogram in increasing or decreasing order of size. If the slope is not clearly dependent on direction, this could be reduced to a two-dimensional graph of maximum slope versus horizontal length.

However, the clutter of items on a scattergram might tend to obscure the underlying pattern, and therefore calls for further abstraction. The single value representing several occurrences at each distance, calculated in geostatistics as just described, is the semi-variance, which can be plotted against distance as a semi-variogram. A candidate for a single value for the variation of slopes of facets of a specific area would be the slope variance (Loudon, 1964[363]). Slope variances could be plotted against area as a slope variogram.

The slope variogram can suggest the form of the uncertainty envelope between known points, within which the real surface is likely to lie (*Figure 38*). The small sample of measurements in a local study is unlikely to define a single smooth variogram. However, the

---

[363] Loudon, T.V., 1964. *Computer analysis of orientation data in structural geology.* Technical Report No. 13 of ONR Task No. 389-135, Northwestern University, Illinois. 138 pp.

work of van Wagoner et al. (2003[364]) suggests that there is a uniformity of shape produced by at least some geological processes operating with common physics over a wide range of scales (see Grain, set and patch 251). It is therefore likely that typical slope variograms might characterise the form of certain object classes within a region, and consequently have just a few representative forms, as do the semi-variograms of geostatistics. The same shape of uncertainty envelope could then be used where the same processes were thought to have operated. The envelope could help to determine where additional observations would be particularly helpful in reducing uncertainty, or where the simple model does not conform to the observations, suggesting where a composite model, such as one introducing faulting, is required. Smoothing uncertainty over the area as a whole would be misleading where the variation is in fact strongly localised.

Mallet (2002, page 443[365]) points to the useful distinction between interpolation and what he terms simulation. Interpolation estimates the value of properties between their observation points, for example, showing the most likely elevation of a surface. Simulation, in this context, is a means of illustrating one of many possible arrangements of values of the property between the observation points – an illustration of a possible form or shape, created by simulating a process. A pattern of tight folding shown in an unexposed area might thus be regarded as a simulation. The geologist might consider that the hidden rocks would follow that pattern, without being able to predict the actual location of individual folds or to plot them accurately on a map. The form or shape of such a feature is invariant under rigid-body transformations, but the location of its component features are not, and their positioning within a visualisation is rather arbitrary. A map obscures the distinction (see Ambiguity and map representation 148). A model might separate estimation (based on observations) from shape simulation (based on expectations of the underlying processes). The simulated features could be generally expected to lie within the probability envelopes.

The geometrical approach throws some light on the choice of methods for interpolation, but implementation is not straightforward. Before exploring its potential further, it may be helpful to review some methods discussed in The geometry of interpolation 243, relating the algebra to these geometrical concepts.

The geometry of interpolation

<<Transforming space 224

Summary: *Geometrically, interpolation functions can be regarded as transformations, but vertical adjustments of shapes generated by combinations of sine waves or power series lack clear geological interpretation. Abstraction (as in variograms) brings together properties unaltered by moving in space or geological time, and gives feedback to control interpolation.*

[364] Van Wagoner, J.C., Hoyal, D.C.J.D., Adair, N.L., Sun, T., Beaubouef, R.T., Deffenbaugh, M., Dunn, P.A., Huh, C., and Li, D, 2003. Energy dissipation and the fundamental shape of siliciclastic sedimentary bodies. *Search and Discovery* Article #40080. http://www.searchanddiscovery.com/

[365] Mallet, J.-L., 2002. *Geomodeling.* Oxford University Press, Oxford. 599pp.

Spatial models should reflect geological interpretation, supporting and complementing human skills through computer visualisation, virtual and augmented reality.

Spatial transformations help to clarify the types of knowledge we can bring to bear on interpolation. The interpolation methods described earlier (see Estimation by interpolation 212) represent the configuration of the surface on which points and lines are known to lie. The form of the surface is determined by adding together varying amounts of basis functions, such as sine waves or elements of a power series, z=ax, z=bx^2, and so on. In transformational terms, the basis functions are scaled or magnified by varying amounts (a and b in this case) before addition. A function such as z=f(x,y) transforms each point by vertical translation. Each (x,y) point on a flat plane through the origin is moved vertically (parallel to the z axis) by an amount calculated from an equation in x and y. Moving average and geostatistical methods likewise bring about vertical displacements from a horizontal plane. This may be obvious to a mathematician, but can seem rather artificial to a geologist. Deposition and compaction may or may not be seen as processes that act vertically, but even so, a formation top, say, is unlikely now to be in its original orientation.

A more flexible transformation is the parametric representation x=f(s,t); y=g(s,t); z=h(s,t), which translates each point (s,t) on a flat plane in the x, y and z directions (see Estimation by interpolation 212). In principle, the original surface (in this case with coordinates s and t) need not necessarily be a flat plane. One could, for example, add another coordinate to represent the elevation of an earlier surface and a third variable in the functions to represent this third dimension, but the complexity of mutual relationships among surfaces makes this approach unattractive for most general applications.

There are other properties of the surface, however, that geologists sometimes use in manual interpolation and generalisation. From their background knowledge and experience, they have a feel for the likely form and shape of the surface, independent of its position in space. By definition, shape is invariant under rigid-body transformations, such as translation or rotation. Abstracting by bringing together geometrical properties from many sources is obviously dependent on this spatial invariance. Because slopes are invariant under translation, and angles under rigid-body transformations, they can be brought together from many locations in one meaningful diagram. Objects that have been formed by a particular set of geological processes may have characteristic shapes. The abstraction process might therefore be able to quantify the shape properties and make them available to guide interpolation.

Few of the surface fitting methods described here (see The geometry of the spatial model 203) take such geological knowledge into account. But the methods of geostatistics and kriging (Isaaks and Strivastava, 1989[366]) have a clear geological significance. The variogram is a method of abstracting general information from specific observations, and providing feedback (see Abstracting from reality to model 131, Benefits of an object-oriented system 199) to guide further investigation. It measures differences in elevation of points at specific distances apart. This is a property analogous to slope. It also is invariant under translation, and can therefore be abstracted from several sources. The variograms represent the

[366] Isaaks, E.H. and Srivastava, R.M., 1989. *Applied Geostatistics*. Oxford University Press, Oxford. 592pp.

expected variability of the surface at different wavelengths (reflecting the effects of combinations of processes at different scales) and make this knowledge available for interpolation.

The example of *Figure 35* (in Drawing the line 209) pointed to one limitation of the geostatistics model. In many areas of geoscience, knowledge of the configuration reflects ideas about surfaces and their deformation that involve slopes and curvatures. Interpolation that is based on assessing the similarity of the elevation at an unknown point to the elevations of nearby points may give an appropriate estimate of the elevation, but not of other spatial properties, such as form or shape (see Shape). Geometrical forms and movements, such as conical folding or listric faulting, do not necessarily fit the basic model of geostatistics. There is always a need to look carefully at the assumptions of the model, and to limit its use to appropriate cases. Stratified sediments, with an emphasis on the geometrical aspects of structural geology and sedimentology, may benefit from a model that handles slopes on the surface (see Spatial variation and uncertainty 241) rather than just lines joining data points.

A significant drawback is the difficulty of relating a mathematical representation in algebraic form to field observations, visualisations or geological processes. When examining, say, the form of a formation top along a cliff exposure, it is not obvious what its variogram would look like, nor is it easy to picture the form of a surface by examining its variogram. The shapes of functions like power series or sine waves (see Estimation by interpolation 212) and their composite surfaces are not obviously related to forms produced by geological processes, and larger breaks in the surface, such as faults or hinge-lines, have to be handled separately and superimposed on the function. As discussed in 'DSIs, FEMs and their geometrical significance 235', discrete smooth interpolation and finite element methods can bring greater flexibility to the relationship between interpolation and form. Local elements can be described in terms of location, slope and curvature. Interpolated values can be based on this geometry. The geometrical features of the elements can be geometrically transformed in response to their surroundings, just as a geologist might envisage geological processes transforming an object to create a particular form within the context of the surrounding geology.

The geometry of the various types of interpolation can be visualised, perhaps throwing some light on their suitability for the conceptual geological processes that the surveyor is trying to reflect. The immanent geological processes (see Abstracting from reality to model 131) that created the various geological objects are invariant under translation in space and time, and the form or shape of their products is likewise invariant. Shape is therefore a potentially powerful tool in identifying processes from observation of the products, and in guiding interpolation procedures to give a realistic result (see Shape 266, Morphometrics 268. It plays a significant part in geological reasoning and can be formulated in terms of geometrical ideas, which are appropriate to observation and visualisation, and match geological thinking. The geometrical aspects may be converted to algebra for computer implementation, but this need not be apparent to the user. Ideally, the approach to handling the spatial information would extend and assist the innate, trained, human abilities of the geologist throughout the process from observation to description, filtering, interpolation, abstraction, feedback, generalisation, visualisation and presentation. In geological surveying, human perception and visual imagination dominate the spatial analysis. It is desirable to select computer methods that support their weaknesses and complement their strengths.

"A visual comparison has a number of strengths and weaknesses. The most obvious strength is the simplicity with which a comparison is completed. Humans can observe, recognise and interpret spatial fields automatically, integrating their background knowledge and understanding of the spatial field being viewed. They can then compare two spatial fields and make a qualitative assessment of their similarity, exploiting the outstanding ability of the human brain to synthesise disparate information. The comparison will involve looking at overall similarity, the similarity of specific features and even the possible similarity of features if they were shifted or altered slightly. Yet, amongst all these strengths emerge the weaknesses with this approach. While the spatial field can be interpreted and observed, the observer can personally bias the interpretation and there are limits to the capacity of the brain to assess multiple images or large spatial extents... It is evident that the human visual system works predominantly with features that command attention due to their intensity, size, shape, location or value" (Wealands et al., 2005[367], page 20).

Computer visualisation techniques link human perception and the spatial model. Accounts of geoscience visualisation are available at, for example, BGS (2010b[368]). They refer to techniques of virtual reality that were developed for diverse applications, many in aerospace, energy, and medicine, and typically require extensive, fixed and dedicated facilities. They make it possible to examine, discuss, reconcile and amend three-dimensional models of geological features. Field geologists may be able to link spatial models to reality through visualisation of models in the field, on screen, with paper prints, or potentially with augmented reality (a composite view of the real scene viewed by the user and a virtual scene generated by the computer) see The field survey model 300.

Regardless of the type of display, computer methods for creating, interpreting and manipulating the spatial models should make sense in geological terms. Geologists should be able to see a direct link between the computer modelling processes and the underlying geological thinking, surveying procedures, and geological processes. Changes in model parameters should have intuitive significance for geologists, and the specification of model behaviour should be geologically meaningful. The focus is moving from individual models to a broader approach linked to human vision in a more comprehensive knowledge system, considered in Seeking shared concepts 247.

Implementation note: Computer-assisted visualisation helps to explore complex three-dimensional structures by interactively combining the unique human abilities of the geologist to interpret the images with the ability of the computer to create, record and display them. Ideally, the underlying mathematics should (but may not) realistically represent the likely results of geological processes.

[367] Wealands, S.R., Grayson, R.B., Walker, J.P., 2005. Quantitative comparison of spatial fields for hydrological model assessment – some promising approaches. *Advances in Water Resources*, **28** (1), 15-32.
[368] BGS, 2010b. Geoscience technologies. http://www.bgs.ac.uk/research/technologies.html

Seeking shared concepts

<<Table of contents 2
<<The future geological map 125
<<Transforming space 224
Seeking shared concepts 247
Zoom 247
Grain, set and patch 251
Scale-space 255
Multiresolution survey 259
Boundaries: discontinuities and zero-crossings 264
Shape 266
Morphometrics 268
Deformable models 274
Reconsidering geological mapping 277
>>Mapping geology into the knowledge system 282

Abstract: Various integrative concepts have emerged within the systems approach of diverse disciplines, such as ecology, landscape diversity, biomedical science and cognitive science. They have been sporadically explored in geoscience, but are not yet part of mainstream geological thinking. There is, however, an obvious case for benefitting from these developments in a solid Earth systems model, in improving its scientific rigour, in sharing work done elsewhere, and in widening the interoperability of geological products. They include criteria for defining boundaries of spatial objects (such as stratigraphical units); classifying the extent of processes and object classes in scale-space; their application in a wide range of interpolation models; statistical shape analysis (morphometrics), including changes as a configuration of objects evolves through time; and the integration of spatial knowledge (including uncertainty) in inhomogeneous deformable models – "the confluence of geometry, physics and approximation theory".

Zoom

<<Seeking shared concepts 247

Summary: *Scale is a critical factor in studying geological phenomena. Map generalisation filters to reduce size, scale, detail and resolution. Legacy maps provide the likely initial content for the spatial model. The visualisation model filters the information again as the user zooms in and out, examining it across a wide range of scales. The metadata can record the appropriate range of scales for visualising each object (such as a 1:50 000 map sheet).*

This is the key to accommodating the wide range of scales at which the field geologist observes spatial entities, and leads to reconsidering the nature of spatial objects.

Scale is a critical factor in studying geological phenomena, as Carey (1962[369], page 100) emphasised. "Our thinking is done with models; concrete models such as a spheroid we can picture as the earth, but also by mathematical models; for when we write down symbols to represent the physical behaviour of the earth these symbols are also models, and however erudite our mathematical operations, the answer applies only to the model, and may have little relation to the behaviour of the real earth." He points out the effects of the great changes of scale in our mental models. Within the fields of rock deformation, structural geology and geotectonics, the linear scale varies through sixteen orders of magnitude, and the scale of time is as wide as the scale of size. "Many behaviour thresholds exist and because the several physical properties involve different powers of length or time, terms which are quite insignificant at one scale may be the dominant ones in others. No one mental or mathematical model has validity over the whole field; hence models must be deliberately selected for the time-size field of thinking or calculation. Nomenclature should change with field so that it signifies the kind of behaviour relevant to the scale" (page 97).

In his Fig. 2 Carey illustrates what has since been referred to as scale-space, his horizontal and vertical axes representing scale in time and in space. The same concepts have now led to computer implementations representing spatial characteristics of geological objects within scale-space (see Scale-space 255). Carey proposed five scale intervals in structural geology and geotectonics, studied respectively with: the electron microscope; the petrographical microscope; hammer, compass, clinometer and tape; regional maps and sections; continental and global maps. In his readable and thought-provoking commentary, he related a range of geological processes and events to these scales and to the scale of geological time.

Geological surveying is concerned with (but by no means limited to) phenomena at the scale of regional maps and sections; embracing structures from 10m to 10km. "The conclusions of our outcrop-scale observations are abstracted and symbolized on these maps, and from them regional structures are induced. Here we are concerned with folds, faults, and plutons, in their many combinations and permutations. Not rocks or beds, but stratigraphical formations are the deformed units. In the upper part of the range are horsts and graben, and the smaller nappes, and geanticlines" (page 102). Warren Carey's insights seem highly relevant as we edge towards a computer-based geoscience knowledge system. Scales of time and space are essential parameters in the classification and interpretation of the objects, processes and models, and in the procedures for their interpolation and visualisation.

Conventionally, one procedure for handling the range of scale-space in geological survey is generalisation of maps to smaller scales while moving up the hierarchy of stratigraphical classification. *Map generalisation* is a process of reducing the volume of information depicted on a map while retaining the most significant elements for the particular geological application. Generalising to produce a smaller-scale map should preserve legibility and the

[369] Carey, S. Warren, 1962. Scale of geotectonic phenomena. *Journal of the Geological Society of India*, **3**, 97-105.

desired appearance of lines and areas, maintain the correct spatial relationships, and clarify the essential character of features and structures by removing inappropriate detail. The derivation of the reduced-scale maps calls for an understanding of the geology and intended application, and requires specialised cartographical skills (Buttenfield and McMaster, 1991[370]). The introduction of digital cartography to geology by no means simplified the methods (Downs and Mackaness, 2002[371]).

The geologist probably associates scale with the ratio of the distance between two points depicted on a map and the distance between their real-world counterparts. The visualisation of a spatial model, however, may be magnified to fit a computer screen, a diagram on a printed page, a hologram, or the wall of a projection room. Map scale is then no longer relevant. Instead a geographical grid superimposed on the image can indicate its size in the real world. "From an absolute perspective, scale corresponds to a standard system, such as cartographic scales and census units, used to partition space into operational spatial units" (Stewart et al., 2004[372]).

Scale can refer to the broader concept of the ratio between the spatial resolution at which a property of an object was observed, and the spatial resolution at which it is displayed. The *resolution* refers to the amount of detail that an image can hold, or that a sensor (such as the human eye) can detect, that is, the smallest distance apart of two objects that can be seen as distinct. The sensor filters information from the real world, which may then be filtered again for representation and recording, and at least once more for visualisation, generalisation and publication. *Filtering* (defined in 'Invariant properties and classification 232' as a process of selectively enhancing or reducing specified components of the information stream) can be seen here as a continuous abstraction process of removing the finer spatial detail to give an image at any coarser resolution.

Conventionally, in order to reduce detail and to place observations and interpretation in their wider context, relevant information is filtered and transferred to maps at predefined smaller scales, to generalise and provide overviews of the broader picture. Geological Survey maps are normally printed at a limited number of scales, filtered by the largely manual procedures of map generalisation. These visualisation filters are selected to give a legible view of the geology on the published map, and inevitably influence what is recorded in the field. In remotely sensed images or in interpolated grids, pixels are usually of the same size throughout the full extent of the image, and the pixel size determines the finest resolution of the image. On a printed map, the acuity of human vision and the printing process determine the finest observable resolution (usually a small fraction of a millimetre) throughout the map.

---

[370] Buttenfield, B.B. and McMaster, R.B. (editors), 1991. *Map generalization: making rules for knowledge representation*. Wiley, New York. 245pp.

[371] Downs, T.C., Mackaness, W.A., 2002. An integrated approach to the generalisation of geological maps. *Cartographic Journal,* 39(2), 137-152. http://www.geos.ed.ac.uk/homes/wam/DownsCartJournal2002.pdf

[372] Stewart, S.A., Hay, G.J., Rosin, P.L., and Wynn, T.J., 2004. Multiscale structure in sedimentary basins. *Basin Research*, **16**, 183-197. doi: 10.1111/j.1365-2117.2004.00228.x http://www.geog.umontreal.ca/gc/PDFs/New_PDFs/2004_stewart_Hay.pdf

Before the visualisation filters are applied, the field geologist handles a wide range of resolutions. With individual observations from separate boreholes, the resolution may be determined by their spacing. But with outcrops, their size and spacing does not determine the resolution of the observations. The geologist might observe at successively finer resolutions, first by looking at the overall landscape, then at an outcrop as a whole, then more closely at interesting parts of the exposure, then perhaps at a smaller part through a hand lens. If more detail is needed, a section of the rock can be examined at finer resolution with an optical or even an electron microscope. The geologist adjusts the resolution of the filter, depending on the amount of detail required, simply by moving the sensor (usually the eye) nearer to or further from the object, as in peering closely or standing back. Of course, the type of information depends on the method of observation, which can change along with the resolution.

The geological map does not differentiate between levels of observational resolution, even when generalised to several smaller scales. Vertical exaggeration of cross-sections can introduce more detail in the vertical direction, but the finer points of many spatial observations must still be relegated to the text and diagrams of the map explanation. Furthermore, the maps are not just records of the sparse observations, but also interpretations of the geology. They may emphasise different aspects of the geology at each map scale. However, the interpreted origins of objects, spatial patterns, and spatial relationships arising from separate processes at different scales are not differentiated on the map. The map combines the results of a wide range of scales of observation and interpretation in a single image of fixed scale. The consequences include obscurity and ambiguity in the map (see Ambiguity and map representation 148, Scale-space 255) and inappropriate handling of the outcome of complex systems (see Complex and emergent systems 159).

With a spatial model, on the other hand, the geologist can observe, record, model, visualise and communicate spatial information about each geological object at any resolution. The scale can be modified with additional filters when required for a specific purpose. This allows a more flexible approach to issues of generalisation. Presentational aspects can be handled at the visualisation stage. Maintaining spatial relationships among objects when changing the scale of visualisation, including the links between geology and topography, is simpler because the relationships can be recorded explicitly. Formal designs to ensure clarity of presentation at predefined map scales are less necessary where the user can zoom in and out to see the desired level of detail.

Initially, the spatial model must depend largely on legacy material that was collected with different presentational procedures in mind. In other words, pre-existing maps and related records are the objects with which the initial model is concerned, rather than the hierarchies of geological objects observed in the field. Digital maps can of course be modified to a more flexible form before representing them as part of a spatial model. The geology may be separated from the topographical base map, and might be converted to three dimensions, perhaps by fitting it to a digital terrain model. With some additional work, structures and entities of geological significance can be identified on the map, and described and represented as separate objects in the model. The objects still relate to scale, however, whether they come from pre-existing map representations or directly from field observations.

The freedom to zoom in and out and to pan around while examining a spatial model is an important benefit. But the content of the model has to change in step with the scale and the area. For example, a visualisation based on a 1:50 000 map might show increasing detail on zooming in, until the display reflects the full resolution of the original map. As the user continues to zoom in, the display might switch to a smoothed view of 1:10 000 maps of the same area, which in turn becomes increasingly detailed until the full resolution of the original is reached. There is thus a limited range of scale (or visualisation resolution) over which each object can appropriately be displayed. This should be recorded, presumably in the metadata for the object, where software can refer to it and switch objects when appropriate.

The spatial model should be designed to accommodate foreseeable developments, in order to reduce later backtracking and to encourage exploration of the system's potential. Metadata should therefore include scale characteristics for the wide range of spatial objects that play a role in the survey spatial model (see Scale-space 255). Objects can then be regarded as occupying a position along a continuous range of scales, opening a new outlook for descriptive procedures in the field; a viewpoint that in related fields has led to reconsideration of the nature of spatial objects (see Grain, set and patch 251).

Implementation note: In practice, most Surveys start to build spatial models on the basis of their legacy of geological maps. A clear strategy for adding value during and after map digitisation is helpful. The design should look ahead to zooming freely through a wide range of scales, reflecting observational resolutions in the field.

Grain, set and patch

<<Seeking shared concepts 247

Summary: *Some long-standing classifications can obscure understanding of emergent processes. Ecologists propose a more flexible structure rooted in complex systems. Typically applied to remote-sensing imagery, the pixel size determines the grain or resolution; relatively homogeneous zones of sets of adjacent pixels are referred to as patches or blobs and may be relevant to interpolation. Patch dynamics studies the form, structure, function and development of the patches and may reflect phenomena of interest.*

The Grid (see Semantic Web and Grid 50) supports a knowledge system that is increasingly global, both geographically and in subject matter. Within this more unified knowledge system, advances in spatial modelling in such active areas as medical imagery, aerospace, computer vision and remote sensing already impinge on geoscience. It therefore seems appropriate in the present scenario, which is neither a prediction nor a recommendation, to explore ideas; consider alternative possibilities; look across fences at work in nearby fields; and report premonitions of possible impediments to mutual understanding.

The top-down view (see The surveyor's holistic view 60, The geometry of interpolation 243, Deformable models 274) of geological surveying suggests that an important task during fieldwork is for the surveyor to identify zones of relative homogeneity, reflecting coherent geological environments within which spatial characteristics can be evaluated. Within each zone, when the observed geological characteristics originally developed, the ambient conditions (such as stress fields, values and gradients of temperature and pressure, or chemical, sedimentational or biological environments) were presumably relatively homogeneous and continuous (see Continuity, fractals, octrees and wavelets 217).

For practical reasons, the results of geological survey have conventionally been presented at a limited range of resolutions, corresponding to map scales. Relatively homogeneous zones on a large scale are defined by such concepts as terranes, subterranes, fault blocks, fold belts, metamorphic zones, and facies. Stratigraphical categories (see The stratigraphical framework 139, Stratigraphical units in space and time 141), as shown on a map, might also define such zones. Large discontinuities that separate the zones, such as major faults and unconformities, are identified and shown on the map; smaller ones are smoothed over by continuous lines and surfaces. Orientations of irregular surfaces are represented by precise measurements made according to local rules of thumb (see Unevenly spaced data 238). Ambiguities are arbitrarily resolved, perhaps by positioning the base of a sandstone formation after a decision that the lower part has many shale beds, or conversely, moving the boundary up by deciding that the underlying shale has sandstone lenses near the top.

Hay et al. (2002[373]) sound a warning to landscape ecologists that has resonance in geoscience: "We assign meaning to these [landscape] patterns, but as it turns out, this meaning may be completely inappropriate for describing the underlying processes, or understanding the 'system' as a whole, because we have been trying to coax from these landscape patterns a hierarchical mirror of our definitional classes." Perhaps geologists likewise try to coax from observed geological patterns a hierarchical mirror of predetermined stratigraphical classes that fit well on the map. Maybe these too are inadequate for elucidating emergent processes or understanding the system driven by their interactions (see Complex and emergent systems 159).

Van Wagoner et al. (2003[374]), point out the need for "a new physics and hydrodynamics-based sedimentology that provides a unifying context for the analysis and interpretation of clastic sedimentary systems, largely independent of depositional environment and scale." They suggest: "a more logical approach may be to develop depositional models that are based on the properties of the decelerating flows responsible for most clastic deposits...the physics of turbulent flow deceleration and sediment transport transcend many depositional environments and scales." Although considered in the context of sand body formation, the search for unifying concepts extends more widely throughout geoscience. The 'common

[373] Hay, G.J., Dubé, P., Bouchard, A., Marceau, D.J., 2002. A scale-space primer for exploring and quantifying complex landscapes. *Ecological Modelling*, **153** (1-2), 27-49. http://www.sciencedirect.com/science/article/pii/S0304380001005002

[374] Van Wagoner, J.C., Hoyal, D.C.J.D., Adair, N.L., Sun, T., Beaubouef, R.T., Deffenbaugh, M., Dunn, P.A., Huh, C., and Li, D, 2003. Energy dissipation and the fundamental shape of siliciclastic sedimentary bodies. *Search and Discovery* Article #40080. http://www.searchanddiscovery.com/

language of stratigraphy' (see The stratigraphical framework 139) does indeed encounter 'discrepant basic principles' as it moves from a map basis to a more wide-ranging model-based system.

Field observation and the concepts of fractals and complex systems (see Complex and emergent systems 159) suggest that there may be a complete spatial range of zones of relative homogeneity in nested progressions from microscopic to regional scale. As geology is a historical science, a relatively homogeneous zone, such as a widespread sandstone deposit resulting from a depositional event, might later be split up by tectonic events into several structural zones, say various fault blocks and an adjoining fold belt. Also, discontinuities arise, not only from the interaction of processes affecting the same rocks at different times, such as sedimentation, intrusion, folding, faulting and weathering, but also from concurrent interactions of the same process at different scales, as with the jet-plume pairs described by van Wagoner et al. (2003[375]). Thus, it is unlikely that homogeneous zones based on reconstruction of processes and events in geological history will fit into one simple classification, or one single nested hierarchy.

Landscape ecologists, less burdened by the complexity of geological time and unevenly spaced observations, have developed a general terminology in the context of studying landscapes from remote-sensing data. The term *grain* refers to "the smallest intervals in an *observation set* ... equivalent to the spatial, spectral, and temporal resolution of the pixels composing an image" while *extent* refers to the range over which observations at a particular grain are made (Hay et al., 2002[376]). To the ecologist, *patches* "represent discrete areas of relatively homogeneous environmental conditions, the definition of which is artificially imposed according to a phenomenon of interest and only meaningful when referenced to a particular scale. ...at a larger scale, [a patch] can be viewed as a mosaic (or landscape) of its own, consisting of smaller patches... represented by a collection of pixels in a remotely sensed image" (Rashed, 2004[377]). This definition of patch is not to be confused with that in computer-aided design, where it refers to a local area fitted by a specific function (see DSIs, FEMs and their geometrical significance 235). The definition of grain could also confuse a geologist. A similar concept to that of a patch arises in computer vision, where a *blob* refers to a region that strongly differs in luminance from the surrounding background (Marr, 1982[378]), not to be confused with the binary large object referred to as a

[375] Van Wagoner, J.C., Hoyal, D.C.J.D., Adair, N.L., Sun, T., Beaubouef, R.T., Deffenbaugh, M., Dunn, P.A., Huh, C., and Li, D, 2003. Energy dissipation and the fundamental shape of siliciclastic sedimentary bodies. *Search and Discovery* Article #40080. http://www.searchanddiscovery.com/

[376] Hay, G.J., Dubé, P., Bouchard, A., Marceau, D.J., 2002. A scale-space primer for exploring and quantifying complex landscapes. *Ecological Modelling*, **153** (1-2), 27-49. http://www.sciencedirect.com/science/article/pii/S0304380001005002

[377] Rashed, T., 2004. Quantifying the ecological patterns of urban densification through multiple end-member spectral mixture analysis, landscape metrics and fuzzy logic. *International Archives of the Photogrammetry, Remote Sensing and Spatial Information Sciences, Vol XXXV-B7:503-508* http://parker.ou.edu/~rashed/Publications/TarekRashed-ISPRS2004.pdf

[378] Marr, D., 1982. *Vision: a computational investigation into the human representation and processing of visual information.* W.H. Freeman, New York. 397pp.

blob in data management. The patch or blob is a particular type of object, and both terms are used in the literature.

"Spatial patchiness is ubiquitous in ecological systems. The theory of patch dynamics, assuming that ecological systems are dynamic patch mosaics, studies the structure, function and dynamics of patchy systems with an emphasis on their emergent properties that arise from interactions at the patch level" (Wu and David, 2002[379]). The aim is to provide a unifying framework to understand the formation, evolution and decay of patches across different systems and scales, and the patterns, mechanisms and consequences of patchiness.

Eiden et al. (2000[380]), for example, describe the technique and explore its potential in a study of landscape diversity for the European Commission. They discuss the significance of landscape metrics, such as patch density, number of classes, ratio of patch perimeter to area, diversity, and juxtaposition. Most work on patch dynamics refers to remote sensing, where images offer coverage that is relatively complete and uniform compared with the scattered outcrops and boreholes of field geology. The term most often refers to two dimensions, and 'patch body' has been used to indicate their form in three-dimensions. Applying patch dynamics concepts to geoscience survey starts from very different data sources and methodology. Nevertheless, the patches identified in studying a present-day landscape are likely to have their counterparts in, for example, the sedimentary record of similar processes operating on the landscapes of geological history.

In a geological spatial model the typical patch body would be defined in three spatial dimensions and perhaps by the geological time-span in which it formed, and is a particular kind of spatial object. The use of the term 'patch' or 'blob' may be useful where the object is a relatively homogeneous region, and where the procedures of patch dynamics and flexible local categories could be helpful. The immediate relevance is that a patch is presumably created by a coherent set of processes, and the gaps within it might therefore be filled by one style of interpolation. Their wider future relevance is suggested by the developments outlined in Scale-space 255 and Multiresolution survey 259.

The patches of interest to the field geologist are likely to be three-dimensional, and based only partly, if at all, on remotely sensed images. Nevertheless, identifying zones of relative homogeneity are an initial means of categorising field observations within a spatial context, regardless of whether they are called stratigraphical units, objects, patches or blobs. The geologist might have expectations of the spatial properties of the patch, based on observation and reasoning, guided by background knowledge of the setting, and by analogous situations, processes and their outcomes elsewhere. Within the local context, the patch might be regarded as a spatial object instance, identified as belonging to an informal object class. It might be interpreted in that light within the configuration defined by its spatial relationships with similar neighbouring objects, and within the hierarchies of spatial

[379] Wu, J. and David, J.L., 2002. A spatially explicit hierarchical approach to modelling complex ecological systems: theory and applications. *Ecological Modelling*, **153** (1-2), 7-26.

[380] Eiden, G., Kayadjanian, M., Vidal, C., 2000. Capturing landscape structures: tools. http://ec.europa.eu/agriculture/publi/landscape/ch1.htm

object classes to which it belongs. The interpretation should define and reconcile the spatial properties within and among the objects – a process dependent on the surveyor's human perception. Patches are of many sizes from microscopic to regional, and smaller patches occur within larger patches, reflecting the range of scales over which geological processes operate. There is therefore an important link from patch dynamics to Scale-space 255 theory.

Implementation note: The literature about new fields that are developing rapidly can be hard to locate. A bibliographical or Web search engine can provide pointers to current developments. Search phrases such as patch dynamics, scale-space theory, multiresolution morphometrics, or landscape ecology metrics lead to many up-to-date references on topics mentioned here.

Scale-space

<<Seeking shared concepts

Summary: *With wide-ranging applications from biomedicine and ecology to 3D seismic analysis, scale-space theory studies the effects of scale change on patches and blobs. It mimics patterns of human perception of the real world at multiple resolutions. The blurring effect of viewing an object from a greater distance can be modelled by convolving the image with a Gaussian kernel (a process of filtering to remove the finer spatial detail). Progressively less detailed images are stacked one above the other, the vertical axis representing scale. The changing properties, and the creation, merging, splitting and annihilation of blobs, are examined within the stack.*

Scale-space theory is of wide application in generalisation, visualisation and computer vision in topics from biomedical imaging to landscape ecology, as Hay, et al., (2002[381]) point out in their primer on the subject. Current applications are for the most part two-dimensional, not because of mathematical limitations, but because of the usual subject matter and to keep the computation manageable. Existing applications generally refer to remotely sensed images, including medical tomography, with evenly spaced data at constant resolution. Geoscience applications have been more in geophysics rather than geology. Stewart et al., (2004[382]) describe applications of scale-space analysis to 3D seismic data. Martin and Stofan (2007[383]) discuss the integration of multiple-scale sensor data from the planet Mars. Multi-scale geological models were the topic of the GSA Penrose Conference in 2006 (see

[381] Hay, G.J., Dubé, P., Bouchard, A., Marceau, D.J., 2002. A scale-space primer for exploring and quantifying complex landscapes. *Ecological Modelling*, **153** (1-2), 27-49. http://www.sciencedirect.com/science/article/pii/S0304380001005002

[382] Stewart, S.A., Hay, G.J., Rosin, P.L., and Wynn, T.J., 2004. Multiscale structure in sedimentary basins. *Basin Research*, **16**, 183-197. doi: 10.1111/j.1365-2117.2004.00228.x http://www.geog.umontreal.ca/gc/PDFs/New_PDFs/2004_stewart_Hay.pdf

[383] Martin, P., Stofan, E.R., 2007. Planetary science: Multiple data sets, multiple scales, and unlocking the third dimension. *Geosphere*, **3**, 435-455. doi: 10:1130/GES00089.1

Wawryzniec et al., 2007[384]). Patch dynamics and scale-space theory will surely be relevant to geological surveying as it moves away from the map-bound conventions of fixed scales towards a more flexible spatial framework of models and objects studied at many resolutions.

Scale-space theory was developed in the context of computer vision (Marr, 1982[385]). His work also throws light on how human beings analyse features of the real world at multiple resolutions, when looking at objects at varying distances from the eye. An important aspect for present purposes is that the computer can mimic patterns of human perception, thus harmonising the visualisations provided by interactive computer support with an evolving geological interpretation driven by human visual skills and background knowledge.

Scale-space theory views scale as the ratio of the resolution of the filters for observational records and for visualisation (see Zoom 247). In the usual application, to remote sensing imagery, the observational filter yields pixels of constant resolution. With complex systems in mind (see Complex and emergent systems 159), ideas from ecology and computer vision are brought together in scale-space analysis, to study the changing form of spatial objects as the scale changes. Spatial patterns, and consequent understanding of the underlying processes, are shaped by the relationship between the observed objects and the scales and resolutions at which we visualise them.

"Conceptually, *scale* represents the 'window of perception', the filter, or measuring tool, with which a system is viewed and quantified; consequently real-world objects only exist as meaningful entities over a specific range of scales" (Hay et al. 2002[386]). "A simple example is the concept of a branch of a tree, which makes sense at a scale from, say, a few centimetres to at most a few meters" (Lindeberg, 1996[387]). The leaf, the branch, the tree, the stand of timber, the forest are regarded as distinct but linked object classes, responding to separate processes and behaving according to different rules. Similarly, a geological spatial object class might be associated with a specific range of scales or resolutions. Lindeberg (1994b[388]) provides a detailed mathematical account of the application of scale-space representation for analysis of image data at the lowest levels in the chain of information processing of a visual system.

The usual visualisation filter selected in scale-space theory is known as a *Gaussian kernel*, which has appropriate mathematical properties (linearity, and no bias for location,

---

[384] Wawrzyniec, T.F., Jones, R.R., McCaffrey, K., Imber, J., Holliman, N., Holdsworth, R.E., 2007. Introduction: Unlocking 3D earth systems, harnessing new digital technologies to revolutionize multi-scale geological models. *Geosphere*, **3**, 406-407. doi:10.1130/GES00156.1

[385] Marr, D., 1982. *Vision: a computational investigation into the human representation and processing of visual information.* W.H. Freeman, New York. 397pp.

[386] Hay, G.J., Dubé, P., Bouchard, A., Marceau, D.J., 2002. A scale-space primer for exploring and quantifying complex landscapes. *Ecological Modelling*, **153** (1-2), 27-49. http://www.sciencedirect.com/science/article/pii/S0304380001005002

[387] Lindeberg, T., 1996. Scale-space: a framework for handling image structure at multiple scales. *In Proc. CERN School of Computing, Egmond aan Zee, The Netherlands, 8-21 September 1996.* http://www.nada.kth.se/cvap/abstracts/lin96-csc.html

[388] Lindeberg, T., 1994b. *Scale-space theory in computer vision*. Kluwer Academic Publishers, Dordrecht, Netherlands. 420pp.

orientation or scale) and is less prone than most filters to introducing spurious patterns during generalisation. According to Marr (1982[389], page 56): "the Gaussian distribution has the desirable characteristic of being smooth and localized in both the spatial and frequency domains and, in a strict sense, being the unique distribution that is simultaneously optimally localized in both domains. And the reason, in turn, why this should be a desirable property of our blurring function is that if the blurring is as smooth as possible, both spatially and in the frequency domain, it is least likely to introduce any changes that were not present in the original image." Stewart (2004[390]) reports on a trial of smoothing seismic data with the Gaussian kernel: "the most smoothed profile is free from the high-frequency artefacts produced by moving average filtering... lateral migration of fold hinges is less pronounced."

Gaussian smoothing is seen as a close analogue of the receptors in human vision. It is based on the bell-shaped curve, familiar in statistics as the normal distribution shown in *Figure 39*. Reflecting its application in statistics, the central point is referred to as the mean, and the root mean square deviation of values from the mean is the standard deviation. For convenience, the mean is generally taken as the origin with coordinates (0,0), and the distance of a point from the mean is measured in standard deviations. The curve is rotated about the vertical axis through its central point, to form a filter that can be applied to circular areas on the image. The value of the curve at any point indicates the amount of information passed by the filter. Thus, the Gaussian filter passes most information at the centre and progressively less towards the edge, which for computation might be set at about 4 standard deviations, as the filter passes little information beyond this boundary (*Figure 39*). The scale is determined by the width of the filter (usually measured as the distance on the image corresponding to one standard deviation in the filter).

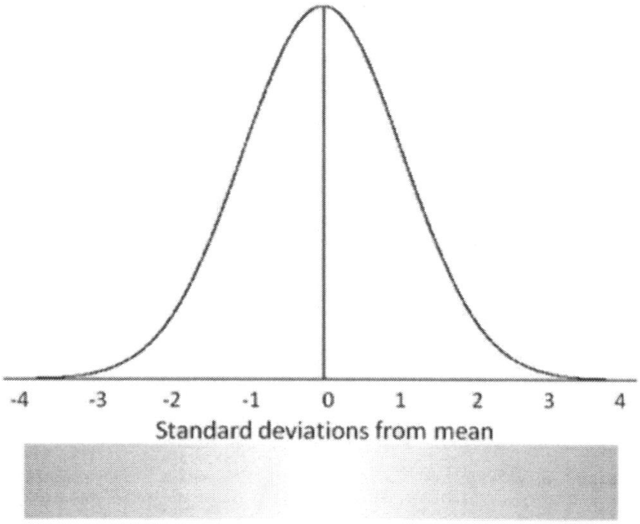

Figure 39: The bell curve used in Gaussian filtering

[389] Marr, D., 1982. *Vision: a computational investigation into the human representation and processing of visual information.* W.H. Freeman, New York. 397pp.

[390] Stewart, S.A., Hay, G.J., Rosin, P.L., and Wynn, T.J., 2004. Multiscale structure in sedimentary basins. *Basin Research*, **16**, 183-197. doi: 10.1111/j.1365-2117.2004.00228.x http://www.geog.umontreal.ca/gc/PDFs/New_PDFs/2004_stewart_Hay.pdf

Human vision presumably collects data simultaneously from each receptor cell in the retina of the eye. But the corresponding computer process, known as convolution, scans the image by moving a window step by step over the image, centred on each pixel in turn. The window or filter in this case is the Gaussian *kernel*, that is, the mathematical function describing the form of the curve, probably represented by a small array of numbers for weighting the appropriate pixels. Each pixel of the new image is calculated by applying the weighting values to the original pixel values within the window. The results are recorded in sequence as pixels at the corresponding points on a new image, which can have the same number of pixels as the original, but now shows the image blurred by the filter.

In scale-space analysis, the original image is scanned repeatedly, transforming the window at each scan by widening it to increase its standard deviation. The effect is similar to viewing the image at increasing distances, blurring the fine detail. The successive images are stacked one above the other, with the coarsest resolution at the top. A typical scale-space or multiresolution diagram might be a view of a remotely sensed scene with east and north as the horizontal coordinate axes. The vertical axis represents the 'scale', or resolution at which the objects are filtered, regarded as an additional dimension. Conceptually, the stack is continuous, but for practical reasons, the scale is incremented in a number of steps, typically about 100. Mathematically, the result of the Gaussian smoothing can be considered as the diffusion gradient of the grey-level intensity of an image as it diffuses up to coarser levels of the stack (the diffusion equation describes the physical process that equilibrates concentration differences without creating or destroying mass).

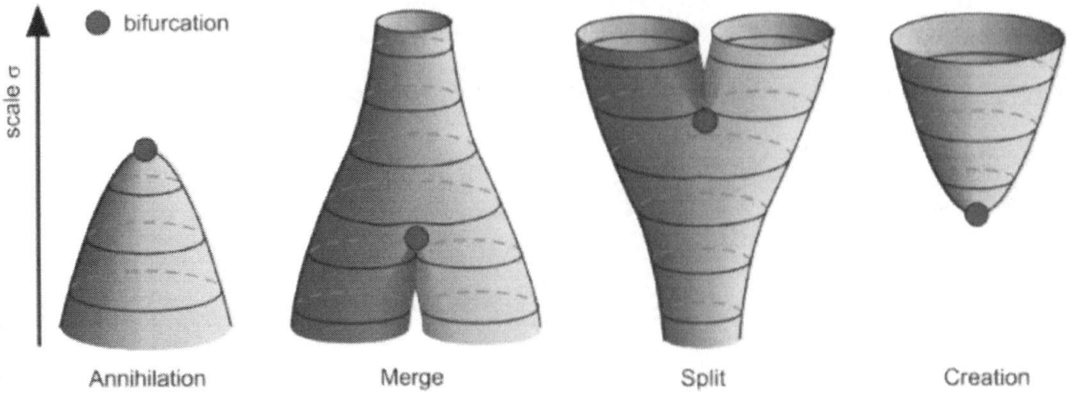

Figure 40: Topological elements in scale-space (From fig.10, Stewart et al., 2004)
(©Blackwell Publishing Ltd, from Basin Research, 16, 183-197.)

The stack shows how structures diffusively persist and change form over a range of scales. Stewart et al. (2004[391]) illustrate the appearance in scale-space (see *Figure 40*) of the topological events affecting blobs: *creation*, as a new blob appears; *annihilation*, as a blob disappears; *merging*, as two blobs merge into one; and *splitting*, as one blob splits into two. Each event occurs as a bifurcation point at a single pixel in scale space. In the earlier example, leaves might merge as a branch, the branches merge as a tree, and so on. The

[391] Stewart, S.A., Hay, G.J., Rosin, P.L., and Wynn, T.J., 2004. Multiscale structure in sedimentary basins. *Basin Research*, **16**, 183-197. doi: 10.1111/j.1365-2117.2004.00228.x
http://www.geog.umontreal.ca/gc/PDFs/New_PDFs/2004_stewart_Hay.pdf

three-dimensional nature of geological investigation and interpretation (with scale-space therefore of four dimensions), the range of sampling resolutions, the diversity of shape of geological objects, and the need to balance observation, interpretation, interpolation and generalisation complicate the application of these ideas to geoscience survey. Nevertheless, the concepts can be applied to generated 2D images and may prove helpful in the multiresolution approach that is typical of geological surveying, and in the spatial models representing the results.

Multiresolution survey

<<Seeking shared concepts 247

Summary: *Despite generalisation to several scales, ambiguities arise because geological maps seldom differentiate resolution from scale. Spatial models are more flexible, and scale-space concepts can clarify multi-resolution observations. The chain of reasoning may identify features of salient importance for visualisation in a particular context, regardless of their size. Interpretation of the outcome of geological processes and their interactions at various scales should determine the interpolation methods. The interpreted results could be located within scale space and generalised where required by Gaussian blurring.*

The model (unlike the map) can address scale-space issues, which are intrinsic in geological thinking and should therefore be considered in the system design. It may be helpful to think about them here in the context of a spatial model, using a scale-space diagram (see Scale-space 255) of the 3D geology, not as a working tool but as an informal concept to clarify ideas. The difficulty of visualising four dimensions can be overcome by thinking of various separate 2D projections of the geology with scale as the third dimension, or by using time to represent the scale dimension and imagining a time sequence of scale changes to the solid model.

Ecologists base their scale-space diagrams on the analysis of evenly spaced data, which are in short supply in geological survey. Initially, therefore, the concept may prove more useful for clarifying and recording the resolution or scale of observations; the range of scales over which a specific object or model is valid (see Zoom 247) and can be generalised; and the levels in scale-space at which alternative objects or models should be substituted. The concept also provides a basis for exploring, generalising, visualising and testing the interpretation. It provides a framework within which interpolation of unevenly spaced information, such as scattered well or borehole data, can be better understood. In due course, experiments with appropriately collected data from, say, thin sections or areas of good exposure, may throw light on how local processes behave and interact at different scales.

Observations during geological surveying vary in scale from microscopic analysis of a thin section to an overview of an entire landscape (see Zoom 247). A huge range of resolutions can therefore be recorded from observations at sparsely distributed rock exposures of varying extent. The importance of an observation does not depend on its extent or resolution. One thin section might provide a vital clue to the level of metamorphism; an ash horizon a few millimetres thick or a small exposure of a major unconformity might be the key to understanding the stratigraphy; a single microfossil might define the environment of deposition. Generalisation by blurring would obscure the regional significance of these salient features. The importance of a feature, however, depends on the objectives of the visualisation (see Abstracting from reality to model 131). For example, a geophysicist matching gravity anomalies to the underlying geology might have little interest in the microfossil. The chain of reasoning (see Microdocuments and the threads of reasoning 190) should therefore record the contexts in which the item has salient importance. References in the metadata could then ensure that, regardless of size, the item would be identified and annotated in visualisations, but only where the context required it.

In scale-space, one might expect to find, say, observations of features of individual grains or crystals and their relationships, mineralogical and then petrographical interpretations, at the most detailed level. Such information is sparse and local, but could be accessible through a spatial index (BGS, 2010a[392]). Conceptually, the items are likely to be related to objects at higher levels, and could be connected to them by hypertext links, such as threads of reasoning. Ascending the stack to levels of coarser detail, these items might give way to lithological properties in sedimentary structures of various sizes, then to complete beds, facies and informal groupings, which, as the filters become increasingly coarse, might fall into the familiar sequence of stratigraphical classification. At some level in this sequence corresponding to the 'survey scale', a space-filling interpretation might be attempted, forming the base level of the main scale-space stack (as opposed to local stacks referring to detail at, say, outcrop level). The objects at various ranges of scale have formed through the operation of distinct scale-dependent processes. The interpretation of each may therefore lead to different methods of interpolation.

Even within a single object, a wide range of spatial properties and interpreted processes may be encountered (see Zoom 247). For example, various processes of erosion and deposition might shape a formation top (regarded as an object) at all scales with features ranging from a river delta to a small ripple mark or a wind-sculpted sand grain. The surface may subsequently be deformed by regional tectonics at a coarse scale, and by microfolds and faults at a more detailed scale. To complicate matters, it may be necessary to study such a surface on the basis of unevenly spaced data points at various resolutions. Slope variograms (see Unevenly spaced data 238) at various resolutions might throw some light on the geometrical characteristics of the formation top, but careful examination of the surface, or similar surfaces, at outcrop, could lead to more convincing conclusions. Coarse and fine scale features do not naturally combine in one mathematical equation. Even with local fitting (see Unevenly spaced data 238) a single method is unlikely to be appropriate with scattered data. In scale-space, therefore, the single object at one resolution may have to be regarded on more detailed examination as a composite of several separate spatial objects (such as sand grains, ripple marks, channels, local depositional forms, formations). The number of separate objects depends on the resolutions of observation, interpretation and visualisation rather than the spacing of the data points. The smaller objects may be observed and

---

[392] BGS, 2010a. GeoIndex. http://www.bgs.ac.uk/geoindex

interpreted only locally and sporadically, probably included in the reasoning, but not represented in maps or visualisations at survey scale.

Geological decisions are needed to guide the procedures for interpolation. The sedimentary features may predate the structural ones. It could be appropriate therefore to interpolate their properties with axes referring to the bedding rather than to a horizontal plane. Features interact in various ways, perhaps reflecting different processes or separate episodes in the time sequence of their formation. For example, the large folds might have refolded the earlier small folds, or both sets might have formed simultaneously under the same stress field. These relationships are probably apparent in the field. They could be recorded as metadata, as their consequences for the model extend beyond specific observations and apply over a much wider area. More generally, metadata about the behaviour, constraints and relationships of object classes, object instances, and processes may result from direct observation in the field. In turn, a better understanding of the characteristics of the products of complex systems may prompt some informative field observations, complementing the results of experiments and subsurface geoscience studies such as 3D seismic surveys.

In general, interpolation should reflect interpretations that are based on geological considerations and should take into account the effects of scale, including the superposition of patterns at different scales. Areas with inadequate data are still affected by fine-scale processes, even if their consequences were not directly observed. This could be made clear in a visualisation by depicting the uncertainty or simulating a possible configuration (see Spatial variation and uncertainty 241, Shape 266). Where information at a required scale is inadequate, the visualisation could indicate the availability of relevant information at other scales. The task of reconstructing the geology at various scales and visualising the results is considered in the section on Deformable models 274.

Map generalisation, by retaining salient features while reducing the scale, can be seen as a special case of abstraction (see Abstracting from reality to model 131). Abstraction leads not only to images at lower resolution, summaries, interpretations and explanations, but may also consider wider areas and provide a general account of the properties and behaviour of objects that can provide feedback for the processes of interpolation. Interpolation, on the other hand, builds a more complete picture (using the feedback gained by abstraction) to fill gaps between the limited observations. Interpolation based on geological interpretation helps to explore the validity and the consequences of the models, and to refine their predictions by testing them against the real world. The processes of interpolation and abstraction are essential aspects of geological field survey and are mutually dependent. They must therefore interface to work together through compatible representations. For example, the abstraction process should lead to metadata that describe the expected behaviour of an object in a standard form that can invoke the appropriate interpolation methods and parameters to create grids or multiresolution octrees (see Mark-up and metadata 175, Making a mesh 207) suited to the visualisation software.

The Gaussian kernel (see Scale-space 255) is widely used to blur images for generalisation. It could also be used for interpolation, as a weighting function in the moving average method (see Estimation by interpolation 212). It explicitly aims to avoid any commitment to a specific interpretation or any creation of spurious patterns. "The theory developed here is rather aimed at describing the principles of the very first stages of low-level processing in an uncommitted visual system aimed at handling a large class of different situations, and in which no or very little a priori information is available. Then, once initial hypotheses about the structure of the world have been generated within this framework, the intention is that it should be possible to invoke more refined processing, which can compensate for this, and adapt to current situation and the task at hand" (Lindeberg, 1994a[393]).

The application of the Gaussian kernel to interpolation is therefore limited to areas where background knowledge is lacking, and its application to generalisation of evenly spaced observational data can seldom be used in geological survey. However, interpolation procedures based on geological interpretation as described earlier (see DSIs, FEMs and their geometrical significance 235, The geometry of interpolation 243) yield values at evenly spaced points. Subsequent generalisation of these interpolated values should avoid additional assumptions or the creation of spurious patterns. The Gaussian kernel therefore has a possible role in changing the scale of an interpretation for visualisation, within the appropriate range of scale-space.

Scale-space theory could throw light on some spatial properties of the interpreted view of an object. The aim is to understand the object's spatial properties and relationships as a whole, not as a single-scale image but across the full range of scales where they apply. For the geologist, there is nothing new in looking, say, at how sand grains combine as small structures that knit together as a bed, which interleaves with others to form a larger unit, how these stratigraphical units form in rhythmic sequence, exhibit onlap, offlap, and so on. However, building the concepts into the framework of an explicit spatial model (which can then be visualised) is less familiar.

Computer methods can detect objects that persist across a range of scales. The 'natural scale', or the scale showing the strongest expression of a multi-scale object, can also be calculated. Methods are available to clarify whether apparent features are likely to be real or to have arisen by chance. Hay et al. (2002[394]) and Lindeberg, (1996[395]) point to details of their implementation. However, the concern of this scenario is merely to consider the potential relevance of the methods to geological survey.

Marr points out (1982[396], page 70) that if objects at different resolutions coincide, they suggest a single physical phenomenon. Blobs and edges that persist over a range of scales are likely to indicate significant objects and, not surprisingly, human vision has evolved to

---

[393] Lindeberg, T., 1994a. Scale-space theory: a basic tool for analysing structures at different scales. *Journal of Applied Statistics*, 21 (2) 225-270. http://www.nada.kth.se/~tony/abstracts/Lin94-SI-abstract.html

[394] Hay, G.J., Dubé, P., Bouchard, A., Marceau, D.J., 2002. A scale-space primer for exploring and quantifying complex landscapes. *Ecological Modelling*, **153** (1-2), 27-49. http://www.sciencedirect.com/science/article/pii/S0304380001005002

[395] Lindeberg, T., 1996. Scale-space: a framework for handling image structure at multiple scales. *In Proc. CERN School of Computing, Egmond aan Zee, The Netherlands, 8-21 September 1996.* http://www.nada.kth.se/cvap/abstracts/lin96-csc.html

[396] Marr, D., 1982. *Vision: a computational investigation into the human representation and processing of visual information.* W.H. Freeman, New York. 397pp.

help us to detect them. Obscuring parts of an image with coarse rectangular pixels therefore confuses the eye, as conflicting information is sent at different scales about the same structure. On nearly closing the eyes, the larger edges are less apparent, revealing more of the underlying image.

The visualisation of a geological spatial model can likewise confuse the eye where patterns from, say, sedimentation and tectonics or conflicting patterns of folding from several episodes at different scales, are inextricably combined in one image. This may be inevitable in a map representation. However, the patterns may have been analysed separately in the field (see The multifaceted model 297), and the confusion arises only when they are later combined as a single image. Spatial models might encourage the development of separate sub-models that can either be viewed separately to understand the geology, or combined to show its disposition on the ground.

The interpretation of each object involves processes that apply over a limited, though maybe wide, range of scales. Generalisation would only be valid within that range, which should therefore be specified in the metadata, along with the likely consequences of passing its boundaries, such as annihilation, splitting, merging, or creation of a new object (see Scale-space 255).

The basic 3D Gaussian kernel has no preferred direction. However, "it can be advantageous to use filters that correspond to different scale values along different directions" (Lindeberg, 1994a). With sedimentary rocks, for example, the properties of the rock are likely to vary most rapidly normal to the bedding. There is therefore a case for analysis in terms of *canonical coordinates*, that is, based on a frame of reference uniquely determined by the object's internal spatial properties.

A geometrical transformation (see Geometrical transformations 228) is the means of moving from geographical coordinates to canonical coordinates and vice versa. The geologist who exaggerates the vertical scale of a cross-section is adopting such an approach. The results of these procedures in geographical space might be flattened blobs, compressed normal to the bedding and more closely resembling the shape of sedimentary bodies. The coarsely defined blobs may have a role in depicting vague forms or rock bodies of uncertain shape. But neither geological analysis nor human vision is limited to multiresolution blobs in determining configuration, shape and form; nor is the family of Gaussian kernels (see Boundaries: discontinuities and zero-crossings 264).

Implementation note: One purpose of this scenario is to consider the potential value of new techniques. Some will extend the scope of computer methods. But for many tasks, well-established methods, such as generalisation by sub-sampling or interpolation by splining, are simpler and may well be fit for purpose.

Boundaries: discontinuities and zero-crossings

<<Seeking shared concepts 247

Summary: *Discontinuities of assemblages of properties, in terms of position, slope and curvature at all scales (such as faults, hinge-lines and stratigraphical or facies boundaries) are fundamental to the interpretation of a geological system. They mark an important type of boundary. More generally, boundaries mark change, and the type of change can be identified and located at specific scales by zero-crossings (where the value of a property changes sign) using derivatives of the Gaussian kernel.*

Discontinuities (see Continuity, fractals, octrees and wavelets 217) in the geological record can result from feedback mechanisms (see Complex and emergent systems 159). For example, if movement along a fault surface created a zone of weakness, the feedback would cause the fault to move subsequently and repeatedly along or near to the same surface, relieving the slow build-up of stress by localised release, rather than dissipating it evenly across the area. Similarly, the initiation of a fold may alter the strength characteristics of the rock, giving positive feedback that results in the fold increasing in size, until negative feedback from larger-scale influences, such as confining rigid beds or a changing stress pattern, intervenes to limit its development. Likewise, feedback effects in the interaction of processes of deposition, consolidation and erosion create discontinuities separating individual structures and beds, and introduce unconformities and disconformities at all scales. In general, processes interact at a wide range of scales within the systems and processes that operated throughout geological history. For example, larger scale patterns of sedimentation interact with processes such as basin subsidence, mountain uplift, erosion and transportation, sea-level changes and dissipation of wave energy. Events, such as an episode of sea-level change or faulting might cause breaks in the depositional process, or might occur later, affecting the deposited sediment and causing discontinuities of another kind.

Discontinuities are widely relevant in geological surveying (Gillespie et al., 2011[397]). Spatial discontinuities, in position, slope or curvature, such as faults and hinge-lines, are fundamental to the interpretation of a geological surface. Discontinuities in composition or properties such as bedding and stratigraphical boundaries are of obvious interest, and lateral discontinuities within a rock body might suggest an ancient shoreline or a facies boundary. However, the boundaries surveyed in the field and shown on geological maps do not necessarily mark discontinuities. They do indicate change. From a systems viewpoint, it is helpful to define the type of change.

The concept of *zero-crossing* helps to detect, identify, represent and analyse points and lines of change in configuration, form and shape (Marr, 1982[398]). When scanning a line across the representation of a property of a surface, points may be encountered where the property

---

[397] Gillespie, M.R., Barnes, R.P., Milodowski, A.E., 2011. British Geological Survey scheme for classifying discontinuities and fillings. *British Geological Survey Research Report,* RR/10/05. 56pp. ISBN 978 85272 674 7 http://nora.nerc.ac.uk/13986/

[398] Marr, D., 1982. *Vision: a computational investigation into the human representation and processing of visual information.* W.H. Freeman, New York. 397pp.

has a value of zero, as the scan crosses from positive values on one side of the point to negative on the other, or vice versa. They are referred to as zero-crossing points. For example, on successive scans across a digital terrain model, points might be detected where the elevation was zero, separating positive values (above sea level) on one side from negative values (below sea level) on the other. The line joining the zero-crossing points marks the coast.

If the elevation model is replaced by its first derivative, it no longer shows the elevation, but instead the rate of change of elevation, namely slope, with a value of zero where the ground is flat. Zero-crossings would occur at ridges or valleys where upward slopes along the scan line give way to downward slopes. The zero-crossings on a folded bed would indicate the highest and lowest points of folds. The second derivative of the surface would show the rate of change of slope (akin to the curvature of the surface). The zero-crossings of scan lines would be the points of inflection where concave and convex parts of the surface meet.

Ramsay (1967[399], page 347), used zero-crossings of second derivatives of elevation in his work on structural geology. On a fold profile "there are generally points of inflection where the rate of change of slope is zero... The points where inflections occur are the limits of individual folds in the cross section, and the lines joining these points in adjacent profiles are lines of inflection delimiting the separate folds in the surface." Stewart et al. (2004) determine the scale of folds delimited by zero-crossings. They state (page 19) that: "The examples presented in this paper strongly indicate that whether a fold is perceived or not depends entirely on the scale of observation, so a measure of fold scale should always accompany classification that is otherwise based on fold style or geometry."

Third and higher order derivatives are seldom considered in analysing surfaces or processes, or in studying vision, as Marr (1982[400]) pointed out. "Choice of the second derivative as the cutoff point rests on the empirical observation of car designers that customers [in the showroom] notice discontinuities in the first and second derivatives of a surface but not in the third." The same presumably applies to geological observations in the field.

One attraction of the Gaussian kernel for detecting and generalising blobs (see Scale-space 255) is its membership of a larger family of functions that can detect other shapes, such as ridges, corners, lines and edges (Lindeberg, 1996[401]). The kernels of interest in the larger family are derivatives of the Gaussian function (*Figure 39*), including its first and second derivatives, and the Laplacian (a second order differential operator which is the sum of the unmixed second order derivatives and is invariant on rotation). The significance of the Laplacian is that, whereas other derivatives detect lines of change in a predetermined orientation, the Laplacian picks out points where there is sudden change in any direction. It thus detects zero-crossings with no bias for location, orientation or scale.

---

399 Ramsay, J.G., 1967. *Folding and fracturing of rocks.* McGraw-Hill, New York. 568pp.

400 Marr, D., 1982. *Vision: a computational investigation into the human representation and processing of visual information.* W.H. Freeman, New York. 397pp.

401 Lindeberg, T., 1996. Scale-space: a framework for handling image structure at multiple scales. *In Proc. CERN School of Computing, Egmond aan Zee, The Netherlands, 8-21 September 1996.* http://www.nada.kth.se/cvap/abstracts/lin96-csc.html

When looking at an image, the human eye, trained by a lifetime's experience, readily detects blobs, edges, corners and ridges. Marr (1982) considers the larger family of Gaussian kernels as basic mechanisms for human vision. He points out that the derivatives can be approximated by the difference between two Gaussian kernels that have a slight lateral displacement between them. He speculated that such a mechanism may be built into the system of optic nerves, allowing them to act as edge-detector cells. Just as receptors may be specialised to enable us to see blobs of different sizes, so they may also detect edges of different lengths and directions, which he saw as the basic elements of human visual recognition. And just as the computer can convolve Gaussian kernels of different sizes to detect blobs at different scales (see Scale-space 255), so it can also convolve Laplacians of different sizes to detect zero-crossing bands at different scales for representation in scale-space (bands, because in geometry lines have no thickness and therefore do not scale).

Boundaries, such as faults or unconformities, can be regarded as objects, things of geological interest in their own right. As with blobs, the scale-space analysis of discontinuities is likely to apply to interpretations based on field observations and to their subsequent generalisation, rather than directly to the observations, as these are unlikely to be evenly spaced (see Multiresolution survey 259). In contrast to applications of these techniques in many other fields, their geological applications focus more on interpretation than on data analysis. The system aims for an interactive approach, in order to gain the advantages of complex human reasoning and background knowledge, and combine the best features of mind and machine. The computer analysis and visualisation should therefore be consistent with the geologists' methods of field observation and interpretation of past processes and configurations. The concept of Shape 266 has an important role in human reasoning and in constructing the digital spatial model.

Shape

<<Seeking shared concepts 247

Summary: *The shape of an object refers to its form or structure, as opposed to its content or significance. It describes invariant spatial properties that are independent of where the object is, how it is positioned, how big it is, or how long it has been there. Statistical analysis of shape may refer to characteristics such as slope or curvature, and may be used to compare, combine and analyse information from many sources.*

In studying, say, the outcrop of a particular formation, field geologists might observe, sample, estimate or measure, and record values of spatial properties, such as location, extent, orientation, curvature, size, shape, texture, pattern, resolution, relationships and arrangement. They might follow rigid, predetermined investigational and sampling designs, perhaps leading to rigorous statistical analysis aiming to confirm a pre-existing hypothesis or interpretation. More likely they would follow a flexible exploratory course, responding to what they learn by continually extending their interpretation, testing hunches, and adjusting their hypotheses and procedures as the investigation proceeds. They might study patterns in the field, subdividing the exposed rock bodies into relatively homogeneous areas or patches (see Grain, set and patch 251). Through examination of their properties, they might picture immanent processes and historical configurations (see Abstracting from reality to model 131) that could have created the outcome they see today. To do this, they might examine spatial characteristics like shape or pattern that originated in the past and persist to the present day as properties unaffected by rigid-body transformations (see Invariant properties and classification 232).

The *shape* of an object may be regarded as an aspect of its form or structure, as opposed to its content (its substance and significance). Shape is an important spatial characteristic in our human perception of the world around us. We take it into account in everyday classification, recognition and comparison (that must be a yam, it's the wrong shape for a potato; I remember your face but not your name; she looks just like you). These subtle skills are carried through to geological surveying where equivalent remarks might be made about, say, rock types, fossils, folds or sedimentary structures.

Shape does not depend on where the object is, how big it is, how it is positioned and oriented, or how long it has been there. This time and space invariance of shape is the basis of much geological spatial reasoning. It enables us to see the present as a key to the past and to compare geological features wherever they occur, however they are positioned, and whatever their size (see Invariance and processes 56). Spatial invariance thus has an essential role in abstracting spatial information from objects as general statements for feedback (see Abstracting from reality to model 131). It leads to wide-ranging insights that create expectations about likely spatial properties and suggest to the surveyor what to look for next.

The spatial insights might lead to comparisons with, say, shape features characteristic of the type section of a formation (for identification or classification); or those recognised at nearby exposures or elsewhere (for stratigraphical correlation); or similarities with features on a larger or smaller scale (for understanding processes and properties). They might link the observations to hypothetical historical configurations of geological entities, events and processes, as previously interpreted from other evidence or as suggested by the observations themselves. Interpretation and observation thus develop together, each throwing light on the other. The insights encompass both object instances and classes (see Object instances and classes 181). Thus features might be regarded as specific to the object instance, or as typical of the object class, or as inherited from object classes at higher hierarchical levels.

Observations in the field intertwine with an underlying geoscience spatial model (see The imperfect model 156, The general geoscience spatial model 293, The solid Earth systems model (sEsm) 71) that represents our understanding of the properties of the rocks, their distribution and relationships, and the processes and configurations of their development and history. Concepts of the form and patterns created by geological processes, and how they relate to what is observed, develop in the geologist's mind, perhaps expressed in field sketches. For example, the pattern and shape of ripple marks on a sandstone bed might bring to mind the effect of waves and wind at the sea margin or in flume or wind tunnel experiments at the present day. This in turn might suggest other characteristics that could be looked for, with implications for other spatial properties and the place of the sandstone bed in the wider configuration indicated by historical geology. Combining such information from a diversity of sources requires abstraction procedures that generalise the information and distil its essence for the feedback procedures that recycle the knowledge gained in the field (see Abstracting from reality to model 131). Shape concepts play a vital part.

If shape properties can be made explicit and included in a computer analysis, then spatial observations might be more rigorously integrated with their interpretation and the associated reasoning. Geologists may take abstraction and feedback for granted, but computer implementation requires a more mechanical view of how spatial knowledge from different sources and separate objects can be combined. The abstraction procedures depend on the mathematical concepts of geometrical transformations (see Geometrical transformations 228, Invariant properties and classification 232) and invariant properties (such as shape) that can be measured, analysed and integrated.

In their rigorous mathematical account, Kendall et al. (1999[402]) explain that shape analysis has its own mathematical procedures because "classical statistical methods are not always adequate or, at least, not clearly appropriate for the statistical analysis of shape and it is necessary to adapt them to work on unfamiliar spaces." Dryden and Mardia (1998[403]), in an account somewhat more accessible to the non-mathematician, also describe statistical methods for shape analysis. They define shape as "all the geometrical information that remains when location, scale and rotational effects are filtered out from an object. So, an object's shape is invariant under the Euclidean similarity transformations of translation, scaling and rotation". Because of this invariance, shape characteristics can be used in comparing analogous features formed at different times and different places, including those from conceptual models or experiment. Shape properties of objects from many sources can be compared, combined and analysed (see Morphometrics 268). In a sense, they are intrinsic features of the objects that may relate to their origin, history and significant properties, and contribute to the reasoning process in geology.

Morphometrics

<< Seeking shared concepts 247

Summary: *Various techniques for spatial interpolation, classification, recognition, comparison, abstraction, reasoning and simulation can be applied to spatial objects in geology. An appropriate next step is the integrated spatial analysis of configurations of objects and their relationships, which involves the concepts of morphometrics. Landmarks (useful markers or spatial reference points) and outlines (that define the forms of boundaries between patches) are a basis for superimposing, comparing and summarising the shape characteristics of two or more objects. Shape interpolation or morphing visualises*

---

402 Kendall, D.G., Barden, D. Carne, T.K., Le, H., 1999. *Shape and Shape Theory*, John Wiley & Sons, New York. 306 pp.

403 Dryden, I.L., and Mardia, K.V., 1998. *Statistical Shape Analysis.* John Wiley & Sons, New York. 347 pp.

intermediate shapes between objects, and can be linked to measures of shape change of entire objects and the finer-scale components within them.

The concepts of scale-space analysis (see Scale-space 255) provide a means of describing the location and form of a wide diversity of spatial objects. Zero-crossings are a means of detecting and defining their boundaries. The spatial relationships between objects are a means of clarifying their arrangement and interactions, such as 'adjacency' to ensure that space is filled appropriately and no voids are accidentally left in the model of the Earth's crust. Individual spatial objects, such as patches or blobs, boundaries, edges, corners and ridges, can be combined as higher level geological objects in a hierarchical sequence. They lead on naturally from analysis of the individual objects to analysis of their configuration. The *configuration* of objects, that is their relative spatial arrangement, relationships and interactions, is a key to understanding the geology. It can integrate the individual objects as a higher level object in the hierarchy.

In d'Arcy Thompson's comparisons of biological form (see Geometrical transformations 228), he retained the topological configuration of the skeletons he compared, and studied the geometrical transformations needed to match the two specimens. His successors in the field of measurement of shape variation and its relationships to other variables, now termed *morphometrics*, usually identify and compare shape characteristics by means of *landmarks* (useful markers or spatial reference points) and outlines (that define the forms of boundaries between patches). These are geometrical structures, namely points or outlines such as zero-crossings (see Boundaries: discontinuities and zero-crossings 264), within the objects being compared, which are thought to correspond in each of the objects. In biological applications, they may be loosely referred to as *homologous*, implying that they have a consistent biological or biomechanical significance in the processes under consideration. They provide a spatial framework or configuration within which shape characteristics can be compared, and variation of shape can be studied.

In geology, discontinuities may define landmarks and outlines that are topologically invariant (see Geometrical transformations 228, Invariance and processes 56) and have a degree of permanence in geological time. For example, the sandstone beds on either side of a fault would not change sides during later tectonic or other processes, and the strata within a fault block would remain within the fault block during subsequent deformation. Discontinuities, such as faults, may define patch boundaries (see Grain, set and patch 251), and may be helpful in defining landmarks and outlines. Both may be required as points and lines of correspondence when comparing different objects and in the analysis of shape (see Shape 266) and the overall interpretation of the configuration. Examples are a comparison of the shapes of a set of sand dunes or of folds, or comparison of the form of a sand body at the present day with its palinspastic reconstruction at the time of deposition, or reconstruction of the form of a sand body from segments that have been split apart by faults and folded separately, or of course the comparison of shapes of a number of fossils that are thought to belong to the same species.

The so-called *Procrustes methods*, which minimise the sum of squares of distances between corresponding landmarks, are widely used for superimposing, comparing and summarising the shape characteristics of two or more objects (Adams et al., 2004[404]). The procedure might be to translate the centroids of the objects being compared to the same origin; scale the objects to the same unit size; and rotate them to minimise the sum of squared distances between corresponding landmarks. Unlike earlier approaches that applied statistical methods to measurements of characteristics such as length or width of selected features, they apply the spatial transformations to the entire object, aiming to preserve the geometrical relationships of the structures as a whole.

Richtsmeier et al. (2002[405]) provide an accessible account of the conceptual issues, stressing the importance of the appropriateness of the selected method and its validity (the ability of the method to find the correct answer). They define their model as "a mathematical construct that attempts to characterize certain aspects of the underlying phenomena (e.g., dimensions, dynamics, properties, interactions)... A model is formulated using statistical expertise and intuition, based on whatever previous experience and knowledge the scientist may have. Once a model is formulated, data are used to determine those parameters of the model that are most compatible with the observations" (page 70).

They see deficiencies in the Procrustes model, and argue that the analysis should start from data that are invariant under the required transformations, rather than filtering out the effects of translation, scaling and rotation during subsequent analysis. "Answers based on information that can be known from the data are of more use to biological inquiry than those based on unjustifiable assumptions" (page 63). They consider that the translation, scaling and rotation of the Procrustes model are *nuisance parameters*, that is, they are not of immediate interest but in this model have to be accounted for in order to analyse the shape parameters. Inconveniently, they cannot be accurately estimated as the data contain random variation. They advocate an alternative model based on a matrix of relative distances between landmarks within each object. The comparisons are made between the matrices, and the transformations applied to them.

Shape interpolation, using *morphing* or *blending* techniques, creates a smooth transition from the shape of an initial object to that of a target object, based on landmarks that correspond in the two objects. Morphing has found applications in such fields as: computer vision; 3D modelling and animation for visual effects in film, television and computer games; and in object reconstruction from medical imagery (for example, in predicting the growth pattern of a child's jaw to guide orthodontic procedures). The shapes can be represented in many ways. One possible approach is using implicit surfaces (Turk and O'Brien, 2002[406]) to represent the boundaries of a solid object. The boundaries are interpolated from a set of points, each known to be on the boundary of the object or to be within or outside it. The interpolated object should contain all the points known to lie inside it, and none of those outside. In addition, orientations can optionally be specified.

---

[404] Adams, D.C., Rohlf, F.J., Slice, D.E., 2004. Geometric morphometrics: ten years of progress following the 'revolution'. *Italian Journal of Zoology, **71**:5-16.* http://life.bio.sunysb.edu/morph/review/review.html

[405] Richtsmeier, J.T., Deleon, V.B. and Lele, S.R., 2002. The promise of geometric morphometrics. *Yearbook of physical anthropology,* 45:63-91, 63-91. www.hopkinsmedicine.org/FAE/JTRVBDLS2002YPA.pdf

[406] Turk, G., O'Brien, J.F., 2002. *Modelling with implicit surfaces that interpolate*. ACM Transactions on Graphics **21** (4), 855-873. http://www.cc.gatech.edu/~turk/pubs.html

Systematic changes of shape are also of interest in *allometry*, which studies biological relationships where shape changes with size. This can lead to a joint study of size-and-shape or *form*, where shape on its own is not invariant under scaling. Alternatively, separate measures of size and shape can be developed and their relationships examined; a similar approach could be relevant to the analysis of self-affine (see Invariant properties and classification 232) geological features in scale-space (see Scale-space 255).

Morphometric methods have seldom been used in geology outside palaeontology and occasionally igneous petrology (Perugini et al., 2002[407]). Potentially, however, they might bring interactive computer support to the field geologist, based on spatial transformations equivalent to visualisation of processes of erosion, transport, deposition, intrusion, folding, and faulting (see Deformable models 274).

Most geological applications take a more intuitive and less rigorous view of shape. Geologists may think of shape properties as showing systematic change, rather than invariance, under translation. They may be concerned with variation of shape in space or time, as for example, in the shape of a fold that changes along its axis from a circular to a more angular cross-section, or a transition from deposition of beds of lenticular shape to younger beds of more uniform thickness. The stereograms of structural geology and crystallography are based on orientation measurements, which meet some invariance criteria and illustrate shape properties, such as symmetry and cylindrical or conical folding, although their invariance on moving through scale-space (see Scale-space 255) is questionable. The semi-variograms of geostatistics (see Estimation by interpolation 212, Spatial variation and uncertainty 241) reflect a shape property, namely the variability of elevation differences between two points on a surface and their relationship to the distance separating the points. Semi-variograms from different surfaces might therefore be compared and possibly combined. As with the Procrustes model, however, it may be better to start from invariant data. Even where these examples make geological sense, their computer analysis requires care. Its assumptions must match those of the data collection.

Feedback procedures can build on the abstraction process to fit characteristic forms to particular configurations, thus combining expectations of the shape with the location of significant points or boundaries defined by survey. Feedback of some shape characteristics may help to estimate elevations at unknown points in surface interpolation. For example, a surface that is thought to slope uniformly might be interpolated or contoured on a different basis from one known to cross a steep-sided reef. More generally, kriging feeds back information from the semi-variogram just mentioned to weight evidence, on the basis of distance from surrounding data points, in estimating an unknown point on a surface. In a sense, kriging is acting like a spatial filter, of a rather different type to the Gaussian filter described in Scale-space 255, and like any filter is liable to introduce unwanted artefacts into the interpolation.

Kriging quantifies the idea that nearby points are likely to have more similar elevations than those farther apart. More subtle control of interpolation might be achieved through the

---

[407] Perugini, D., Poli, G., Prosperini, N., 2002. Morphometric analysis of magmatic enclaves: a tool for understanding magma vesiculation and ascent. *Lithos*, **61,** 225-235.

geometrical representation of finite elements (see Unevenly spaced data 238), where slope and curvature estimates at known points could be weighted to estimate their values at an unknown point (see Drawing the line 209) or could be constrained by, say, principal directions detected by zero-crossings (see Boundaries: discontinuities and zero-crossings 264) or other aspects of the differential geometry of the configuration (see Representing wider knowledge 283). Such methods were initially based on simple shape characteristics such as slopes or curvatures, which can be summarised statistically (Watson, 1966[408]). Strebelle (2002[409]) shows how multiple-point statistics, based on three or more data points together (rather than just two as in the traditional variogram) can be carried through to the simulation model and anchored to the hard data. Similar procedures might be taken for granted in hand contouring, but as usual the computer methods call for more explicit reasoning (Strebelle and Levy, 2008[410]). The development of high-precision Global Positioning System field surveying (Pearce et al., 2006[411], McCaffrey et al., 2005[412], McCaffrey et al., 2008[413]) and 3D seismic subsurface data can provide detailed three-dimensional information at many scales (Stewart and Podolski, 1998[414]). Methods based on differential geometry can provide a framework for analysing the number of folds in an area, their arrangement and type based on curvature characteristics (Lisle and Toimil, 2007[415]). Such procedures call attention to the somewhat fuzzy boundary between interpolation and simulation.

Interpolation aims to calculate the likely values of a variable between known data points; simulation aims to show possible outcomes of partly random processes. If simple shape characteristics like the slope and curvature of a surface are known to vary little over appropriate areas, this fact could be helpful for interpolation (see Drawing the line 209). But more complicated patterns on the surface may reflect processes where random variations build on one another, as in emergent systems (see Complex and emergent systems 159). Although the resulting shapes may be predictable, the location of individual features may be

---

408 Watson, G.S., 1966. The statistics of orientation data. Journal of Geology, 74, 786-797.

409 Strebelle, S., 2002. Conditional simulation of complex geological structures using multiple-point statistics. *Mathematical geology*, **34** (1), 1-22. doi: 10.1023/A:1014009426274 http://www.springerlink.com/content/8g2meagu5k0u07pk/

410 Strebelle, S., Levy, M., 2008. Using multiple-point statistics to build geologically realistic reservoir models: the MPS/FDM workflow. Geological Society, London, Special Publication, **309**, 67-74. doi: 10.1144/SP309.5

411 Pearce, M.A., Jones, R.R., Smith, S.A.F., McCaffery, K.J.W., Clegg, P., 2006. Numerical analysis of fold curvature using data acquired by high-precision GPS. *Journal of Structural Geology*, **28**, pp. 1640-1646.

412 McCaffrey, K.J.W., Jones, R.R., Holdsworth, R.E., Wilson, R.W., Clegg, P., Imber, J., Holliman,N., Trinks, I., 2005. Unlocking the spatial dimension: Digital technologies and the future of geoscience fieldwork. Geological Society (London) Journal, **162**, 927-938. doi: 10.1144/0016-764905-017

413 McCaffrey, K.J.W., Feely, M., Hennessy, R., Thompson, J., 2008. Visualisation of folding in marble outcrops, Connemara, westen Ireland: an application of virtual outcrop technology. Geosphere, 4, 588-599. doi:10.1130/GES00147.1

414 Stewart, S.A., Podolski, R., 1998. Curvature analysis of gridded geological surfaces. *In* Coward, M.P., Dalbatan, T.S., Johnson, H., (eds) *Structural geology in reservoir characterisation*. Geological Society, London, Special Publications, **127**, pp. 133-147.

415 Lisle, R.J., Toimil, N.C., 2007. Defining folds on three-dimensional surfaces. *Geology*, **35** (6), pp. 519-522. doi: 10.1130/G23207A.1 http://geology.geoscienceworld.org/cgi/reprint/35/6/519

largely unknown. However, where the overall shape characteristics of a surface are known, it may be possible to simulate its form and anchor that representation to known data points.

The procedures are akin to the simulation methods described by Mallet (2002[416]), see Spatial variation and uncertainty 241. The simulation may illustrate important spatial properties and provide realistic views of some aspects of the geology, such as the likely form, variability and appearance of a surface. It might call attention to shape features of interest, such as the possible occurrence of anticlines of an interesting size in a particular area. The shape characteristics may also help in estimating the probability envelope (see Spatial variation and uncertainty 241). But if the true locations of the features showing shape characteristics are unknown, the simulation is inappropriate for estimating elevations at unknown points. As always, the appropriate visualisation model depends on the users' objectives.

Many geological applications rely on human perceptions of shape that cannot readily be fitted into a rigorous mathematical structure. Landmarks may prove difficult to define or relate between geological objects, and some shape properties, such as the subtle features of dendritic stream patterns, are difficult to quantify geometrically, although Zhang et al. (2005[417]) show how they can be simulated. The human mind is skilled at visualising pattern, and adjusting it to fill gaps between observations. In these circumstances, computer methods might be employed, not to derive a solution, but as a tool to assist geologists in manipulating and visualising the results of their interpretation.

Geologists, however, are not just concerned with shape, but draw on all types of spatial information, including location, scale, and configuration, to reconstruct and represent their interpretation of the geology. They think in terms, not just of surfaces, but also of three-dimensional objects at many scales. For example, the rocks within a fault block might at one level be considered as a single object that was folded as a unit. At a more detailed level, smaller objects such as individual beds within the fault block might be subject to folding on a smaller scale. Computer methods, stemming from morphometrics, can incorporate the interpolation of boundary surfaces and subsidiary internal surfaces into the transformation of the geometric properties of the object as a whole (see Morphometrics 268). The same approach can link spatial objects between sub-models of the geology (see The multifaceted model 297), for example, by carrying an object created in a depositional model through to a structural model to analyse the effects of folding and faulting.

---

416 Mallet, J.-L., 2002. *Geomodeling.* Oxford University Press, Oxford. 599pp.

417 Zhang, T., Switzer, P., Journel, A., 2005. Merging prior structural interpretation and local data: the Bayes updating of multiple-point statistics. *Proceedings of IAMG '05: GIS and spatial analysis,* **1** 615-620.

Deformable models

<< Seeking shared concepts 247

Summary: *Spatial modelling is a global activity, sharing developments from fields such as biomedical image analysis, including deformable models. These can offer an intuitive, interactive, probabilistic approach to multiresolution spatial analysis of objects and their evolution through time. The model quantifies changes in location, size, and shape of objects in terms of spatial transformations of an inhomogeneous, elastic object responding to external forces. In a geological context, deformable models could bring together ideas from interpolation, generalisation, shape and multiresolution analysis, as an aid to geological visualisation and reasoning.*

Digital cartography became a practical and economic success in geological surveys only when effective general-purpose cartographical systems became commercially available. These were designed to meet a worldwide requirement in fields such as topographical mapping and remote sensing. Geological applications, built on the foundation of years of painstaking experimental work that helped to determine how analogous systems applied in the geological context, could then be adapted to fit the more general solution. Similar comments apply to the development of geological databases. Now, in other fields with an extensive literature unnoticed by many geologists, major developments in spatial modelling may point the way ahead for global systems that will include geoscience spatial models.

Non-invasive medical techniques, such as magnetic resonance imaging and positron emission tomography (MRI and PET), drove advances in the processing of biomedical images. Three-dimensional models were developed, based on earlier work in computer graphics, remote sensing imagery, computer vision, pattern recognition and morphometrics. Models of biomedical systems and sub-systems adopt a different terminology from their geological counterparts, but they have much in common.

Both are concerned with the configuration, connectivity and spatial relationships of the objects involved; the movement of fluids; erosion, transportation and deposition of solid material; heat flow; changes in configuration through time; the description of characteristic shapes and forms; and deviations from them. Background knowledge plays a large part, and information from many different sources is reconciled and combined in order to reconstruct the objects and the system. Categories of frequently occurring assemblages of properties, structures and growth patterns are classified, identified and named. Interpretations of characteristics or deviations from them are sought in terms of function, origin, and developmental history, and abnormalities or anomalies explained in terms of provenance, environment and relationships with other objects.

In one promising approach to representing such systems, spatial transformations (see Geometrical transformations 228) act on deformable spatial models, where the geometric transformations relate to processes that can be readily visualised and understood. Transformations can be summarised geometrically and statistically. In such a prolific and fast evolving field, up-to-date work may be found through Web search engines or in the appropriate sections of the extensive bibliography of Price (2010[418]). The following examples may at least suggest some useful phrases for searching.

Atlases of the brain that show its detailed anatomy are apparently widely used in planning and performing surgical operations, providing detailed background information to compare with the patient's magnetic resonance imagery (MR) scan. Ganser et al. (2004[419]) prepared a version of a well-established brain atlas as a set of digital spatial models. They also provided software to modify the models. The image of each type specimen from the brain atlas is regarded as a deformable model, which can be adjusted to match the scans of an individual brain that might, for example, contain a tumour.

"Because there are differences between an atlas book and a MR image concerning dimensionality (2D plates versus 3D volume dataset), medium of representation (printed on paper versus digital display), and anatomic shape (standard anatomy versus individual brain), the information transfer from the atlas to the MR image has to happen solely in the mind of the surgeon. It is evident that this procedure stresses the physician's 3D imaginative capability very much, and it requires a long time experience to gain success. The usage of a book in the aseptic environment of the operating theatre is an additional problem which limits the application of an atlas to the preoperative planning stage. Another disadvantage of a printed atlas book is its finality, i.e. there is no way to correct errors and include new insights or further information... To overcome these problems, as neurosurgeons demand, we decided to develop a computerized atlas system...

• It shall include 3D reconstructed surface models of brain structures as well as the original atlas plates,
• it shall provide an easy-to-handle matching feature to adapt the atlas nonrigidly to individual brain images,
• it shall offer a powerful visualization to display atlas and MRI in joint views,
• it shall include additional information which extends the contents of the Talairach atlas; moreover it shall be open to further extensions, and
• it shall offer an interface to a navigation system in order to provide the atlas information intraoperatively." (Ganser et al., 2004, pages 3, 4)

Similarly, field geologists may have to work in a difficult environment, and struggle to relate the field evidence to idealised concepts of the outcome of geological processes set in their historical configuration and to visualise the full 3D consequences of tentative interpretative models. They might refer to textbook illustrations of typical examples of various geological objects and structures in the office, but have problems consulting them in the field. At the abstract level of geometry, the techniques of manipulating the biomedical spatial models seem widely relevant. Similar features available to a fully connected digital field notebook could therefore be helpful.

---

[418] Price, K., 2010. Annotated Computer Vision Bibliography. http://iris.usc.edu/Vision-Notes/bibliography/contents.html

[419] Ganser, K.A., Dickhaus, H., Metzner, R, Wirtz, C.R., 2004. A deformable digital brain atlas system according to Talairach and Tournoux. *Medical Image Analysis*, vol **8**(1), pp 3-22.

Methods of processing deformable models, stemming from computer vision and computer graphics, are widely used in medical image analysis. Montagnat et al. (2001[420]) review the mathematical aspects of deformable surfaces. Davatzikos (2001[421]) addresses a problem of comparing anatomical shapes using shape transformations. The unit of shape is a template (analogous to a geologist's type example). The differences of individual shapes from the template quantify the characteristics of the shape with respect to the template, and are measured by the shape transformations that map the template to the individual shape. Although different brains have different shapes, their underlying structure is similar and homologous points on the brain surface (which in geology might be defined by the zero-crossings of Boundaries: discontinuities and zero-crossings 264) are first brought into register.

Shape change, such as tumour growth or tissue loss due to ageing or disease, can occur anywhere in the interior of the structure (Zacharaki et al., 2008[422]). The transformation therefore treats the images as inhomogeneous elastic objects, and deforms them by external force fields until they are in registration with one another. The elastic properties vary from one region to another, allowing some regions to deform more readily than others, just as a sandstone bed may deform less than the surrounding shale, in response to, say, tectonic folding, igneous intrusion or consolidation. The elastic transformations tend to preserve the relative positions of anatomical structures, while being flexible enough to allow for variability between individuals. The key point for present purposes is that the method deforms the object as a whole, rather than as separate lines or surfaces. External forces (which might relate to adjacent objects) control the deformation. The internal variation of properties is described by, and is reflected in, the deformation.

In their comprehensive survey, McInerney and Terzopoulos (1996[423], page 92) state that: "The mathematical foundations of deformable models represent the confluence of geometry, physics, and approximation theory. Geometry serves to represent shape, physics imposes constraints on how the shape may vary over space and time, and optimal approximation provides the formal underpinnings of mechanisms for fitting the models to measured data... The physical interpretation views deformable models as elastic bodies which respond naturally to applied forces and constraints." They point to the ability of deformable models to combine top-down knowledge of the structures with bottom-up information from images; to their ability to accommodate considerable variability through time and between individuals; and to their support for highly interactive mechanisms that readily bring to bear the expertise of the human specialists.

---

[420] Montagnat, J., Delingette, H., Ayache, N., 2001. A review of deformable surfaces: topology, geometry and deformation. *Image and Computing Vision*, **19**, 1023-1040.

[421] Davatzikos, C., 2001. Measuring biological shape using geometry-based shape transformations. *Image and Vision Computing*, **19** (1-2), 63-74.

[422] Zacharaki, E.I., Hogea, C.S, Biros, G., Davatzikos, C., 2008. A Comparative Study of Biomechanical Simulators in Deformable Registration of Brain Tumor Images. *IEEE Transactions on Biomedical Engineering*, Volume 55, Issue 3, March 2008, pages 1233-1236.
http://repository.upenn.edu/cgi/viewcontent.cgi?article=1146&context=meam_papers

[423] McInerney, T., Terzopoulos, D., 1996. Deformable models in medical image analysis: a survey. *Medical Image Analysis*, **1** (2), 91-108. http://mrl.nyu.edu/~dt/papers/mia96/mia96.pdf

They describe how dynamic, deformable models can represent a broad range of shapes and are useful in a wide range of applications. They quantify not just static shape, but also shape evolution through time. They can accommodate uncertainty. They link to finite element methods. A multi-resolution strategy can proceed from coarse to fine detail, improving local similarity and global coherence. And they yield an accurate description of the object and quantitative information about it, in an intuitive, convenient form. Their properties make deformable models a possible candidate for the task of reconstructing the geology, which can be taken into account in Reconsidering geological mapping 277.

Reconsidering geological mapping

<< Seeking shared concepts 247

Summary: *Geological surveyors, by careful observations informed by background knowledge, can reconstruct a coherent view of local geology, covering many aspects, each with its own strand of reasoning, and all interacting. Computer methods introduce new methods, such as object-orientation, complex systems, multi-resolution analysis, and a more rigorous mathematical framework linking many sub-models. The methods augment, but should not displace, the relevant background knowledge of the experienced geologist. The mathematics and algorithms must therefore make sense in terms of the geology and human perceptions, while offering computer assistance in analysis, visualisation, modification and manipulation. The end-product should be an integrated view from which a range of representations can be generated by users to meet their diverse objectives.*

The conventional delivered products of geological surveying are a map and text explanation representing an interpretation of the geology of an area of interest. Earlier comments on conventional field mapping (see The surveyor's holistic view 60, Estimation by interpolation 212) refer to its top-down approach, stressing the importance of background knowledge in arriving at an accurate and useful geological map that is internally consistent, and consistent with the surrounding areas and with the global context. That background knowledge refers to the metadata that define our classification and terminology of geoscience objects; the processes that formed the objects and determined their properties; the behaviour of object classes and their specific instances in the processes in which they take part (see Object instances and classes 181); the historical configurations of objects, processes and events through geological time; as well as familiarity with the surface expressions of the geology and the techniques for studying it.

This implies constant interaction between the field survey model and the dynamic stratigraphy model along with its process and palinspastic models. The surveying procedure depends on background knowledge, but as it proceeds, more is learned of the local (and possibly global) dynamic stratigraphy. There is thus a constant two-way flow of knowledge between every aspect, amending models and metadata through the complementary processes of abstraction and feedback (see Abstracting from reality to model 131).

Geologists in the field simultaneously observe many different aspects of the geology, such as stratigraphy, sedimentation, structure, petrography, palaeontology, and igneous and metamorphic features. They have to keep in mind the interactions among all aspects but, in the field records, each aspect may be distinct, with its own strands of reasoning (see The multifaceted model 297). They are all subsequently assembled together as parts of a small set of published maps and related documents. An important objective of the geological survey of a particular area is thus, by careful observation informed by a wide range of background knowledge, to reconstruct a coherent view of the geology as a whole.

The rationale for the geological map extends naturally to the creation of the geoscience spatial model. The human mind is remarkably accomplished at comparing the spatial characteristics of two objects seen side-by-side, making full use of short-term and spatial memory (Loudon, 2000[424], part I, section 4). But mental comparison is much less effective for observations separated in time or space. The surveyor therefore describes, sketches and measures relevant features of the objects as field records for a geological map or a computer database, thereby transferring the salient points from short-term memory to an external record. The external records capture selected information for deferred side-by-side comparison, at a convenient scale for visualisation. Abstractions from many observations are thus available to the short-term memories of the originator or other users, as and when they are required.

Although the rationale remains the same, changing the emphasis – from publishing an interpretation as maps, cross-sections and map explanations to representing it within a computer-based knowledge system – has far-reaching consequences. Computer support within a knowledge-based system can enhance methods of reconstructing the geology. Ideas and operations implicit in geologists' thinking, such as abstraction, reasoning and feedback, can be made explicit by representation within the spatial model. Observations and interpretations can thereby be more widely integrated, shared, and opened to wider scrutiny, evaluation and enhancement. Because computer-aided methods can manipulate large amounts of information quickly and accurately, they can extend the geologists' model by introducing ideas such as object-oriented analysis (see The object-oriented perspective 179), complex systems (see Complex and emergent systems 159), and multiresolution analysis (see Multiresolution survey 259), which may augment field studies and lead to a better understanding of the definition, behaviour and relationships of the geological objects.

This implies a more rigorous mathematical framework, which can only be achieved by means of computer support, effective and reliable systems, and widespread adoption of suitable standards, all of which take time to develop. It is suggested (see Mapping geology into the knowledge system 282) that the geoscience knowledge system will evolve towards a set of interacting spatial sub-models that can be regarded as facets of the general geoscience model. The same geological objects participate in various facets of the general model, and in various metric spaces defined by the axes or dimensions along which aspects of their description are measured.

---

[424] Loudon, T.V., 2000. Geoscience after IT: a view of the present and future impact of information technology on geoscience. Elsevier, Oxford. 142 pp. Also available as *Computers & Geosciences*, Special Issue, **26** (3A), A1-A142 http://nora.nerc.ac.uk/2404/1/Part_I.pdf

A *metric space* is a collection of objects with a defined means of measuring the distance between them. The *distance* from object A to object B must be a real number, never negative, and zero only if the two objects coincide. The distance from A to B must be the same as that from B to A. The distance from A to C plus that from C to B cannot be less than that from A to B, regardless of the location of C. Distance may have its usual geographical sense (in which case the objects between which distances are measured would usually represent points) or may simply be a value measuring the dissimilarity or difference between the objects. Different spaces may have different properties that determine which mathematical methods, and therefore which computer processes, are appropriate.

The spaces of interest in the present context include: present-day two-dimensional space, as on a map; 3D geographical space, as in a spatial model; scale-space (see Scale-space 255), where an additional dimension refers to the resolution, showing the range of scales over which specific objects exist and processes operate; the spaces of dynamic stratigraphy that include geological time as a dimension; spaces in which the absolute locations of objects may be unknown, but their relative locations are defined by their spatial or time relationships; shape-space (see Shape 266) where objects and processes are unchanged by rigid-body transformations and uniform scaling; and the spaces of geo-reasoning, where geological reasoning links together ideas or memes (see Abstracting from reality to model 131) to develop explanations. The same object may be represented in several such spaces. For example: a sandstone body might be represented on a map and in a spatial model; its scaling and shape characteristics might be analysed in appropriate models; it might be placed in its stratigraphical setting by its relationships to other objects; and it might be connected to historical processes by the reasoning in a text explanation.

The objects within these spaces are also of various types. Some objects like blobs and patches (see Grain, set and patch 251) may refer to solid bodies; others like zero-crossings (see Boundaries: discontinuities and zero-crossings 264) refer to lines and surfaces, which may or may not define the boundaries of objects. All potentially refer to composite, multiresolution objects, made up of various component objects linked by their spatial relationships. Spatial relationships between the objects may locate their positions relative to other objects without quantitative measurement. For example, the relative position of formations in a stratigraphical sequence may be known (in space or geological time) through relationships such as above, below, before, after, without knowledge of their absolute ages or geographical coordinates.

Geologists without computer support seem to have few problems in handling these objects and spaces simultaneously in their minds, and give little thought to the complexity of their reasoning process. The human brain has unique abilities (see Stages of concept development (geological thinking) 56) to reconcile and synthesise disparate information, and an experienced geologist has relevant background knowledge beyond the potential of any machine. Computer methods must therefore be seen as augmenting, rather than displacing, the skills of the geologist. To achieve this, the mathematics, its visualisation, and the computer algorithms that implement them must make geological sense. They must follow the geologist's pattern of thought, be consistent with the geoscience reasoning process, and help to clarify that process and the consequences of adjusting aspects of the reasoning.

The geological interpretation is based on spatial concepts, which lead naturally to the more rigorous mathematical framework of geometry (see The geometry of interpolation 243). Computer processes translate the geometrical operations into algebra and statistics, but, for the geologist, the operations must still make sense in spatial (geometrical) terms, and should be readily visualised, with computer help as appropriate. The visualisation must match the pattern of human perception, in this case reflecting and extending the imagery of geological field observations and sketches and the more formalised visual representation of geological maps.

Ideally, the abstraction and feedback processes in the spatial model should also rest on geometrical summaries and generalisations that can be visualised, and preferably recognised and estimated visually in the field. Geologists would then be able to sketch an example of a generalisation or point out its characteristics in a suitable exposure. Examples are conical and cylindrical folding, and the typical shape of the results of a particular type of depositional process, such as a deltaic deposit. The computer could display a type example for comparison with field data, modifiable to match the local situation and specific observations. The system must be interactive to combine the strengths of the human brain and computer processing. In the field or office, the computer working model should respond interactively as geologists add new observations or amend their interpretation, enabling them to weigh up the consequences of each change, and decide how the survey can best proceed.

With such an approach, computer methods can offer greater flexibility in representing and manipulating the geologists' ideas. They should offer a dynamic multi-dimensional representation that can be manipulated mathematically, communicated instantly, modified interactively, and displayed for visualisation. This contrasts with current computer methods, which generally apply to selected, isolated, components of the surveying activity. A more comprehensive system will provide the flexibility to acquire information once, and adapt and reuse it as required, throughout the procedures of data collection, interpretation and presentation. In the longer run, the procedures of survey and the communication of the results will be transformed to become an active component of a global knowledge base. In the light of these requirements, we need to look again at The geometry of interpolation 243.

Widely used methods of interpolating geological surfaces tend to refer to single isolated surfaces. Their results are difficult to relate to geological processes such as erosion, transport, deposition, subsidence, melting, intrusion, faulting, folding, metamorphism and so on. Deformable models (see Morphometrics 268) based on spatial transformations might overcome these issues, but have drawbacks of their own. They compare an initial template with a target instance of an object, and geological type examples of object classes are seldom available to act as templates. The comparison generally relies on homologous points and again these might be difficult to recognise and correlate in geological examples. The methods tend to assume evenly spaced data, which are unlikely to be available from field survey. These drawbacks, however, regard the solution as an inverse model, which, for reasons given earlier, is unlikely to be a realistic scenario (see Forward and inverse models 154).

For shape interpolation in geological survey, the geologist would have to define corresponding features on the initial and target objects. Examples might be a sketch, or a simulated model, of the expected outcome of the processes and configuration of the environment of deposition of the sandstone bed. The tasks of the computer system might be to interpolate the complete boundaries of the objects from the point data; to present appropriate visualisations of the two objects and of a sequence of intermediate stages of morphing (see Shape 266) between the initial and target objects; maybe to maintain defined spatial relationships with adjacent objects; and maybe to calculate a material budget that ensured that mass was preserved during deformation. This could establish an explicit link between, say, the depositional model and the field survey model (see The field survey model 300).

To develop a computer system that helps to reconstruct the geology of a particular area from the available evidence, the system must have access to information on the constraints that apply to each object, relationship, model and space. Detailed analysis is essential to create process modules with algorithms based on the appropriate mathematics. They must be well defined to ensure that they can be reused and linked as appropriate to meet many different situations and requirements. The many different objects and models, with which they may be used, must also be well defined to ensure that their properties are compatible with the procedures for their analysis. The methods should in due course provide field support to geologists by recording observations and assembling a three-dimensional interpretation through interpolation of the geological objects and their relationships.

Geological interpretations in the future knowledge system may be represented as a set of related but separate models. Within these, various sub-models might be handled separately to respond to different objectives and modes of observation and data collection. The system will invoke analytical procedures that are not specific to geoscience, and were probably developed in other fields. As the focus moves from the geological map to the geological model, the opportunity arises of providing a range of products to meet many diverse objectives.

The infrastructure (see Overview of the infrastructure model 37) that is now being developed worldwide to support a global knowledge base (the so-called advanced cyberinfrastructure or Semantic Grid) offers the potential for considerable flexibility, dependent on appropriate recording of information. This requires the concepts to be defined and structured as knowledge base components, the structure and definitions being represented as an ontology (see Overview of the geological framework model 35). Careful bookkeeping will be needed to ensure that appropriate constraints are applied during processing to allow for the complex interactions of the various objects, relationships, models and spaces in the geoscience part of the knowledge base. Geologists must ensure that any relevant structure and content is appropriate for their purposes. A formal structure of sub-models is therefore required (see The multifaceted model 297, Overview of the geological cyberenvironment 28).

Mapping geology into the knowledge system

<<Table of contents 1
<<The future geological map 125
<< Seeking shared concepts 247
Mapping geology into the knowledge system 282
Representing wider knowledge 283
The role of the dynamic model 287
Broadening the framework 291
The general geoscience spatial model 293
The multifaceted model 297
The field survey model 300
The digital geoscience spatial index 305
The conceptual model 307
The system framework 310
Conclusions on mapping to the knowledge system 313
>>References 314

Abstract: The dynamic spatial model (how things got there) contains much of the geoscience reasoning underpinning the static spatial model (where things are) as shown on a map. Both models are essential components of the geoscience knowledge system. Top-down geological survey would benefit from access to The geological cyberenvironment (gce) 85 during field and office work, with the ability to capture and adjust the evolving interpretation. The solid Earth systems model (sEsm) 71, deals conceptually with the history and three-dimensional distribution of geoscience objects, and provides a framework[425] that links individual fragments, filtered and projected to fewer dimensions, which are all that we can observe, record in the field, and carry forward as facets of an interpretation. By formally identifying the objects and models, the system could record a reasoning process in the field, which cannot be disentangled from a completed map. Users of the sparse and varied spatial information require access to the sEsm and the Web generally, supported by a spatial index and object store, with browser[426], database[427], GIS[428], and visualisation[429] facilities. A

[425] Framework: A logical structure and guidelines giving a broad overview for classifying and organizing complex information, within which detail can be added as required.

[426] Browser: A software application that assists the user in searching for, retrieving and presenting relevant information, typically from the internet or World Wide Web

[427] Database management system: a set of computer programs to create, maintain and use a database (an organised collection of digital data), typically providing facilities to select and retrieve items meeting criteria, such as values equal to, less than or more than values specified by the user for each of a set of variables.

[428] Geographic Information System (GIS): An integrated system for the capture, storage, management, retrieval, analysis, manipulation and display of geographically referenced spatial data and its attributes.

[429] Visualisation: Transforming quantitative data (including the results of interpolation) into sensory information – images that the eye and brain can interpret and visualise.

comprehensive, flexible and extensible framework is needed to interlink sub-models representing fragments of the general model structure, to evaluate information, and to guide searching and browsing. Designing The sEsm as a predictive machine 76 should enable the solid Earth systems model to infer likely properties where they have not been directly observed. The same framework of systems[430], metadata[431], models[432], objects[433], attributes and relationships should guide procedures of survey, interpretation, and use of the resulting spatial model, following familiar procedures in the geological cyberenvironment[434].

Representing wider knowledge

<<Mapping geology into the knowledge system 282

Summary: *Achieving a realistic representation of the geology calls on, and contributes to, wide-ranging background knowledge, from many sources and many facets of the general model. The abstraction and feedback procedures of surface and subsurface geology overlap, leading to a more consistent view. The benefits of well-documented data and reasoning supported by the cyberinfrastructure, should lead to flexible presentation of more rigorous results. The systems approach supports the complex connections and interactions.*

A system is a set of interacting parts that function as a whole, and the systems approach involves study of the linkages among the component activities. For example, the Earth systems science approach is a means of understanding the connections between the solid Earth, the water cycle, living things and the atmosphere. It is part of the changing emphasis in the work of many Geological Surveys: away from delivering a restricted range of published end products, towards maintaining integrated information resources from which end users can select services that give a flexible response to their specific needs.

Even authors addressing narrower topics see a need to place them in their wider context. "The 'winds of change' refer to much more than just anisotropic rocks whose seismic anisotropy provide insight into permeability anisotropy. The way we view our planet is changing: the planet is being studied as the interaction of interlocking systems. The four systems are: the hydrosphere, the atmosphere, the lithosphere, and the biosphere, as all of

---

[430] System: A set of interacting parts that function as a whole. The systems approach involves study of linkages or interfaces between the component activities.

[431] Metadata: Metadata is a description of data (often used in a broad sense of representations of information or knowledge) and its context. It is structured to assist the user or computer to find, manage, control and understand the data.

[432] Model: A formalised representation giving a simplified view of aspects of the real (or of an imaginary) world relevant to the purposes in hand.

[433] Objects: Representations of real-world or conceptual things or entities of interest in a particular context.
Object class: An abstraction giving a general description of the expected properties and behaviour of the objects belonging to that class. They are a means of categorising object instances within larger groupings.
Object instances: Representations of specific, identified, real-world or hypothetical objects.

[434] Cyberenvironment: Aspects of the cyberinfrastructure assembled to meet requirements relevant to a particular field of enquiry, aiming to maintain global compatibility while providing access through interfaces that match users' working practices.

us absorb energy from our sun. The planet and all its subsystems, like geology or the weather, are explicitly evaluated as inter-dependent and interacting..." Lynn (2004[435]). Some consequences of a more comprehensive and more fully connected spatial framework (see Overview of the geological framework model 35) and geoscience knowledge system (see The emerging geoscience knowledge system 42) are discussed next, starting from the relevance of many aspects of geoscience knowledge to geological survey.

The discussion in The imperfect map 145 quoted the advice of Tearpock and Bischke (2003[436]), who emphasised that a geological map is not 'correct', but should aim 'to develop the most reasonable and realistic interpretation'. If strata are known to be folded along, say, east-west axes, it would presumably not be realistic to show north-south folding on the map simply to give a better mathematical fit to sparse elevation data. It might be unrealistic to show the irregularities of a formation boundary spread evenly across an area where it was known that the folding and faulting were highly localised. A realistic geological map is presumably drawn to conform to expectations of the objects' behaviour, but there remain questions of balance between interpretation and observation, and how far preconceptions should impose on the surveying procedure. Compared to the map, spatial models can be more flexible in depicting the interpretation. Interpolation and generalisation procedures will improve to do more justice to the geologist's thought processes.

At one extreme, the survey might start from a clean sheet and observe and record data on specified properties and relationships, following well-established predetermined procedures. This approach is effective with many geophysical and geochemical surveys or for a rapid survey of relatively unknown geology, and should yield a consistent and informative dataset. A straightforward interpolation method, such as splining (see Estimation by interpolation 212), which makes no commitment to a specific geological interpretation and resembles familiar manual methods, might then be appropriate. At the other extreme, the survey of an area might start with clear ideas about the geology, and aim to relate these ideas to what can be seen on the ground. It could benefit from a more comprehensive systems framework for geological knowledge and cyberinfrastructure support.

Between these extremes, geological surveying may in practice combine a top-down view (see The surveyor's holistic view 60)[437] of the evolving framework (within which the surveyor operates) with the bottom-up collection of specific detail following predetermined procedures (which progressively clarify the top-down view). In these circumstances, the Geological Survey map is far from being a mechanical selection of observations; the features recorded on the map are a considered selection from the innumerable observations that

---

[435] Lynn, H.B., 2004. The winds of change: anisotropic rocks – their preferred direction of fluid flow and their associated seismic signatures. CSEG Recorder, October 2004, 5-11. http://www.cseg.ca/publications/recorder/2004/10oct/10oct-winds-of-change.pdf

[436] Tearpock, D.J. and Bischke, R.E., 2003. *Applied Subsurface Geological Mapping*. 2nd ed. Prentice-Hall, Upper Saddle River, NJ. 822pp.

[437] Holistic: A view of a system that emphasises its properties and interrelationships acting as a whole, as opposed to the reductionist approach of studying its components in isolation as distinct entities.

could be made. The map is not just an interpolation[438] or a tool for predicting what formation underlies a specific location; it attempts to develop the interpretation of the geology, taking into account, testing and extending a wide range of background knowledge. Many different geological models might guide the interpolation of surfaces, and the final product must reconcile information from many sources.

The knowledge system should be able to support this two-way approach. In order to combine the strengths of human reasoning and computer processing, the computer analysis should assist and extend the visualisation and thought processes of geologists. It should relate to the reasoning procedures of geological surveying, which, considered in the light of related developments in other fields, might help to develop an appropriate spatial framework for a geoscience knowledge system. Interpolation and geological interpretation, for example, could be linked in geometrical terms (see The geometry of interpolation 243).

Geological Surveys generally obtain most of their information from surface or near-surface observation. However, most applications of computer interpolation are deeper subsurface, in oil or mining investigations. They study the same geological objects and processes, although the differences in accessibility to observation influence the methods. The field geologist may emphasise the two-dimensional view of the Earth's surface as presented on a map, based on surface rock exposures that are laterally more extensive, but less complete vertically, than down-hole measurements.

Field geology provides more opportunities than subsurface geology to observe and collect geometrical information relevant to interpolation, as aspects can be examined in three dimensions at a wide range of scales, along with their interactions and spatial relationships. This may enable geologists to match their results more closely to the underlying geological reasoning and to the vagaries of real-world geology. Field and subsurface geology are complementary, contributing to various aspects of abstraction and feedback in geological investigation (see Abstracting from reality to model 131). More comprehensive computer models could lead to greater overlap and sharing of insights and reasoning between subsurface and field studies.

A Geological Survey could attempt to provide comprehensive information to meet a diversity of needs such as those listed in Diverse objectives and products 150. With traditional methods, that would come at a high price and limitations on resources mean that priority goes to more limited solutions. Comprehensive computer support, however, would alter the balance of costs and benefits. Reusable field information could be recorded and stored much more readily than with manual procedures, while bringing the benefits of computer reworking to meet varied objectives. A record of the supporting reasoning and its consequences is required to clarify the extent of the information's relevance.

Many complex decisions are made during geological survey, but the final map conceals much of the process. Spatial models[439] enable users to visualise geological structures of

---

[438] Interpolation: The estimation of values, for example at a point or along a line or surface, in order to predict a value or complete a visualisation.

[439] Spatial model: A model interpreting spatial data and relationships to clarify and understand spatial forms and processes and thereby predict real world attributes.

greater complexity, based on several information sources, as illustrated by the examples in Mallet (2002[440]). More explicitly than the map, the visualisations may incorporate many geological decisions. A comprehensive spatial model of this kind, with countrywide coverage, would require an explicit record of the reasoning behind all aspects of the interpretation, in order to justify, explain and assess the results, and to inform subsequent modifications (see A framework for the reasoning 135). The surveyors know more about these things than the users of the information. Users of the survey, on the other hand, are likely to know more than the surveyors about their own objectives (see Diverse objectives and products 150). They might accept the default visualisation. Alternatively, they might wish to specify a different form of visualisation for their own purposes, or to retrieve a dataset for analysis alongside their own data. To meet these needs, the surveyors could record metadata[441] based on their background knowledge, constraining each object's behaviour when it is subjected to procedures like interpolation or generalisation. That should help users to evaluate the information in their own context, but they also require access to the reasoning that underlay the interpretation.

In principle, a model could track each confirmed decision, and record its justification. Experts familiar with the geology of the area and the surveying procedures might be able to add some retrospective interpretation to existing products. But significant aspects of the reasoning should ideally be recorded during surveying, as the context of background thinking is otherwise lost. The conclusions can be amended if necessary as the survey proceeds.

There is a need to record some vague impressions, explore their consequences, and elucidate or discard them as more is learned. Initially an observation may be of dubious significance. For example, a tentative decision during survey might ascribe an observed pattern to concealed faulting rather than folding. A landscape feature might look vaguely like the topographical expression of a fault. As a recorded observation this would carry little weight, and might eventually be discarded. But when the detailed structure is studied, it might turn out to be a useful clue to the fault pattern. Casual observations of uncertain value must be stored in such a way that they do not obscure more important aspects, but where they can be found and retrieved if required.

Mallet's methods of Discrete Smooth Interpolation are based on a node and link structure, which, as suggested in Rationale 224, could extend naturally to a structure of objects and the relationships between them (see Relationships between objects 183). Within their spatial hierarchy, an object might be a single observation, or a composite object such as an interpolated surface, or a complete spatial model for a topic within a particular area. A relationship, where appropriate, could take the form of a URI (a uniform resource identifier, such as the URLs widely used on the Web). It might reference predetermined codes for software instructions to invoke or constrain computing procedures, or an explanatory

[440] Mallet, J.-L., 2002. *Geomodeling.* Oxford University Press, Oxford. 599pp.

[441] Metadata: Metadata is a description of data (often used in a broad sense of representations of information or knowledge) and its context. It is structured to assist the user or computer to find, manage, control and understand the data.

microdocument[442] (such as a field note), or both, in a mark-up[443] language (see Mark-up and metadata 175). Microdocuments can record detailed level of reasoning, and could of course also be referenced from higher levels of explanation.

Decisions about interpolation or generalisation could also be recorded as part of the reasoning process. In presenting his convincing case for Discrete Spatial Interpolation methods, Mallet (2002) shows how spatial models can build on the concept of balanced cross-sections to ensure that conservation of mass is taken into account when simulated geological processes transform or move material within the model. Process models might also consider aspects like mass conservation in sediment budgets. He also shows how the concepts of principal directions studied in differential geometry (familiar to structural geologists as fold axes and axial planes) can be studied and used in surface interpolation, throwing light on past stress distribution (see also Morphometrics 268) . And he shows how the computer-aided design concept of a developable surface (in which a flat sheet undergoes complex deformation without shearing or plastic deformation) can constrain the interpolation of a folded surface.

These examples illustrate the need to extend the scope of the model and its underlying geoscience reasoning; to move beyond interpolating individual surface segments towards a broader scientific context. “A considerable improvement in the integrity of geological models can be achieved if information about the order of surfaces in the stratigraphical succession, and about the position of unconformities, faults, surface topography and the bedrock surface, can be considered simultaneously in the modelling process” (Hughes, 1989)[444]. In a realistic model, each surface relates to its neighbours and to intersecting surfaces such as faults or fold axial planes. The movement pattern of faults and folds should be mechanically and geologically feasible. The role of the dynamic model 287 must be considered.

The role of the dynamic model

<<Mapping geology into the knowledge system 282

Summary: *The task of geological survey is not just to draw lines joining similar rock types. Rather, it is to illustrate a current interpretation, matching observations with preconceptions based on training and experience, while searching for meaningful patterns to test and adjust the interpretation. Many features and characteristics are observed that are not shown on the map, but contribute to the dynamic model (how things got there) that underpins the interpretation in the geologist’s mind and in turn interacts with the static model (where*

---

[442] Microdocument: A short note or module of information, typically referenced by a URI and seen here as a system component, such as a minimum revisable unit.

[443] Mark-up: Symbols inserted in a document in a mark-up language such as SGML, HTML or XML, to tag the beginning and end of character sequences that can be interpreted by machine, and can be omitted for displaying to the user.

[444] Hughes, J.D., 1989. A multiple-layer strategy for analysis of geological data in layered sequences. *Geological Survey of Canada, Paper* **89-9**, 571.

things are) illustrated on the map. The static and dynamic models are essential components of integrated geoscience survey, and therefore of a supporting knowledge system.

Computers can of course draw maps by mathematical interpolation from the available data, and geologists can edit them to match their intuitive preferences and traditional rules of thumb. But to take full advantage of the opportunities of the cyberinfrastructure, we need to look beyond traditional map making[445]. Spatial modelling should offer more rigorous procedures within the broader rationale, methods and implementations of geoscience survey. The main issue is not the detail of algebraic manipulation, but the relationship of mathematical reasoning to the concepts of geoscience. Computer methods should build on existing geological insights, not blind us to their value. With this in mind, we look again at some features of the general model that are implicit in geoscience survey but hidden in conventional maps.

Consider geologists mapping conventionally in the field. Their task is not just to draw lines connecting similar rock types. Years of training would not be needed for that. Rather, from background knowledge and the small amount of available evidence, their task is to build and record an interpretation of the observed objects showing their likely form and behaviour within their regional setting. In an unfamiliar area, the interpretation starts as a vague outline. Maybe there are thick beds of flat-lying sandstone and conglomerate, and perhaps basalt that might be a lava flow or part of a volcanic vent. The initial impressions suggest sketchy ideas about the possible geometry and what to look for next – ideas that develop as the survey proceeds. There is a top-down iterative process (see The surveyor's holistic view 60), starting from general background knowledge: look, interpret; look again more closely and widely to test, revise, reconcile, refine and extend the interpretation; repeat as necessary (see The geological investigation model 98). Does the basalt extend over only a small area, cutting across the sediment like a vent, or is it in the form of widespread sheets parallel to the bedding of sandstone below? Is the sandstone uniform or does it terminate against the thick conglomerates? How were the rocks folded and faulted? Can the depositional and structural effects on the geometry be separated? Does the interpretation match the landforms? What pattern of processes lies behind the observed distribution? The early views may establish some ideas of the possible nature, size and shape of the objects, a guide to what to look for as the survey proceeds. They are ephemeral concepts, perhaps only in the geologist's mind, but they provoke questions, and "most exposures provide answers only to questions that are put to them" (Gilluly, quoted approvingly by Mackin, 1963[446], page 161).

Observations that answer some questions are recorded on the field map, clarifying the conceptual model and raising further questions. The answers tie the loose geometry of the

---

[445] Mapping (Geological): Conventionally, geological mapping leads to a graphical depiction, usually on a flat surface, of spatial relationships and forms of geological features or properties in a selected area of the Earth's surface or subsurface. As defined in mathematics, mapping relates the elements of one set to those of another. A broader definition of geological mapping could be 'relating elements of geological observation or interpretation of the solid Earth to corresponding elements in appropriate models in the geoscience knowledge system'.

[446] Mackin, J.H., 1963. Rational and empirical methods of investigation in geology, in Albritton, C.C. (editor), 1963. *The fabric of geology*. Freeman, Cooper & Co, Stanford, pages 135-163.

tentative interpretation to locations on the ground and thence to the rigid geometry of geographical co-ordinates. The map face records the precise position on the ground of surveyed points, lines and orientation measurements. But for the most part, direct geological evidence may be missing, inaccessible or concealed by overburden. The gaps are filled to give a 'realistic' view illustrating the current interpretation, matching observations with preconceptions based on training and experience, while searching for meaningful patterns to test and refine the interpretation. Geologists observe many features that are too small to show on the map. They may be recorded in notebooks and eventually in the map explanation, and help to confirm or refute evidence from other scales. Observations are not restricted to the position of objects, but extend to many other features and characteristics (structural patterns, petrography, fossil content, sedimentary structures and so on) that throw light on the processes, the objects' properties and their historical configurations. They contribute to the dynamic model in the geologist's mind and through it to the static model shown on the map.

While surveying the geometrical characteristics and spatial relationships of instances of objects in the field, geologists may also formulate and develop their more abstract ideas. These refer to the classes of object to which the instances[447] (see Object instances and classes 181) belong, the processes that affected them, their likely historical behaviour in response to these processes and the configuration[448] of past objects and events. Extensive observation of instances may lead to general impressions about the geometrical properties of the object class, impressions that might be matched with the results of hypothetical or experimental processes (see Rationale 224). These ideas might be visualised in the geologist's mind or depicted as sketches in a field notebook or scratches in the sand. Even if the impressions are somewhat vague and seldom recorded, they influence the surveyor's expectations of the geometry of the object instances illustrated on the map face. In a spatial model, expectations of the form of interpolated surfaces (based on geological interpretation) could be explicitly invoked and recorded.

As the impressions take form, testable ideas emerge concerning the main features of the mapped geology, reflecting its development through past configurations. Earlier arguments suggested that in part the configuration would have evolved unpredictably, as an emergent[449] system (see Complex and emergent systems 159) that incorporated the consequences of historical and idiopathic[450] incidents as they occurred. In part it would have behaved predictably, the predictions being based on scientific reasoning and analogies with

---

[447] Objects: Representations of real-world or conceptual things or entities of interest in a particular context. Object class: An abstraction giving a general description of the expected properties and behaviour of the objects belonging to that class. They are a means of categorising object instances within larger groupings. Object instances: Representations of specific, identified, real-world or hypothetical objects.

[448] Configuration: The spatial arrangement, pattern, form and shape of objects or their properties. Used by Simpson (1963) in contrasting 'The actual state of the universe or any part of it at a given time, its configuration, is not immanent and is constantly changing' with 'The unchanging properties of matter and energy and the likewise unchanging properties and principles arising therefrom are immanent in the material universe.'

[449] Complex system: A complex, emergent system has many adjacent parts that may interact according to simple rules without central control. Feedback mechanisms may result in effect not being proportional to cause and the linear equations of physics may not apply.

[450] Idiopathic: [Of a disease] arising spontaneously from an unknown cause.

present-day experiments, as the system moved towards appropriate attractors[451] (Complex and emergent systems 159) in accordance with similar instances and immanent[452] laws (see Abstracting from reality to model 131). Geologists are likely to modify their ideas as more is learned, probably by adjusting the arbitrary elements of their view of the historical configuration of events rather than the inevitable elements of the immanent laws that control the outcome, though not the location, of geological processes. A considered interpretation should give a consistent and credible view of the configuration and its development within the regional setting.

The static model of the present-day disposition of the geology (where things are) underpins the cartography: the dynamic model (see The imperfect model 156) of processes and configurations evolving through geological time (how they got there) underpins the interpretation. The static and dynamic models must interact in the geologist's mind, because the dynamic model encompasses much of the scientific reasoning. It integrates concepts and provides the basis for filling gaps between observations to complete the static model, where ideas are constrained by regional knowledge based on widespread observations and their interpretation by many workers – and can be tested against the real world (see SEED, 2011)[453].

The surveying task, particularly in an area like the UK with well-known geology and easy access, is not so much to collect data and fill gaps by mechanical interpolation, as to seek and record significant observations as a guide to adjusting interpretations and placing the objects that reflect them. After all, the final product is an interpretation that fills space by observation and prediction (see The sEsm as a predictive machine 76) at defined and constant scales, rather than just a set of scattered observations of varying resolution. This implies that the static and dynamic models are essential and interrelated components of geoscience survey, Broadening the framework 291 of comprehensive computer support for a knowledge-based system.

Implementation note: In field mapping, The surveyor's holistic view 60 demands that dynamic models of past events, processes and configurations (even if only in the geologist's mind) should support the static models of the present-day configuration that result from survey. Both can potentially be represented as computer models. To encourage future integration in a knowledge-based system, they could both with advantage take their place now in the ontologies of the framework model.

---

[451] Attractor: In complex systems, a preferred position in state space, such that if the system starts from another state, it tends to evolve towards an attractor.

[452] Immanent: Something naturally inherent and intrinsic within and throughout its domain.

[453] SEED, 2011. CyberGeologist. https://www.planetseed.com/node/15211

Broadening the framework

<<Mapping geology into the knowledge system 282

Summary: *Computer methods lack the human insight to build on the analogies from which much geoscience reasoning is constructed. But with computer support, we can formalise our ability to record, share, analyse and integrate spatial knowledge of different geometrical types from various sources, thus overcoming an obstacle to a more comprehensive system. Top-down survey, building on an object-oriented approach, would benefit from access to a full knowledge base in the field as well as at the desktop. This would include dynamic models, and the ability to link in information sources, and capture and adjust the threads of reasoning, while the survey is in progress. Short-term developments can be guided more effectively if long-term aims are allowed for in the project plan.*

It was suggested (At the interface 133) that at present computer-based expert systems are inappropriate for a survey model dependent on human insights and nebulous analogies. Instead, it "must focus initially on interactive visualisations of spatial patterns and relationships, supported by explanatory text, dominated by legacy information and unrecorded experience." This suggests that an important barrier to progress is the difficulty of comprehensively representing and communicating spatial knowledge. However, computer methods should in due course extend our ability to record, share, analyse and integrate spatial knowledge of different geometrical types from various sources (see A wish list for integrated geometry 221), thus overcoming a major obstacle to a more comprehensive system. Computer graphics can provide the flexible geometry needed to represent and visualise the dynamic models. By quantifying the geometry, a more rigorous statistical approach to spatial description and interpretation is possible. In turn, this could lead to quantifying the expected spatial behaviour of object classes, and utilising this global information for local interpolation and generalisation.

Geoscience surveying builds a model step by step (see The dialectic model 129)[454] by juxtaposing and extending interpretations and observations through interactions between them ("exposures provide answers only to questions that are put to them"). In general, surveying is a top-down procedure going from the general to the specific, starting from what is already known or believed and aiming to test and add to that prior knowledge, usually at a more detailed level. A surveyor is obviously concerned with spatial information, but also requires access to non-spatial aspects of the general model, because properties of spatial objects may reflect spatially-independent processes. Full access to the knowledge base is therefore desirable in the field as well as at the desktop. The surveyors and users of the resulting models both need access to a wide range of relevant background information, which they could appropriately access through a browser system. The BGS GeoIndex (see The field survey model 300 and BGS, 2010a[455]) is an example of an index linked to a browser[456]. When mapping a new area, surveyors actively search for sources of additional

---

[454] Dialectic: The theory and practice of weighing and reconciling juxtaposed or contradictory arguments for the purpose of arriving at the truth.

[455] BGS, 2010a. GeoIndex. http://www.bgs.ac.uk/geoindex

[456] Browser: A software application that assists the user in searching for, retrieving and presenting relevant information, typically from the internet or World Wide Web.

information in their area of interest, and in so doing could extend the Index by linking in references to their findings as the survey proceeds. Conceptually, the geological investigation model is thus embedded in The general geoscience spatial model 293, and this in turn is embedded in the all-purpose hypermedia knowledge repository of cyberspace (see The conceptual model 307, Workflows, collaborative networks and Linked Data 53).

In this broader context, the digital cartography approach is only of limited and short-term value, because it starts from the product (the map) not the process (the survey). It may be a useful learning step and a route to carry forward the customer base and legacy information. The developments described in The future geological map 125 take us beyond this to spatial modelling and a process-based approach. When these steps are secured, they are a possible basis for further developments within a broader framework that could bring greater long-term gains.

In the longer term, cyberinfrastructure-based systems should link into dynamic models (see The role of the dynamic model 287) to explain features of the static model and clarify some of the ambiguities of the map (see Ambiguity and map representation 148). This implies a need to capture the threads of the thought process (see The multifaceted model 297, Mechanisms 45), many of which emerge while the survey is in progress but are untraceable in the completed conventional map. Such systems would therefore have to offer comprehensive support to the entire process from desk survey to fieldwork to final conclusions. They would recognise that geologists arrive in the field with extensive experience, a detailed theoretical background and the ability to augment it by consulting appropriate texts and expert specialists. Once in the field, the surveyors modify and extend their existing knowledge by directed observations, in a sequence that can be recorded as a workflow. At some future time, access to a network of spatial models from various sources might be available to guide the interpretations and observations of geoscientists in the field, laboratory and office, enabling the scientists to record, filter, visualise, test, reconcile[457], clarify, focus, refine, extend, evaluate and amend their shared models. A comprehensive knowledge base of the most recent information would thus be generally available as and when required.

Short-term developments should take longer-term ambitions into account, directing the tension between past and future. On the one hand, working systems must be based on well-established methods: on the other hand, their framework[458], at a time of a rapidly changing infrastructure, should look ahead to encourage future development. On the one hand, spatial models must use tried and tested technology to enable geoscientists to communicate ideas cost-effectively in new representations: on the other, their modelling framework has ramifications and consequences that can be determined only by decades of trial and error. We will need this conceptual framework to guide consistent development and efficient

---

[457] Reconciliation: Kent (1978, pp. 202-203) points out that people have different views of reality, and that these change with time. But the views overlap and so can be *reconciled* with varying degrees of success to serve different purposes. "By reconciliation, I mean a state in which the parties involved have negligible differences in that portion of their world views which is relevant to the purpose at hand."

[458] Framework: A logical structure and guidelines giving a broad overview for classifying and organizing complex information, within which detail can be added as required.

interaction of the activities. It must build on established techniques, while being flexible and extensible. It must evolve as we learn, and look ahead with tentative links to experimental research models to ensure that they can be incorporated if and when they prove their worth.

The feasibility of computer implementations of aspects of the dynamic model will emerge from exploratory projects – research tasks, perhaps undertaken within an academic environment, justified not by cost-effectiveness but by the usual research criteria. As their likely gestation periods match those of the solid Earth systems model, their immediate conception would be timely. The framework should encourage future evolution towards compatible geometrical representations and towards a knowledge-based system in which cooperating dynamic and static models share the central role. Such a framework should take into account the diversity of spatial representations and the complex interplay of ideas during the survey (see The surveyor's holistic view 60, The general geoscience spatial model 293).

Implementation note: To reduce later backtracking, Geological Surveys and other information communities with a long-term interest in a comprehensive view of geoscience might plan an adjustable and extensible framework for comprehensive support, including aspects where implementation may not be feasible for some considerable time. For short-term studies, the overhead is probably not justified.

The general geoscience spatial model

<<Mapping geology into the knowledge system 282

Summary: *Conventional survey employs a variety of representations of the geometry: on the map face, in the marginalia and in text explanations. They are linked through their relationships to the same implied general geoscience spatial model. This refers at all levels of detail to the three-dimensional disposition and configuration of the present-day observable objects of geoscience, to their observed and interpreted properties and composition, and also to their history throughout geological time including the processes that created and altered them and are crucial to their interpretation. But only fragments are available for study, analysis and visualisation.*

Geoscience survey is a complex operation calling on a vast amount of existing recorded and unrecorded knowledge. The complexity carries through to the task of providing it with comprehensive computer support, involving spatial models embedded in a knowledge base. The framework (see The geological framework model 105) should help to make sense of the complexity, by setting out the component parts and establishing the relationships between them. To clarify the requirement, consider the conventional geological map once more.

A geological map shows a two-dimensional projection of the land-surface geology on the map face. In the marginalia, the map key records the classification in geological-time sequence of the surface-bounded stratigraphical objects depicted on the map face, essentially a filtered extract from the metadata[459]. Structural events, igneous activity, and metamorphic events may be recorded in the same time sequence, for even the initial interpretation must consider the historical development. The map margin may also record a geological column or generalised vertical section (GVS) summarising the thickness variations and relationships of rock units within the map area. Although much of its content may be derived from the map face, the GVS also takes field observations of measured sections into account and may represent thickness and spatial relationships between stratigraphical units (such as merges with, lies unconformably on, offlaps) more precisely than the map face. Like the map face, it provides an interpretation, in this case referring to the thickness of beds in a sequence corrected for post-depositional structural deformations. Other cross-sections may show an interpretation of the vertical configuration[460], based on field mapping and perhaps on information from subsurface studies. The projections imply an underlying three-dimensional model of the geology, at least in the minds of the mapmakers and users.

A map explanation, usually a separate publication, provides additional spatial information as text descriptions and graphical illustrations. It may include orientation measurements, possibly displayed as stereograms. It is likely to concentrate on explaining and justifying the map interpretation and describing details (such as a sketch of the geometry of an exposed unconformity) that are not readily shown on the map. Where the historical interpretation is not clear from the time sequence, additional sketches might illustrate past events or configurations, along with short text explanations. There may be explanations of the relationships to other maps, perhaps showing geophysical or geochemical properties, seen as a different expression of the same underlying conceptual model. The map explanations may include detailed local maps and sections, and refer to features too small to show at the map scale. The maps may also be generalised for presentation at less detailed scales. Supporting information, such as field notebooks and borehole logs, may be archived but not necessarily made accessible to the map users. Map explanations make it clear that the conceptual model extends back in time to include the development of the observed configuration through geological time, and of explanatory ideas through successive interpretations.

Conventional survey thus employs a variety of representations of the geometry, many of them on a single map. The framework model should identify the various facets and indicate how we think they relate to one another. The diverse facets are linked, and some can only be linked, through their relationships to the same implicit general geoscience spatial model (see The imperfect model 156), represented by a framework model (see The geological framework model 105). This refers at all levels of detail to the three-dimensional disposition and configuration of the present-day observable objects of geoscience, to their observed and interpreted properties and composition, and also to their history throughout geological time, including the processes that created them and are crucial to their interpretation, and the conceptual objects that preceded them. It aims to answer the structured equivalent of

[459] Metadata: Metadata is a description of data (often used in a broad sense of representations of information or knowledge) and its context. It is structured to assist the user or computer to find, manage, control and understand the data.

[460] Configuration: The spatial arrangement, pattern, form and shape of objects or their properties.

the questions a young child might ask on looking into a dark cupboard: What is in there? What is it called? Where is it? How is it arranged? What does it look like? What is it made of? What does it do? Where did it come from? How did it get there? How do I know? (Loudon, 2000, part M[461]). The same conceptual framework is relevant to the computer implementation (see The solid Earth systems metamodel (sEsmm) 110)[462], providing a structure to relate and reconcile the small and scattered fragments of the model for which we have direct evidence. As an abstract mental concept, it provides the basis for relating and reconciling (see Reconciliation 186) the various facets of its representation, and enables us to form views of the overall scene.

Extracts, filtered from the model, transformed and projected as images, are all that we can visualise on paper or screen, and perhaps all that we can readily imagine. They are *filtered* (a process of selectively enhancing or reducing specified components of the information stream) to select the relevant and available topics, and *generalised* to remove unwanted detail for visualisation at the chosen scale. They are *transformed* (adjusted geometrically, see Geometrical transformations 228) for clearer visualisation and *projected* (a process of reducing the number of dimensions) for visible representation in one, two or occasionally three spatial dimensions or as images moving through the time dimension. The geoscience map is a set of views of the general model, seen through a specific set of filters and projected onto a two-dimensional plane. The map face, cross-sections, GVS, and map key are filtered, transformed and projected differently onto the same map sheet. Spatial relationships are usually implied rather than stated, although some may be explicit in the GVS. Results are conventionally communicated through printed maps and related documents of a standard pattern, and the procedures of geoscience survey have inevitably developed to match this rigid pattern of visualisation. The general geoscience spatial model (see also The imperfect model 156, The solid Earth systems metamodel (sEsmm) 110) in full detail has no concrete existence, for its size would be vast beyond knowing and beyond representation. We can build on fragmentary observational data and background knowledge by predicting what has not been observed (see The sEsm as a predictive machine 76) to give a more comprehensive view, but only with representations that yield generalised information at relatively coarse resolution.

If the system had access to full detail, then extracts could be filtered out from the knowledge base to meet viewers' requirements while maintaining internal consistency. The filters, on the basis of metadata, might select from such aspects as:

- the area and stratigraphy of interest
- the business setting, range of topics and levels in the object hierarchy requested by the user
- the user's objectives (see Diverse objectives and products 150) and the emphasis placed on particular properties

---

[461] Loudon, T.V., 2000. Geoscience after IT: a view of the present and future impact of information technology on geoscience. Elsevier, Oxford. 142 pp. Also available as *Computers & Geosciences*, Special Issue, **26** (3A), A1-A142 http://nora.nerc.ac.uk/2408/1/Part_M.pdf

[462] Metamodel: A metamodel is a description of the organisation and function of a model, to assist the user or computer to find, manage, control and understand its contents.

- the sources and evaluation or quality assessment of the information
- the levels of resolution and generalisation that match the scale of visualisation
- the levels of confidence, ranging from confirmed observation to tentative explanation
- the spatial relationships valid at the selected levels
- the appropriate spatial properties of objects, such as existence, location, slope, form, texture, arrangement, relationships, behaviour
- the chains of reasoning and the supporting models of immanent[463] processes and historical configurations[464]

Visualisation of the filtered extracts would then involve geometrical transformations including projection.

It is clear that, like conventional methods, the solid Earth knowledge system can only base its predictions on limited information, and much of this can only be at coarse granularity (level of detail). It might, however, be able to construct a knowledge representation structure that conforms to the current view of the general model, its components and their relationships. This could help to define the scope of the survey model, and provide it with an initial framework. Within the framework, we could locate those diverse fragments of the general model that we think we know something about, and visualise them as filtered extracts like those listed above. The surveying procedures of observation and abstraction (including interpretation, analysis and explanation) provide the content (objects, metadata and models) that could be positioned within the container (The geological framework model 105). We must be able to amend and extend the framework as new topics enlarge its scope. However, as The multifaceted model 297 indicates, combining isolated fragments of knowledge in a system framework (The solid Earth systems model (sEsm) 71) is not straightforward.

Implementation note: Ideally, the entire process of geoscience surveying should operate as an integrated whole and include, for example, local process and palinspastic models that interface with their regional and global counterparts. Initially, components of the framework, such as the metadata model, survey models, palinspastic models and process models (structural, depositional, igneous, metamorphic and so on) can be linked only through interaction with the background knowledge of their creators and users. However, the shared space-time framework should be designed to encourage compatible representations of hitherto unrepresented knowledge. Where feasible, the system should capture this knowledge, including the threads of reasoning by which the diverse components and their relationships were reconciled and explained.

---

[463] Immanent: Something naturally inherent and intrinsic within and throughout its domain.

[464] Configuration: The spatial arrangement, pattern, form and shape of objects or their properties. Used by Simpson (1963) in contrasting 'The actual state of the universe or any part of it at a given time, its configuration, is not immanent and is constantly changing' with 'The unchanging properties of matter and energy and the likewise unchanging properties and principles arising therefrom are immanent in the material universe.'

The multifaceted model

<<Mapping geology into the knowledge system 282

Summary: *Various facets of the general model are surveyed together, and may be presented on a single map. Human users can separate the numerous fragments and reconcile them in their minds, up to a point. But a computer knowledge system requires a framework that identifies and keeps track of models, objects and their behaviour and relationships. It would ideally be invoked during survey to clarify the threads of reasoning, which can seldom be disentangled from a completed map.*

The most comprehensive abstraction[465] in geoscience may be the unified general geoscience model (see The general geoscience spatial model 293, The solid Earth systems model (sEsm) 71) but our observations can reach only isolated fragments. The procedures of geoscience survey invoke many separate models, which abstract ideas and information from diverse features of the real world during observation and recording, not just during the final interpretation (see Abstracting from reality to model 131). During a single visit to one outcrop, geologists use a range of procedures and are well able to collect, say, micropalaeontological samples for later examination, orientation measurements for summary and analysis on stereograms, and stratigraphical information to depict on the map face. Surveying has many facets (see Microdocuments and the threads of reasoning 190, Object-oriented survey 195), addressing different issues, maybe leading to objects[466] derived from separate models[467]. Observations related to the various models may be mapped at the same time and may or may not be distinguished on the map. Experienced users are well aware of these issues when they study a map, and probably compensate for them without conscious thought. They connect the fragments by separating, manipulating and reconciling the implications of individual parts of the image in their minds and generally take the methodology for granted.

Computers lack human insight and process information differently. They require a more formal identification of objects and their behaviour. Surveyed objects are not always compatible, but, if the surveying activities had full computer support, the objects or their underlying models might be reconciled[468] (see Reconciliation 186) at an appropriate level. During survey, each reconciliation procedure could be identified and explained (though not at present automated). By thus making the process of reasoning and reconciliation more explicit, the results could be more rigorous. Alternative models could be created for different interpretations, such as revisions or multiple hypotheses. By identifying the separate models, we could clarify the relationships between them, and the points at which they conflicted or were reconciled.

---

[465] Abstraction: reducing the information content of a concept or an observable phenomenon, typically in order to retain only salient information, relevant for a particular purpose.

[466] Objects: Representations of real-world or conceptual things or entities of interest in a particular context.

[467] Model: A formalised representation giving a simplified view of aspects of the real (or of an imaginary) world relevant to the purposes in hand.

[468] Reconciliation: Kent (1978, pp. 202-203) points out that people have different views of reality, and that these change with time. But the views overlap and so can be *reconciled* with varying degrees of success to serve different purposes. "By reconciliation, I mean a state in which the parties involved have negligible differences in that portion of their world views which is relevant to the purpose at hand."

The advantages of diverse sources of information (see Forward and inverse models 154) are apparent in their contribution to geological interpretation. Obvious examples include topographical, geophysical, geochemical, litho-, chrono-, and bio-stratigraphical models. They generate distinct objects reflecting different properties of the same real-world entity and illuminating different aspects of the general geoscience model. Perhaps less obvious are the disparate sub-models that lurk behind the objects on a single geological map, their significance apparent only through background knowledge. For example, bedding orientation measurements might stem from one sub-model referring to areas the size of a notebook, greatly enlarged as symbols visible on the map, but mismatched to the scale of a sub-model for shaping the lines of formation boundaries (see Continuity, fractals, octrees and wavelets 217, Grain, set and patch 251). The set of orientation measurements might in turn break down into a hierarchy of subordinate models. There might be a subset of orientation measurements collected to determine fold shape, sampled to give each part of the fold an equal chance of being represented. Another subset might have been collected to illustrate a broad view of the spatial structure, in which each area should have an equal chance of being represented. The two sub-models require different data collection procedures and lead to different geometrical representations – stereogram and map.

In areas with continuous river or cliff exposures, vertical sections might be drawn in the field, compensating for tilting and deformation by procedures that are widely understood but seldom recorded. Thus a generalised vertical section (GVS) might be developed in the field, following different procedures from the mapping of formation boundaries. Nevertheless, the GVS might significantly influence the general pattern shown on the map. For instance, bands of evenly spaced formations might be shown snaking across a hillside where they were not exposed, to give a 'realistic' view based on indirect evidence. In other areas the reasoning might follow a different route. Small, scattered outcrops might mean that no continuous vertical sections could be observed. An interpretation of the structural geology based on evidence from orientation measurements could then play a major role in estimating the thickness of beds (from the geometry of the map and cross-sections) subsequently fed through to the GVS. Perhaps most areas mix the two approaches.

The complex interplay between the GVS sub-model and its map-face counterpart may be impossible to disentangle retrospectively from a completed map. Abstraction begins during survey, so that rather than creating a full-blown general model and projecting out a GVS, it is created as a primary record, the top-down abstraction being performed in the field and only the results recorded. There can be no inverse model[469] (see Forward and inverse models 154) to enable us to fully reconstruct the original thought processes from a completed geological map. An expert on the local geology may have few apparent problems in handling the diversity of incompatible objects within one composite map, but relies on training and experience that many map users lack. On the other hand, the computer, if provided with appropriate information, could deal more rigorously with the various facets, treating them as distinct sub-models. The sub-models, as in this case, may refer to the same real-world entity but give rise to objects representing separate aspects of its geometry. The objects

[469] Inverse model: A model addressing the 'inverse problem' of looking at the outcome of past events, and trying to work back to understand the processes by which it came about.

belong to separate object classes and are based on different premises and surveying procedures, which determine the level at which the geometry can be reconciled.

The surveyors may follow a top-down approach (see The surveyor's holistic view 60) in determining which sub-models are relevant and how they are to be reconciled. Deciding to invoke a subsidiary model, however, determines a set of procedures that should be described in the metadata and followed as the sub-model is extended into the surrounding area. In other words, bottom-up methods rule within that limited context. The framework represents the top-down view, and must therefore include a mechanism for keeping track of the sub-models as they are invoked, and introducing new models as they arise (see Workflows, collaborative networks and Linked Data 53). There is a case for standardising the framework[470] and the procedures behind the sub-models, regionally or even globally, to support widespread comparisons of individual facets of the overall model, for example, to display vertical sections as fence diagrams, or stereograms that map strain symmetry, on a regional basis.

Separating the various facets, and identifying the levels at which they are reconciled, should help to clarify the reasoning that led to the overall interpretation. This information could be captured naturally as part of the survey process (see The surveyor's holistic view 60)[471] but is difficult for even the surveyor to recreate from the finished map. For this reason, among many others, long-term development of comprehensive computer systems to support field survey is desirable (see The geological investigation model 98), with the ability to retain records of the threads of reasoning as they develop. On the other hand, much ephemeral material should be discarded to a 'deleted items' folder at an early stage to avoid confusion, just as alterations, errors and superseded ideas are discarded during the preparation of a document by word processing, and probably destroyed at the end of the project.

The question then is how subsidiary models can be organised as facets of an overall framework that enables them to be built side by side and interpreted together, to interact and be filtered, transformed and projected collectively to give shared, meaningful visualisations. Through its ability to untangle implicit links within the geological map, The solid Earth systems model (sEsm) 71 could be the key to a geoscience survey component of a comprehensive computer-based knowledge system. Geoscience survey modelling should be able to cope with the study of complex systems in the field. This and other aspects of object-oriented field survey require systems support provided by The field survey model 300.

Implementation note: Comprehensive geoscience information should ideally be available for surveyors to develop their interpretations in the field. At the time of writing, widespread, broadband, wireless computer field support is not economically available. In the meantime, development of the framework in office-based systems and experience with simpler field systems will help to develop new facilities and position surveyors to take advantage of them as they become available.

---

[470] Framework: A logical structure and guidelines giving a broad overview for classifying and organizing complex information, within which detail can be added as required.

[471] Holistic: A view of a system that emphasises its properties and interrelationships acting as a whole, as opposed to the reductionist approach of studying its components in isolation as distinct entities.

The field survey model

<<Mapping geology into the knowledge system 282

Summary: *The spatial model can be fully understood by its users only if they have some perception of the surveying procedures that created it. Systems support in the field is desirable to record the surveying process and is essential for the study of complex systems. Field surveys have specific characteristics, such as unedited records, wide-ranging uncertainties and the importance of tacit knowledge, that suggest the need for a separate sub-model, to be reconciled in due course with the solid Earth systems model. Systems support could include metadata about object-class behaviour and mechanisms for feedback. The framework could include survey procedures and interpretation to guide use of the resulting spatial model.*

Many desirable features of a geoscience knowledge system may ultimately depend on field surveyors having extensive access to system support in the field (see The geological cyberenvironment (gce) 85). Leaving the snags aside for the moment, this is perhaps best described from the surveyors' viewpoint, because users must, in their minds, be able to follow the surveyors' journey for a full understanding of how the surveyors arrived at their results (see The geological investigation model 98).

Sooner or later, broadband wireless access to a comprehensive computer-based geoscience knowledge system from an electronic notebook in the field will be technically and economically feasible. For example, even in the most remote and unfamiliar area, a spatial index to existing knowledge could surely at some scale provide a topographical map or satellite imagery. It could offer metadata providing a statement of the current framework for geological survey, such as relevant stratigraphical tables, and procedures for survey of sub-models, such as depositional, structural, igneous, and metamorphic sub-models together with the relationships among them, or accounts of geological processes with advice on how to recognise their outcome and the consequences for the specific survey.

The screen could offer a top-down view of procedures matching the geologist's reasoning process. From this the surveyor could select sets of interacting sub-models and relate them to the observations, thereby refining the interpretation. Observations might be made according to the standard procedures recorded in the knowledge base, or deviations from these might be recorded and explained. The knowledge base might contain descriptions of object classes of, say, fossils, lithologies and rock types, as templates[472] for comparison with the object instances that were actually observed, again with the option to record deviations from the type specimens. The survey could thus develop along lines of reasoning in harmony with the full knowledge system, by constant interactive feedback between model and observation, as each amends the other.

Support for surveying could also come from special-purpose tools that are specific to the surveying procedures or to users studying the results in the field. These might include support from the Global Positioning System (GPS) and electronic surveying. Augmented

---

[472] Template: A file (or paper form) providing repetitive elements that guide completion of documents.

reality[473] (Wikipedia, 2010[474], Houser and Hartado, 2010[475]) might help the surveyor to compare developing spatial models with landscape features, for example, by viewing the landscape through binoculars or a camera that superimposed its intersections with spatial models of the geology. More generally, surveyors could receive software assistance in the field to assist them in the process of abstraction from the real world to the spatial model, and to filter, transform, project and visualise the results.

This hypothetical picture, however, may clash with reality. Marr (1982[476]), in the context of the human representation and processing of visual information, wrote: "The problem again was that people became so entranced by the mechanisms for doing something that they erroneously thought they understood it well enough to build machinery for it." His work led him to consider the initial processes of human vision where the eye captures huge volumes of information. The initial representation is a flood of intensity values detected by photoreceptors in the retina. Vision is the process that produces, from these images of the external world, a description that is useful to the viewer and not cluttered with irrelevant information (Marr, 1982, page 31). Marr suggested that the eye and brain initially capture key features and tokens that mark and relate points of interest in the visual images, in what he termed a 'primal sketch'. This then contributed aspects to a different, more compact, representation of the 3D spatial entities being organised in somewhat longer-term memory in the brain.

The sheer volume of information is not the only issue. Psychologists have identified separate areas of brain activity supporting two types of navigation (McNamara and Shelton, 2003[477]). The first involves following a familiar route, such as visiting the fridge during a TV commercial, or driving to work. It may track a specific sequence of landmarks and turning points. The second involves wayfinding, the navigation involved in selecting a new route through a known area. This relies on a cognitive map, that is, a three-dimensional representation in spatial memory of where features are located relative to one another. Presumably, the mind constructs the cognitive map as it gains familiarity with the area by exploration involving the first type of navigation.

The processes of vision and navigation may have counterparts in geological survey. At one extreme, the geology of an area may already be well known, and the survey may aim to collect quite specific information following a predetermined pattern, for a well-defined purpose. The data collection might then consist of filling boxes on a form, following rigidly defined procedures that allow the data to be analysed in a wider context than the initial survey. At the other extreme, the field geologist may be faced with a flood of information in

---

[473] Augmented reality: A means of combining in real time images of the actual world with an overlay of computer-generated images registered in three dimensions.
[474] Wikipedia, 2010. Augmented reality http://en.wikipedia.org/wiki/Augmented_reality
[475] Houser P., and Hartado J.M., 2010. Advanced applications of mobile computing and augmented reality for field geology. 2010 GSA Denver Annual Meeting (31 Oct –3 Nov 2010), Denver, Colorado. http://gsa.confex.com/gsa/2010AM/finalprogram/abstract_180524.htm
[476] Marr, D., 1982. *Vision: a computational investigation into the human representation and processing of visual information.* W.H. Freeman, New York. 397pp.
[477] McNamara, T.P., Shelton, A.L., 2003. Cognitive maps and the hippocampus. *Trends in Cognitive Science*, **7** (8), 333-335. http://www.psy.vanderbilt.edu/faculty/mcnamara/lab/CogMap.pdf

which it is at first unclear what is significant. The geologist might find suitable outcrops and ask them questions, but is unlikely to obtain immediate answers relating to predetermined models of the geology appearing on the screen of a field notebook. At this stage, analogous to exploratory navigation and the primal sketch, decisions on what to observe and record depend on past experience and rules of thumb.

Exploratory fieldwork might initially involve gathering vague impressions by following a sequence of clues, in the form of informative observations, to build an initial view of the situation. This might be accompanied by the emergence of a cognitive map of the geology, on paper or in the geologist's mind, locating the results of the observations, perhaps with tentative geological interpretations, in three dimensional space. The exploratory survey might generate field records that are meaningful only within the local context of the survey, and perhaps understood only by the original surveyors. Their function may be as an intermediate record as the survey proceeds, helping the surveyors to build a formal interpretation for later, wider consideration.

In reality, a survey is likely to fall between the two extremes, with more or less systematic collection of certain types of data, selected through experience of similar situations where observed syndromes may point to explanations. A physician notes the patient's symptoms, as well as the diagnosis, and may find them essential for reviewing the case at a later date. In the same way, for their basic observations, geologists use objective, descriptive terminology to ensure that they remain valid as the interpretation develops and changes. The field survey deals with various levels of certainty, from reproducible observations to tentative hypotheses. The degree of certainty could usefully be noted in the field records.

Systematic survey requires careful examination of the rocks, which may well lead to unexpected discoveries, perhaps of totally different features. They in turn may be followed up in exploratory mode, or the systematic survey may be extended to include them as an additional aspect. Thus the linear development of the observations branches out into other lines of investigation, probably linked to other threads of reasoning. A hypertext[478] representation is well suited to organising this non-linearity.

Data collection must yield notes and records, but also important is the development of tacit[479] knowledge. Perhaps some observations cannot be fully described, although they might be demonstrated in the field to experienced colleagues, for example by helping them to 'get their eye in' for identifying the cleavage planes that relate to local folding or the features that help in correlating the local stratigraphy, or to 'get a feel for' the overall structure. Digital photographs and explanatory field sketches, perhaps as overlays to photographs, could help later investigators to retrace the tacit thinking and re-examine the salient observations.

[478] Hypertext: As on the World Wide Web, information units organised as a network, through which the reader can navigate by following links embedded in it by the author. It typically represents a body of written, pictorial and other material interconnected in a complex way that cannot be represented on paper, implemented in html (HyperText Markup Language).

[479] Tacit knowledge: Knowledge which is acquired through practice and is not or cannot be articulated explicitly.

The early stages of collecting information in the field, particularly in a reconnaissance survey of an unmapped area, may inextricably combine ideas that refer to various sub-models. The observations refer to the products of interaction of various geological processes, and initially it may not be desirable, or even possible, to distinguish the effects of facets (see The multifaceted model 297) of the overall model. As the survey proceeds, however, the relevance of previously noted features may become clearer. Areas of clearer exposure, or of less weathering, metamorphism or structural deformation, may clarify the nature of earlier observations of, say, the stratigraphical sequence. Field notes might then be supplemented by tentative explanations, and reviewed in their light. A word processor makes it easier to organise a complicated structure of field notes.

The process of survey, like vision, should produce, from the glimpses of the real-world geology, a description that is useful to the viewer and not cluttered with irrelevant information. It may be only at a late stage in the investigation that ideas can be rearranged under clear topic headings, such as stratigraphy or structural geology, and communicated more widely. Yet even in the final explanation the topics are closely interwoven, and the account of the structural geology, for instance, may contain many references to the stratigraphy, without which it could not be understood.

In these circumstances, there is a case for treating the field survey as a sub-model with which the authorised survey model will be reconciled (see Reconciliation 186), but which also has permanent value in its own right. When the survey is complete, the edited field model, tracing the threads of investigation, might be required by some subsequent users of the information, for validation, resurvey or extension for a specific purpose, such as site investigation. Seeing the end product as a set of models, rather than a map and memoir, could nevertheless have a major impact on procedures of field surveying.

The field survey model is likely to be the most detailed account of the interface between the interpretation and the real world (see At the interface 133). An object-oriented model makes it easier to express the uncertainty, ambiguity and overlap inherent in fieldwork. Multi-resolution models[480] (see Scale-space 255) support general statements of the geographical location and extent of objects, combined with precise locations where these are known. Overlap, for example where there is a transition from a sandstone unit, first to sandstone with intercalated shale beds, then to shale with sand lenses, and finally to a shale unit, could be modelled in coarse detail as two objects (one of sandstone, one of shale), overlapping at the boundary. This could be supplemented, where the boundary had been examined carefully, by a representation of the intercalation in finer detail. The interpretation might indicate whether a similar pattern of transition would or would not apply to the boundary as a whole. Compared with a map, a spatial model can thus provide a clearer view of the indeterminate boundaries to be expected from complex systems[481] (see Complex and emergent systems 159).

---

[480] Multiresolution model: a model representing an object or process at various levels of detail, in order to display selected scales on request or to investigate interactions of phenomena at different scales.

[481] Complex system: A complex, emergent system has many adjacent parts that may interact according to simple rules without central control. Feedback mechanisms may result in effect not being proportional to cause and the linear equations of physics may not apply.

Field surveying in terms of coarse-resolution objects might show more than one object occupying the same space. This might seem physically impossible, but could indicate intercalation that would be apparent at a finer resolution, or genuine uncertainty, for example, where the pattern but not the location was known. It could then be expressed in terms of Bayesian probabilities (see At the interface 133), just as a bookmaker can offer odds on a win for any horse entering a race, although only one can be the winner. Precise identification of an object is not always possible, and its spatial relationships or properties, including its form, may be uncertain. The field survey sub-model should be able to handle, more effectively than a field map, uncertainties and alternative interpretations. An aim of the survey is to clarify them as the work proceeds, in part by reconciliation with other sources of information. But inevitably some will persist in the final interpretation. By default, visualisation of the completed model would show the most likely situation, with an indication that uncertainty envelopes and less likely alternative interpretations could also be viewed if need be.

In juxtaposing model and reality, much is likely to be known from other surveys, and from background knowledge of dynamic stratigraphy, geological processes and the geological history of the region (see The digital geoscience spatial index 305). One aim is to adjust and refine such a spatial interpretation by connecting it to detailed observations of the area. Another is to amend, clarify and extend the local background knowledge. Local metadata describing the likely spatial behaviour of an object class within the area of interest might come from statistical analysis of the spatial properties and relationships of known instances, from knowledge of the processes that created the object, or both. More general metadata might include information on the properties of spatial object classes, their relationships in space and time, constraints on their behaviour within process-response models, and constraints on filtering[482], morphing[483], projection and visualisation. It might enable local views to be built within regional and global models and metadata. This could lead to a clearer view of the geometry within the sub-models and thence to their reconciliation, which involves a large human input.

Implementation note: A geologist, and more so a Geological Survey, cannot make a radical switch to a cyberinfrastructure-based system for which no appropriate software system is available. The information technology industry cannot develop such systems without a clear specification and demand from geologists, who cannot evaluate the methods until the system exists. Thought experiments and small research projects provide a low-cost starting point. Longer term, methodical clarification of future developments, examination of the underpinning concepts, and development of pathfinder standards, may encourage speculative exploration of new systems linking geoscience and information technology.

[482] Filtering: A process that selectively enhances or reduces specified components of the information stream.

[483] Morphing: A seamless transition that changes one image into another by, for example, applying geometrical transformations to adjust the images so that corresponding points coincide.

The digital geoscience spatial index

<<Mapping geology into the knowledge system 282

Summary: *A geological survey organisation is likely to be concerned with specific areas within the general geoscience spatial model, dealing with lithostratigraphical, structural, igneous, metamorphic and other features of geoscience and their location ,represented by spatial and other models, authorised and supported by the Geological Survey. The user interface requires wider links to the Web, supplemented with GIS, database and visualisation facilities supporting a digital geoscience spatial index that guides users as they search and display the sparse and varied information of the spatial models.*

The phrase 'digital geoscience spatial model' indicates a change of mindset away from the constraints of the environment of pen, paper and printing press in which the traditional geological map developed. The object-oriented approach leads to a viewpoint of geoscience survey based on models from e-science rather than cartography. But information technology also affects geoscience as a whole and many conventions of traditional publication will change as electronic communication gains ground, its unpredictable future form shaped by those who grasp the commercial and academic opportunities.

The user of the cyberinfrastructure has access through the internet to the World Wide Web and cyberspace – the wide variety of linked objects in the hypermedia knowledge repository that extends far beyond spatial models and geoscience. Some objects are refereed publications secure in a long-term archive, but most are ephemeral. They come from a multitude of contributors, may or may not carry their own metadata, and each must be taken on its own terms. The general geoscience model might be thought of as a conceptual region within cyberspace loosely defined by its content (see Broadening the framework 291). Although it is only sparsely populated by information, the general structure of geoscience, and thus of the geoscience model, has frequently been described in conventional publications. Within this again is The solid Earth systems model (sEsm) 71, a more disciplined environment giving a coherent view of the basic geoscience of a defined region, probably dealing at least with its lithostratigraphical, structural, igneous and metamorphic features, and their location on the ground. It is approved, authorised, securely archived and given long-term support by an appropriate information community, normally a Geological Survey. However, digital geoscience spatial models from many sources dealing with various fragments of the general geoscience model are likely to be distributed throughout the knowledge base. The sEsm can be seen as a stable subsystem supported by a well-defined business model (see Projects and information communities 167, Stratigraphical units in space and time 141).

The user interface to access the wide scope of cyberspace must be provided by the current software for Web browsing. Most geoscientists are familiar with the associated search engines, display features and Web links, including links to the scientific literature as catalogues, summaries or full text. They are relevant to the sEsm because much geoscience, including spatial modelling, depends on processes, procedures and techniques that are not fixed in space; on scientific methods of general application from, say, mathematics, physics, chemistry or biology; and on the business context. None of these is specific to geoscience, but all are cited in many geoscience documents. Furthermore, most geoscience research and publication originates outside Geological Surveys, but may be referred to and assessed as part of the survey. Users of the sEsm must be able to access and link into this wider hypermedia knowledge repository through an appropriate framework model and user interface, preferably developed within a wide context.

Geoscientists creating or using spatial models are likely to study a wide range of local information, and can therefore benefit from and contribute to an index of the relevant material. Much of this information is significant because of its location. Therefore the desktop browser for geoscience requires additional features for spatial search and display (including specific visualisation facilities) embedded within its general facilities. For example, it should be able to support a search, within a defined area, for appropriate map sheets and their content and availability, for fossil records within a particular formation, for borehole records within a mile of a fault, or for areas where proposed building developments overlie old mine workings. This requires software with GIS functions beyond the range of the basic Web browser, referred to as Web mapping or Web GIS[484]. The index requires detailed spatial information such as digitised locations and boundaries, supplemented by non-spatial characteristics held in a geoscience database. This should make it possible, for example, to select fossil records referring to a particular species, or boreholes drilled after a specified year, for a particular purpose, to at least a specified depth, and to display the results of the retrieval in their correct positions on a map or other visualisation.

The BGS GeoIndex (BGS, 2010a[485]) illustrates the objectives and benefits of a spatial Web-based index. It refers to published material and also to externally contributed records, such as borehole descriptions, which may or may not be archived by BGS but are not subject to their quality assessment, and indeed may conflict with the BGS view expressed in a map or spatial model. It might be regarded as a prototype Geological Survey index supported by an object store, which could be closely linked to The surveyor's holistic view 60 (see also Broadening the framework 291). An index is vital because of the sparse distribution of information, including incomplete digital coverage during the lengthy transition to systems like the sEsm. When the requested information is not available, the index should at least be able to indicate possible substitutes, such as paper maps and memoirs or their scanned equivalents.

The Survey index is one potential gateway to the sEsm (see The solid Earth systems model (sEsm) 71), and should provide database and spatial search, filtering and display facilities within the software and user interfaces just mentioned. The software (see The geological cyberenvironment (gce) 85) should make it possible to visualise and analyse the outcome of the spatial models in conjunction with other items selected by the user from the wide range of indexed information. From that viewpoint, spatial models might be regarded as depicting part of the general geoscience model, built by reconciling separate views of fragments of it, to give limited but consistent content over a wide region – just like the geological map. It is controlled within the region by a designated surveying organisation participating in worldwide negotiation of standards and metadata. It is a testable interpretation, in which

[484] Web GIS: Web mapping is the process of creating and delivering maps on the World Wide Web. Web GIS adds GIS procedures to analyse and process geographical enquiries and data.
[485] BGS, 2010a. GeoIndex. http://www.bgs.ac.uk/geoindex

models are tied to entities and locations in the real world through the procedures of geoscience survey (see The conceptual model 307).

Implementation note: The user interface to the geoscience knowledge base requires the hypermedia capabilities of a Web browser and should provide a spatial index with database, GIS and associated functions for easy access to the geoscience content and a wide range of supporting material.

The conceptual model

<<Mapping geology into the knowledge system 282

Summary: *The conceptual model specifies the various objects, geological process models, and models of human procedures that interact as a system to represent the results of geoscience survey and the underlying explanations. The benefits, such as more rigorous analysis and wider communication, depend on placing their specifications within a systems framework to supplement the less formal approach of conventional methods.*

An object may contain models, and models necessarily involve objects. The terms 'object' and 'model' thus overlap. An object is a thing, which may have its properties altered, or might be transformed into another object, by a process; a model is more concerned with the sequence of actions or processes that may transform the objects, and the responses of the objects to them. The geoscience survey map takes a static view of present-day objects, whereas dynamic stratigraphy takes a dynamic view of models of geological processes. Geological processes differ from the human activities (see The dialectic model 129) that are involved in geoscience investigations, referred to here as *procedures,* such as observing, recording, interpreting and visualising. In addition to the process models and the procedural models, there are conceptual models that represent a view of a situation, generally in terms of objects, process models, and their behaviour and relationships. A framework model of the solid Earth systems model, seen as part of a knowledge-based system, might attempt to pull together a view of relevant aspects of static objects, dynamic geoscience models and survey procedures.

These models relate geological thinking to a pattern that the cyberinfrastructure can handle. To be useful the concepts must reflect a valid structure for understanding geology, and it is therefore reassuring to think that similar ideas pervade current modes of geological thought. Geological explanations are written in sentences – structures requiring nouns and verbs, probably with modifiers (adjectives and adverbs) and prepositions. These word-types refer respectively to objects, processes, attributes and relationships. Just as an object may embed process models, so a noun, such as 'erosion', may be concerned with processes. The ideas represented in a sentence are assembled into paragraphs and larger sections, building up descriptions of static objects and their dynamic transformations by processes, or, conversely, describing processes and the events that they bring about.

Words, sentences and ideas gain meaning from their context. A description of the lithology of a particular bed gains scientific value from knowledge of its location in space and the stratigraphical sequence. The description may have little value until its relationships to lateral counterparts and the sequence of comparable descriptions of adjacent beds are established by locating its position on a map or by explicit statements in the explanation. Some accounts are rather abstract, dealing with object classes[486], conceptual models[487] and the framework[488] they define. Some are concrete and specific, associating attributes with object instances and processes with specific events. The conceptual model can formalise these lines of thought.

During conventional geological survey, the map explanation obviously has close links to the map, and corresponding objects are described in both. The map shows the static objects of the present-day geology and their spatial location, pattern and relationships. They are shown in two dimensions, but aim to help the user to understand the spatial configuration in three dimensions. Diagrams, probably in the map explanation, may give static views of past configurations, which can help the reader to visualise the dynamic spatial transformations that shaped the present-day configuration. By careful use of colour and ornament, grouped in the map key, the geological map can show the level of objects within hierarchical classifications, for example, of rock types and stratigraphical units. Maps at different scales can depict various levels of spatial detail.

The map explanation contains descriptions of properties, accounts of hypotheses, and the counterpoint between them. It cannot record in text the precise detail of the spatial configuration (that is the role of the map) but it can describe and illustrate configurations from dynamic processes and events. The underlying hierarchies make it possible to refer by a word or phrase to a set of ideas that are described in greater detail elsewhere, not necessarily in the same document and possibly referring to the parallel operation of generalising to maps at a smaller scale.

Conventionally, a memoir or map explanation may deal with many informal and unidentified models, even if neither author nor reader is thinking explicitly in these terms. At one point the text might be concerned with a depositional process-response model, at another a simplified interpolation model representing the form of a formation top, at another a structural model of the deformation and perhaps a process-response model of the stress and resulting strain. Procedural models might describe, say, the design and sequence of the field investigation, or the methods of drafting a map. Models of past configurations might be illustrated by palinspastic maps and diagrams illustrating the written account by depicting the geologists' view of the position and form of objects, such as instances of basins, currents, sources and sinks. The model of the reasoning process might proceed from generalities concerning classes of objects and processes to specific instances observed in the field.

---

[486] Object class: An abstraction giving a general description of the expected properties and behaviour of the objects belonging to that class. They are a means of categorising object instances within larger groupings.

[487] Conceptual model: Descriptive model of a system clarifying the meaning of the terms and concepts, their interrelationships, and system boundaries, independently of the design and implementation of their computer representation.

[488] Framework: A logical structure and guidelines giving a broad overview for classifying and organizing complex information, within which detail can be added as required.

In a conventional explanation, a formal statement of the conceptual structure is seldom required, because the reader can grasp the lines of thinking and reasoning from background knowledge and context. The framework for a computer-based knowledge system, however, will require the various sub-models of geoscience survey to be disentangled and identified by marking them up with appropriate tags. The advantages are a more rigorous structure with explicit links that can be followed by users or by computer software.

The cyberinfrastructure brings new possibilities to geoscience survey, including wider access to more rigorous results and flexible presentation in a shared environment of worldwide and interdisciplinary communication. The system can, and must, support existing forms of map and map explanation and so incorporate legacy information. But equally, it can support quantitative measurements, their statistical analysis, and links to new measuring devices. It can support metadata or ontologies that set out rules that impose consistency on the usage and meaning of words and symbols. It can support hierarchies of objects and processes, with different levels of generality. It can abstract from instances to classes, define the expected behaviour of object classes within specific models, and handle inheritance of properties from one hierarchical level to another. It can store threads of reasoning. It can support computer programs to manipulate, transform and visualise spatial objects of any dimensionality. As geoscience survey moves to the computer environment, the manipulation of images as cartoon models in the mind can be supplemented by more rigorous mathematical and graphical representation. Dynamic models can be brought to bear on spatial objects, with testable results.

A computer-based system will introduce models that cannot be readily handled by conventional means. For example, few geologists would doubt that the systems they deal with are complicated, and many would accept that they are complex (see Complex and emergent systems 159). "What distinguishes a complex system from a merely complicated one is that some behaviors and patterns emerge in complex systems as a result of the pattern of relationship between the elements" (Wikipedia, 2005, 'complex systems'[489]). The methodology for studying complex systems is relevant to explanation of the geometry of rock bodies, and therefore to geoscience survey. This is not apparent on the geological map, which illustrates only the results of the reasoning, leaving explanation of the reasoning process to text-based accounts and to the background knowledge of the users.

The mechanical responses of the computer and the wider and more detailed sharing of information call for a formal structural framework. In other words, we must create a descriptive framework for a conceptual model of the objects, processes, procedures and relationships involved in the systems of geoscience survey. It must be able to deal with the repetition and overlap of ideas across the various sub-models, and to follow threads of thought through the complex structure. It must be capable of containing legacy information but also be open-ended, extending to accommodate structures required by unfamiliar models (see The system framework 310).

[489] Wikipedia, 2010. Complex systems. http://en.wikipedia.org/wiki/Complex_systems

The system framework

<<Mapping geology into the knowledge system 282

Summary: *Not all information is of equal value, and much that is collected in the field is discarded or access is restricted. The more valuable information requires a comprehensive, flexible and extensible framework to structure it, to signal its existence and relevance, to spot the gaps and overlaps, and to guide searching and browsing with indexes. This index must extend beyond the authoritative solid Earth systems model (sEsm) to relevant fragments of the general geoscience model and the hypermedia knowledge repository that enclose it like Russian dolls. The sEsm could include reconciled sub-models and spatial objects along with metadata describing their relationships and behaviour (including legitimate filters and transformations) enabling the browser to initiate scripts for their analysis and presentation.*

This section pulls together, from earlier discussions, some ideas bearing on the system framework[490]. The conventional flow of geological survey information follows the procedures of abstraction[491] (see Abstracting from reality to model 131), emphasising the products, from field notes to edited fair copies to published maps and map explanations to generalised maps at smaller scales and associated descriptions. To some extent, the publication of maps, publication of text accounts, and curation of samples, specimens and detailed records follow separate, parallel paths, with different representations of the information. The more generalised material tends to be communicated more widely. The detailed preliminary material may be of interest to (and understood by) only a few specialists. If it is available at all outside the Survey, it may be only through local experts or intermediaries who can explain the context, thus avoiding the costs of editing and presentation for a wider audience that may not exist.

The system could be structured around the procedures rather than the products. Efficiency demands that undue effort is not devoted to perfecting information of narrow relevance, but it may not be possible to assess its ultimate significance during the initial stages of survey. Procedures for reworking the information are therefore required, discarding many preliminary records, and retaining others in a tentative form with limited access, as they may be comprehensible only with advice from the originator or the survey team. As with conventional methods, the information must be evaluated with the intended audiences in mind, and levels of presentation and explanation adjusted accordingly. The importance of information, however, is not likely to be determined by the map scale and form of its presentation. Its relevance to a particular application is related to its position in the system framework, and the indexes (see The digital geoscience spatial index 305, The solid Earth systems metamodel (sEsmm) 110) should indicate its availability.

A comprehensive, flexible and extensible framework is needed to describe the growing content of the knowledge system, so that we can store items and their relationships where

---

[490] Framework: A logical structure and guidelines giving a broad overview for classifying and organizing complex information, within which detail can be added as required.

[491] Abstraction: reducing the information content of a concept or an observable phenomenon, typically in order to retain only salient information, relevant for a particular purpose.

we can later find and retrieve them and identify missing information. The browser menu and indexes may reflect the framework, or at least be related to it. The general geoscience model may be vast, but the overall structure and features of geoscience are described in many textbooks and are implied in bibliographical databases and thesauri (such as GeoRef, see American Geological Institute, 2010[492]), and database metadata (see Some related initiatives 341), thereby giving form to geoscience and its general model. They provide classes and categories that could be formalised to inform the indexing, linking and cataloguing of geoscience information within a computer-based geoscience knowledge system, as proposed in the The solid Earth systems metamodel (sEsmm) 110.

Within The general geoscience spatial model 293, we might identify the essential elements for the core of the dynamic and static models. Relevant extracts from them are the basis for The geological framework model 105, which supports the representation and explanation of the disposition and configuration of the rock units within a defined area (including their properties, composition, relationships and history). Distinct sub-models could be separately identified within this framework. Their products (object instances) may be inextricably combined in conventional maps, but could be traced back in an appropriately structured map explanation or spatial model. The model as a whole could enable the user to view survey procedures as a process, not just as a product. The structure must be flexible to accommodate the variety of routes that surveying procedures can follow, and the levels of detail to which they refer. For example, the framework must be able to accept existing visualisations as input in order to benefit from the legacy of maps, reworked or otherwise. It must support evolution by modification and replacement of items in the intertwined sets of sub-models and objects.

The boundaries of the authorised framework model may or may not include all the relevant sub-models. For example, irreconcilable external descriptions of boreholes may be excluded from the sEsm, even though they may be indexed and securely archived by the Survey as they could contain information of value in other contexts. The indexes (see The digital geoscience spatial index 305) and framework must cope with conflicting and fragmentary information. It should offer a coherent framework that extends beyond the sEsm into the wider knowledge base, indicating responsibilities and status for external items. The sEsm must make reference to external items, many of which may be unpublished. If the original documents are ephemeral or not widely accessible, then extensive attributed quotations from them may be necessary as notes (microdocuments) within the Survey's object store. Information on which the authorised interpretation depends must be available and held in safekeeping, perhaps in a Geological Survey archive.

An interactive browsing system could guide users (who know most about their own requirements) through the available relevant material. The system must distinguish the core sEsm from peripheral information, and approved quality-assessed information from external contributions, and must help with its filtering, projection and visualisation. Guided by metadata, the browser could initiate scripts for appropriate analysis and presentation of the information.

---

[492] American Geological Institute, 2010. GeoRef Information Services. http://www.agiweb.org/georef/index.html

Special-purpose software may have to be designed for specific filters, such as those mentioned in Invariant properties and classification 232. The software should ensure that its output of authorised material is valid by referring to the metadata. The metadata should therefore record information that determines the validity of combining objects as a composite object for display. Each spatial object and relationship, whether atomic or composite, has its own set of legitimate filters, transformations and projections. These are among the properties that determine the behaviour of the object, and should be recorded in the description of the object instance if they differ from the properties it inherits from the object class. Conventionally, verbal descriptions, chains of reasoning and process models are separate from the map in their various formats. But in the model they refer to objects placed by the metadata in a rigorous framework, and potentially can be spatially filtered and presented along with their links to the graphical visualisations.

There seems to be no prospect of direct translation from, say, an existing memoir into an artefact conforming to a formal systems framework, and without analysing memoirs we cannot arrive at an appropriate framework. The solution seems to be to create an extensible structure that is usable at an early stage and can evolve along with (or slightly ahead of) the system as a whole. Two aspects of the framework are the ability to find information, and to cope with new methods of analysis. The framework can be built step by step, as in moving from geological map to spatial model (see From map to digital model 145), extending the static frameworks to deal with dynamic models and system-supported survey (see Deformable models 274).

In thinking beyond the basic spatial models towards a comprehensive knowledge-based system, the dynamic model of how geological objects originated is central to the reasoning process. It can be seen as part of The general geoscience spatial model 293, and implemented in The solid Earth systems model (sEsm) 71. The contents of the model are inevitably incomplete, but its structure can be relatively comprehensive. Surveying procedures follow separate sub-models of various facets (see The multifaceted model 297) of the overall model, generating objects that may be incompatible. For maximum flexibility during interpretation and revision, the fragments would be captured while the survey is in progress and stored as distinct items. These sub-models would then be reconciled within the sEsm framework, referenced to the enclosing general geoscience (or Earth systems) model, which is embedded in turn in the general hypermedia knowledge base of the Web of Data (see Workflows, collaborative networks and Linked Data 53).

Conclusions on mapping to the knowledge system

<<Mapping geology into the knowledge system 282

The fragmentary information gained from geological surveying and other investigations is structured by and contributes to a holistic[493] knowledge system. This reflects the users' objectives, and relies on their background knowledge of the local situation and a more widely shared general understanding of geology. The emphasis of geological surveying is moving away from mapping geological observations and interpretations on two-dimensional paper base-maps supported by separate text-based documents, towards mapping the information into a comprehensive knowledge system that supports multi-dimensional, multi-resolution models of solid Earth systems and their consequences. An aim of systems geology is to represent this knowledge more fully within a flexible framework of generally accepted ideas, to support more rigorous analytical methods, and to provide convenient access to the information.

The future geological map 125 discusses the changing system from a geological viewpoint, emphasising the methods of geological surveying. The nature of the supporting system and the mechanisms for achieving it are considered in The emerging geoscience knowledge system 42, which offers a scenario[494] proposing longer-term objectives for systems geology. All these aspects are summarised in the Overview 11. The underlying concepts are discussed in the Stages of concept development (summary) 44. The two major components proposed are The solid Earth systems model (sEsm) 71 and its supporting infrastructure The geological cyberenvironment (gce) 85. The system[495] design should match the ways in which geologists think about their science, and provide a coherent structure of subsystems that enable the various components to work together. The geological objectives are the concern of The geological business model 93. The methods of investigation are discussed in The geological investigation model 98. The proposed structure is described in The geological framework model 105 and The geological infrastructure model 117. The requirements for each of these models and for the cyberenvironment arise partly from considerations discussed in The future geological map 125 and are listed in the sections on design requirements for each of these four models.

Implementation note: The implementation should avoid quirky systems that do not meet wider industry standards and protocols, and should provide a straightforward user interface for accessing, processing and using the information. Surveying organisations, producing standardised formal maps, are a likely source of draft frameworks for geoscience knowledge-based systems. International exploration of provisional standards for this purpose may be timely. Anticipatory pathfinder standards can help to focus both theory and solutions, even where a stable system seems a distant prospect. Topics where broad agreements can be reached may justify collaborative investment and shared standards to enhance global understanding of Earth systems.

---

[493] Holistic: A view of a system that emphasises its properties and interrelationships acting as a whole, as opposed to the reductionist approach of studying its components in isolation as distinct entities.

[494] Scenario: A description of a plausible, though uncertain, outcome.

[495] System: A set of interacting parts that function as a whole. The systems approach involves study of linkages or interfaces between the component activities.

References

<<Table of contents 1
<<Introduction and overview 1
<<The emerging geoscience knowledge system 42
<<The future geological map 125
>>Appendices 327
>>Index 359

Adams, D.C., Rohlf, F.J., Slice, D.E., 2004. Geometric morphometrics: ten years of progress following the 'revolution'. *Italian Journal of Zoology, **71**, 5-16.* http://life.bio.sunysb.edu/morph/review/review.html

Adlam, K.A.McL., Clayton, A.R., Kelk, B., 1988. A 'demonstrator' for the National Geosciences Data Index. *International Journal of Geographical Information Systems*, **2**(2), 161-170.

American Geological Institute, 2010. GeoRef Information Services. http://www.agiweb.org/georef/index.html

Anteneodo, C. , Da Lux, M.G.E. 2010. Complex dynamics of life at different scales: from genomic to global environmental issues. Phil. Trans. R. Soc. A vol. 368, no.1933, 5561-5568 doi: 10.1098/rsta.2010.0286 http://rsta.royalsocietypublishing.org/content/368/1933/5561.abstract

Asch, K., 2003. The 1:5 Million International Geological Map of Europe and Adjacent Areas: Development and Implementation of a GIS-enabled Concept. *Geologisches Jahrbuch,* SA 3, BGR, Hannover, 190 p. ISBN: 3-510-95903-5

Ashby, W.R., 1964. *Introduction to cybernetics*. London, Methuen.

BBSRC, 2007. Systems biology (UK Biotechnical and Biological Sciences Research Council). http://www.bbsrc.ac.uk/publications/corporate/systems-biology.aspx

BGS, 2010a. GeoIndex. http://www.bgs.ac.uk/geoindex

BGS, 2010b. Geoscience technologies. http://www.bgs.ac.uk/research/technologies.html

BGS, 2010c. The BGS rock classification scheme. http://www.bgs.ac.uk/bgsrcs/home.html

BGS, 2010d. Geoscience archives, William Smith. http://www.bgs.ac.uk/discoveringGeology/geologyOfBritain/archives/williamsmith/home.html

Baas, A.C.W., 2002. Chaos, fractals and self-organisation in coastal morphology: simulating dune landscapes in vegetated environments. *Geomorphology*, **48**, 309-328.

Bar-Yam, Y., 2002. "Significant points" in the study of complex systems. http://www.necsi.org/projects/yaneer/points.html

Bates, R.L., Jackson, J.A. (eds.), 1980. *Glossary of Geology*, 2nd edn. American Geological Institute, Falls Church, VA.

Beer, S., 1967. *Cybernetics and Management*, 2nd edn. English Universities Press, London, 240pp.

Berman, F., Hey, A.J.G, and Fox, G., 2003. *Grid Computing: Making the Global Infrastructure a Reality*. John Wiley & Sons, 1080pp. ISBN: 0470853190.

Berners-Lee, T., 2007. The Giant Global Graph (blog post) http://dig.csail.mit.edu/breadcrumbs/node/215

Berners-Lee, T., Hendler, J. and Lassila, O., 2001 (May). The Semantic Web. *Scientific American*, **284** (3). http://www.scientificamerican.com/article.cfm?id=the-semantic-web

Bizer, C., Heath, T., Berners-Lee, T., 2009. Linked data – the story so far. *International Journal on Semantic Web and Information Systems (IJSWIS)*, **5**(3), 1-22. DOI: 10.4018/ijswis.2009070101. http://tomheath.com/papers/bizer-heath-berners-lee-ijswis-linked-data.pdf

Blackmore, S., 2000. The power of memes. *Scientific American*, October 2000, **283**(4), 52-61.

Blackmore, S., Dawkins, R., 2000. *The meme machine.* Oxford University Press, Oxford. 264 pp.

Blaschke, T., Lang, S., Hay, G., (editors), 2008. *Object-Based Image Analysis: Spatial Concepts for Knowledge-Driven Remote Sensing Applications*. Springer-Verlag, New York. 818pp. ISBN-13: 9783540770572

Bleistein,N., Cohen, J.K., Stockwell, J.W., 2001. *Mathematics of multidimensional seismic imaging, migration and inversion*. Springer-Verlag, New York. 510pp.

Bonham-Carter, G.F., 1994. *Geographic Information Systems for Geoscientists: modelling with GIS*. Elsevier, Oxford. 398 pp.

Bradley, W.H., 1963. Geologic laws, in Albritton, C.C. (editor), 1963. *The fabric of geology*. Freeman, Cooper & Co, Stanford, pp.12-23.

Brodaric, B., Fox, P., McGuinness, D.L., 2009. Geoscience knowledge representation in cyberinfrastructure. *Computers & Geoscience,* **35**(4), 697-699.

Brodaric, B., 2004. The design of GSC Field*Log*: ontology-based software for computer aided geological field mapping. *Computers & Geosciences*, **30**(1), 5-20

Brodaric, B., and Gahegan, M., 2006. Representing geoscientific knowledge in cyberinfrastructure: some challenges, approaches and implementations. *GSA Special Paper 397*, pp.1-20.

Budd, T., 2000. *Understanding object-oriented programming with Java.* Addison-Wesley, Reading. 420pp.

Burrough, P.A., McDonnell, R.A., 1998. *Principles of Geographical Information Systems*. Oxford University Press, Oxford. 333pp.

Bush, V., 1945. As we may think. *The Atlantic Monthly* (July). http://www.theatlantic.com/doc/print/194507/bush

Buttenfield, B.B. and McMaster, R.B. (editors), 1991. *Map generalization: making rules for knowledge representation*. Wiley, New York. 245pp.

CSDMS, 2008a. Community Surface Dynamics Modeling System Project, Strategic Plan 2008-2013. http://csdms.colorado.edu/mediawiki/images/CSDMS_Strategic_Planv3F-48-op.pdf

CSDMS, 2008b. Community Surface Dynamics Modeling System Project. http://csdms.colorado.edu/wiki/Introduction

Carey, S. Warren, 1962. Scale of geotectonic phenomena. *Journal of the Geological Society of India*, **3**, 97-105.

CGI (Commission for the Management and Application of Geoscience Information), 2009. GeoSciML. http://www.cgi-iugs.org/tech_collaboration/geosciml.html

Chamberlin, T.C., 1897. The method of multiple working hypotheses. Journal of Geology. Reprinted in 1995, *Journal of Geology* **103**, 349-354.

Clapworthy, G,. Viceconti, V., Coveney, P.V., Kohl, P., 2008. Editorial. *Phil. Trans. R. Soc.* A 366, 2975-2978. doi: 10.1098/rsta.2008.0103 http://rsta.royalsocietypublishing.org/content/366/1878/2975.full.pdf+html

Coad, P. and Yourdon, E., 1991. *Object-oriented design.* Yourdon Press, Englewood Cliffs, NJ. 197pp.

Codd, E.F., 1970. "A Relational Model of Data for Large Shared Data Banks". *Communications of the ACM*, 13(6), 377–387. doi:10.1145/362384.362685.

Cornell University, 2006. Department of Earth and Atmospheric Sciences: Strategic Plan. http://www.geo.cornell.edu/lms82/StrategicPlan_Dec.pdf

Cosgrove, J.W., 1998. The role of structural geology in reservoir characterisation. *In* Coward, M.P., Daltaban, T.S., Johnson, H., (eds) *Structural geology in reservoir characterization.* Geological Society, London, Special Publications, **127**, pp. 1-13.

Cross, T.A., and Lessenger, M.A., 1999. Construction and application of a stratigraphic inverse model. In Harbaugh, J.W. et al (editors). *Numerical experiments in stratigraphy: recent advances in stratigraphic and sedimentologic computer simulations.* SEPM (Society for Sedimentary Geology) Special Publication No. 62, Tulsa, pp. 69-84.

Curtis, A., Wood, R. (editors), 2004. *Geological prior information: informing science and engineering.* Geological Society, London, Special Publications, **239**.

Daiches, D., Jones, P., Jones, J. (editors), 1996. *The Scottish Enlightenment, 1730-1790: A hotbed of genius.* The Saltire Society, Edinburgh.

Davatzikos, C., 2001. Measuring biological shape using geometry-based shape transformations. *Image and Vision Computing*, **19** (1-2), 63-74.

Dawkins, R., *The selfish gene*. Oxford University Press, ISBN 019857519X. 224pp.

de Kemp, E.A., 2000. 3-D visualization of structural field data: examples from the Archean Caopatina Formation, Abitibi greenstone belt, Québec, Canada. *Computers & Geosciences*, 26 (5), 509–530.

De Roure, D., Goble, C. and Stevens, R., 2009. The Design and Realisation of the myExperiment Virtual Research Environment for Social Sharing of Workflows. *Future Generation Computer Systems* **25** (5), 561-567. doi.org/10.1016/j.future.2008.06.010 (Preprint at http://eprints.ecs.soton.ac.uk/15709/)

De Roure, D., Jennings, N. R. and Shadbolt, N. R., 2003. The evolution of the Grid, *In* Berman, F., Hey, A.J.G, and Fox, G., (editors), 2003. *Grid Computing: Making the Global Infrastructure a Reality*, pp 65-100. John Wiley & Sons, 1080 pp. ISBN: 0470853190. http://www.semanticgrid.org/documents/evolution/evolution.pdf

De Roure, D., Jennings, N. R. and Shadbolt, N. R., 2005. The Semantic Grid: Past, Present and Future. *Proceedings of the IEEE*, 93 (3). pp. 669-681. http://eprints.ecs.soton.ac.uk/9976/

Downs, T.C., Mackaness, W.A., 2002. An integrated approach to the generalisation of geological maps. *Cartographic Journal,* 39(2), 137-152. http://www.geos.ed.ac.uk/homes/wam/DownsCartJournal2002.pdf

Dryden, I.L., and Mardia, K.V., 1998. *Statistical Shape Analysis.* John Wiley & Sons, New York. 347pp.

Edwards, R.A., Scrivener, R.C., Forster, A., 1987. Applied geological mapping, Southampton area. *BGS Research report ICSO/87/2.* British Geological Survey, Keyworth. 69pp.

Eiden, G., Kayadjanian, M., Vidal, C., 2000. Capturing landscape structures: tools. http://ec.europa.eu/agriculture/publi/landscape/ch1.htm

Emmott, S., et al., 2006. *Towards 2020 Science*. Microsoft Corporation. http://research.microsoft.com/en-us/um/cambridge/projects/towards2020science/downloads.htm

Enge, H.D., Buckley, S.J., Rotevatn, A., Howell, J.A., 2007. From outcrop to reservoir simulation model: Workflows and procedures. *Geosphere*, **3**, 469-490. doi:10.1130/GES00099.1

European Science Foundation, 2007. Systems Biology: A grand challenge for Europe. http://www.esf.org/publications/medical-sciences.html

Fennick, J. R., Keith, J. B., Leonard, R. H., Truong T. N., Lewis J. P., 2008. A cyberenvironment for crystallography and materials science and an integrated user interface to the Crystallography Open Database and Predicted Crystallography Open Database *J. Appl. Cryst.* (2008). **41**, 471-475 doi:10.1107/S0021889808000381

Finholt,T., Van Briesen, J., 2007. WATERS network cyberinfrastructure plan. http://www.watersnet.org/docs/CyberinfrastructurePlan.pdf

Foley, J.D., van Dam, A., 1995. *Computer graphics: principles and practice in C*, 2nd ed. Addison-Wesley, Reading. 1200pp. ISBN-10: 0201848406

Foster, I., Kesselman, C., Nick, J.M., Tuecke, S., 2002. The Physiology of the Grid: An Open Grid Services Architecture for Distributed Systems Integration. http://www.globus.org/alliance/publications/papers/ogsa.pdf

Foster, I., Kesselman, C. (eds), 2003. *The Grid: Blueprint for a new computing infrastructure,* 2nd ed. Morgan Kaufmann, San Francisco. 748pp.

Fowler, C.M.R., 2005. *The solid Earth: an introduction to global geophysics*, 2nd ed. Cambridge University Press, Cambridge, UK. 685pp. ISBN 0 521 89307 0

Frodeman, R., 1995. Geological reasoning: geology as an interpretive and historical science. *GSA Bulletin*, **107**(8), 960-968.

Ganser, K.A., Dickhaus, H., Metzner, R, Wirtz, C.R., 2004. A deformable digital brain atlas system according to Talairach and Tournoux. *Medical Image Analysis*, **8**(1), 3-22.

Gillespie, M.R., Barnes, R.P., Milodowski, A.E., 2011. British Geological Survey scheme for classifying discontinuities and fillings. *British Geological Survey Research Report,* RR/10/05. 56pp. ISBN 978 85272 674 7 http://nora.nerc.ac.uk/13986/

Goodchild, M. F. (1992) Geographical information science. *International Journal of Geographical Information Systems* **6** (1) , 31-45.

Gorbachev, Y.I., 1995. *Well logging: fundamentals of methods*. John Wiley & Sons Ltd., Chichester. 324pp.

Graps, A., 1995. An introduction to wavelets. *IEEE Computational Science and Engineering,* **2** (2), pp 50-61. http://www.amara.com/IEEEwave/IEEEwavelet.html

Griffith, D.A. and McKinnon, R.D. (eds), 1981. *Dynamic spatial models*. Plenum Press, New York. 443pp.

Gruber, T. R., 1995. "Toward Principles for the Design of Ontologies Used for Knowledge Sharing". In: *International Journal Human-Computer Studies*, **43,** (5-6), 907-928.

Harbaugh, J.W., Watney, W.L., Rankey, E.C., Slingerland, R., Goldstein, R.H., Franseen, E.K., 1999. *Numerical experiments in stratigraphy: recent advances in stratigraphic and sedimentologic computer simulations.* SEPM (Society for Sedimentary Geology) Special Publication 62. SEPM, Tulsa.

Harrison, J.M., 1963. Nature and significance of geological maps, in Albritton, C.C. (editor), 1963. *The fabric of geology*. Freeman, Cooper & Co, Stanford, pp. 225-232.

Hay, G.J., Dubé, P., Bouchard, A., Marceau, D.J., 2002. A scale-space primer for exploring and quantifying complex landscapes. *Ecological Modelling*, **153**(1-2), 27-49. http://www.sciencedirect.com/science/article/pii/S0304380001005002

Heap, B. (Chairman), et al. 2004. Peer review and the acceptance of new scientific ideas. *Sense about Science*, London. ISBN 0-9547974-0-X http://www.senseaboutscience.org.uk/pdf/PeerReview.pdf

Hedberg, H.D. (editor), 1976. International stratigraphic guide: a guide to stratigraphic classification, terminology and procedure. Wiley-Interscience, New York.

Heylighen, F., 2001. *The science of self-organization and adaptivity*, in: The Encyclopedia of Life Support Systems. EOLSS Publishers Co. Ltd. (in press). http://pespmc1.vub.ac.be/Papers/EOLSS-Self-Organiz.pdf

Holden, D.J., Archibald, N.J., Boschetti, F., Jessell, M.W., 2000. Inferring Geological Structures Using Wavelet-Based Multiscale Edge Analysis and Forward Models. *Exploration Geophysics*, **31**(4), 67-71

Houlding, S.W., 2001. XML – an opportunity for meaningful data standards in the geosciences. *Computers & Geosciences*, **27**, 839-849.

Houser P., Hartado J.M**.,** 2010. Advanced applications of mobile computing and augmented reality for field geology. 2010 GSA Denver Annual Meeting (31 Oct –3 Nov 2010), Denver, Colorado. http://gsa.confex.com/gsa/2010AM/finalprogram/index.html

Howard A.S., Hatton B., Reitsma, F., Lawrie, K.I.G., 2009. Developing a geoscience knowledge framework for a national geological survey organisation. Computers & Geosciences, **35**, 820–835. doi:10.1016/j.cageo.2008.06.004 http://nora.nerc.ac.uk/7128/1/Author_final_Howard2009C%26G.pdf

Howson, C., Urbach, P., 2006. *Scientific reasoning: the Bayesian approach*. 3rd ed. Open Court Publishing Company, Chicago. 352pp. ISBN 0-8126-9578-X

Hubbard, B.B., 1996. *The world according to wavelets.* A.K. Peters, Wellesley MA. 264pp.

Hughes, J.D., 1989. A multiple-layer strategy for analysis of geological data in layered sequences. *Geological Survey of Canada, Paper* **89-9**, 571.

Isaaks, E.H. and Srivastava, R.M., 1989. *Applied Geostatistics*. Oxford University Press, Oxford. 592pp.

Jain, A.K., Ross, A., Prabhakar, S., 2004. An introduction to biometric recognition. *IEEE Transactions on circuits and systems for video technology. Special issue on Image- and Video-Based Biometrics*. **14** (1), 4-20. http://www.csee.wvu.edu/~ross/pubs/RossBioIntro_CSVT2004.pdf

Kendall, D.G., Barden, D. Carne, T.K., Le, H., 1999. *Shape and Shape Theory*, John Wiley & Sons, New York. 306 pp.

Kent, W., 1978. *Data and reality*. North-Holland Publishing Company, Amsterdam. 211pp.

Kessler, H., Campbell, D., Ford, J., Giles, J., Hughes, A., Jackson, I., Peach, D., Price, S., Sobisch, H-G., Terrington, R., Wood, B., 2009a. Building on geological models : the vision of an environmental modelling platform. In: *Geological Society of America Annual Meeting 2009, Illinois, USA, 18-21 Oct 2009*. Illinois, USA, Geological Society of America, pp. 24-30. http://nora.nerc.ac.uk/8423/

Kessler, H., Mathers, S., Sobisch, H.-G., 2009b. The capture and dissemination of integrated 3D geospatial knowledge at the British Geological Survey using GSI3D software and methodology. *Computers & Geosciences,* **35**(6), 1311-1321**.** http://nora.nerc.ac.uk/7207/1/Kessler_CG_GSI3D_article_final.pdf

Koch, G.S., Link, R.F., 2002. *Statistical analysis of geological data.* Dover Publications, 832 pp. ISBN: 0486495124

Kraak, M-J., van Driel, R., 1997. Principles of hypermaps. *Computers & Geosciences*, **23**(4), 457-464.

Kuhn, T.S., 1962. The structure of scientific revolutions. The University of Chicago Press, Chicago, 172 pp.

Laszlo, E., 1972. The Systems View of the World. Braziller, New York, 131pp.

Lee, S., Morris, W., 2000. Making friends with e-journals. *The Library Association Record* **1002**(7), 392-393.

Leow, A., Yu, C.L., Lee, S.J., Huang, S.C., Protas, H., Nicolson, R., Hayashi, K.M., Toga, A.W., Thompson, P.M., 2005. Brain structural mapping using a novel hybrid implicit/explicit framework based on the level-set method. *NeuroImage*, **24**(3), 910-927.

Lewis, G.M., Lovejoy, S., Schertzer, D., and Peeknold, S., 1999. The scale invariant generator technique for quantifying anisotropic scale invariance. *Computers & Geosciences*, **25**(9), 963-978.

Licklider, J.C.R., 1960. Man-computer symbiosis. *IRE Transactions on Human Factors in Electronics,* vol HFE-1, p. 4–11 (March 1960). http://memex.org/licklider.pdf

Licklider, J.C.R., 1968. The computer as a communications device. *Science andTechnology,* (April 1968). http://memex.org/licklider.pdf

Lindeberg, T., 1994a. Scale-space theory: a basic tool for analysing structures at different scales. *Journal of Applied Statistics*, **21**(2), 225-270. http://www.nada.kth.se/~tony/abstracts/Lin94-SI-abstract.html

Lindeberg, T., 1994b. *Scale-space theory in computer vision*. Kluwer Academic Publishers, Dordrecht, Netherlands. 420pp.

Lindeberg, T., 1996. Scale-space: a framework for handling image structure at multiple scales. *In Proc. CERN School of Computing, Egmond aan Zee, The Netherlands, 8-21 September 1996.* http://www.nada.kth.se/cvap/abstracts/lin96-csc.html

Lisle, R.J., Toimil, N.C., 2007. Defining folds on three-dimensional surfaces. *Geology*, **35** (6), 519-522. doi: 10.1130/G23207A.1 http://geology.geoscienceworld.org/cgi/reprint/35/6/519

Liu, Y., Myers, J., Minsker, B., Futrelle, J., 2007. Leveraging Web 2.0 technologies in a Cyberenvironment for Observatory-centric Environmental Research. http://www.semanticgrid.org/OGF/ogf19/Liu.pdf

Loudon, T.V., 1964. Computer analysis of orientation data in structural geology. *Technical Report* No. 13 of ONR Task No. 389-135, Northwestern University, Illinois. 138 pp.

Loudon, T.V., 1982. The case for computer-based spatial models in geology. *BGS Project 22A, Progress Report.*

Loudon, T.V., 1996. Commentary on a British Geological Survey computing archive 1965-85 : *British Geological Survey, Information & Data Resources Series, Technical Report* WO/96/3 - [i],19p (1996). ISBN: X780907719

Loudon, T.V., 2000. *Geoscience after IT: a view of the present and future impact of information technology on geoscience*. Elsevier, Oxford. 142 pp. Also available as *Computers & Geosciences*, Special Issue, **26** (3A), A1-A142 http://nora.nerc.ac.uk/2395/ to http://nora.nerc.ac.uk/2411/

Loudon, T.V., 2009 Four interacting aspects of a geological survey knowledge system. *Computers & Geosciences*, **35**(4), 700-705. 10.1016/j.cageo.2007.12.009

Loudon, T.V. and Laxton, J.L., 2007. Steps towards Grid-based Geological Survey: suggestions for a systems framework of models, ontologies and workflows. *Geosphere*, **3**(5), 319-336. http://nora.nerc.ac.uk/1084/

Loudon, T.V. and Mennim, K.C., 1987. Mapping techniques using computer storage and presentation for applied geological mapping of the Southampton area: *British Geological Survey, Information & Data Resources, Research Report* WO/87/3 / ISBN X78015852X.

Ludascher, B., Lin, K., Bowers,S., Jaeger-Frank, E., Brodaric, B., and Baru, C., 2006. Managing scientific data: From data integration to scientific workflows. *Geological Society of America Special paper* 397, p.109-129, doi: 10.1130/2006.2397(08). http://users.sdsc.edu/~ludaesch/Paper/gsa-sms.pdf

Lynn, H.B., 2004. The winds of change: anisotropic rocks – their preferred direction of fluid flow and their associated seismic signatures. CSEG Recorder, October 2004, 5-11. http://www.cseg.ca/publications/recorder/2004/10oct/10oct-winds-of-change.pdf

MacEachren, A.M., Kraak, M-J., 1997. Exploratory cartographic visualization: advancing the sagenda. *Computers & Geosciences*, **23**(4) 335-343.

Mackin, J.H., 1963. Rational and empirical methods of investigation in geology, in Albritton, C.C. (editor), 1963. *The fabric of geology*. Freeman, Cooper & Co, Stanford, pages 135-163.

Mandelbrot, B.B., 1982. *The fractal geometry of nature*. Freeman, San Francisco, 460pp.

Mallet, J.-L., 2002. *Geomodeling.* Oxford University Press, Oxford. 599pp.

Mardia, K.V., 1972. *Statistics of directional data*. Academic Press, London. 355pp.

Marr, D., 1982. *Vision: a computational investigation into the human representation and processing of visual information.* W.H. Freeman, New York. 397pp.

Martin, P., Stofan, E.R., 2007. Planetary science: Multiple data sets, multiple scales, and unlocking the third dimension. *Geosphere*, **3**, 435-455. doi: 10:1130/GES00089.1

McCaffrey, K.J.W., Jones, R.R., Holdsworth, R.E., Wilson, R.W., Clegg, P., Imber, J., Holliman,N., Trinks, I., 2005. Unlocking the spatial dimension: Digital technologies and the future of geoscience fieldwork. *Geological Society (London) Journal*, **162**, 927-938. doi:10.1144/0016-764905-017

McCaffrey, K.J.W., Feely, M., Hennessy, R., Thompson, J., 2008. Visualisation of folding in marble outcrops, Connemara, westen Ireland: an application of virtual outcrop technology. *Geosphere*, 4, 588-599. doi:10.1130/GES00147.1

McInerney, T., Terzopoulos, D., 1996. Deformable models in medical image analysis: a survey. *Medical Image Analysis*, **1**(2), 91-108. http://mrl.nyu.edu/~dt/papers/mia96/mia96.pdf

McNamara, T.P., Shelton, A.L., 2003. Cognitive maps and the hippocampus. *Trends in Cognitive Science*, **7**(8), 333-335. http://www.psy.vanderbilt.edu/faculty/mcnamara/lab/CogMap.pdf

Mennis, J.L., Peuquet, D.J., Qian, L., 2000. A conceptual framework for incorporating cognitive principles into geographical database representation. *International Journal of Geographical Information Science*, **14** (6), 501-520.

Minsky, M., 1981. A framework for representing knowledge. Reprinted, pages 95-128 in Haugeland, J. (editor), *Mind Design*. MIT Press, Cambridge. 368pp.

Montagnat, J., Delingette, H., Ayache, N., 2001. A review of deformable surfaces: topology, geometry and deformation. *Image and Computing Vision*, **19**, 1023-1040.

Mossop, G.D., Shetsen, I (comp.), 1994. Geological atlas of the Western Canada Sedimentary Basin. *Canadian Society of Petroleum Geologists and Alberta Research Council*, Special Report 4L http://www.ags.gov.ab.ca/publications/wcsb_atlas/atlas.html

NACSM (North American Commission on Stratigraphic Nomenclature), 1983. North American Stratigraphic Code. *The American Association of Petroleum Geologists Bulletin*, **67**(5), 841-875.

NERC, 2007. Next Generation Science for Planet Earth, NERC Strategy 2007-2012. http://www.nerc.ac.uk/publications/strategicplan/documents/strategy07.pdf

NORA, 2010. The NERC Open Research Archive. http://nora.nerc.ac.uk

National Research Council (U.S.), 1993. *Solid-Earth Sciences and Society*. National Academy of Sciences, Washington, DC, 346pp.

Next Generation Grids Expert Group, 2006. Future for European Grids: Grids and Service-Oriented Knowledge Utilities: Vision and research directions 2010 and beyond. Report 3 for the European Commission. ftp://ftp.cordis.europa.eu/pub/ist/docs/grids/ngg3-report_en.pdf

Nicolis, G., Prigogine, I., 1989. *Exploring uncertainty.* W.H. Freeman, New York. 313pp.

Noble, D., 2010. Biophysics and systems biology, *Phil. Trans. Roy. Soc. A,* **368**, 1125-1139.

OMG, 2010. The Object Management Group (OMG.) http://www.omg.org

OmniMark, 2010. A high-performance content processing platform. http://www.stilo.com/Products/OmniMark/tabid/57/Default.aspx

OneGeology, 2010. OneGeology. http://www.onegeology.org/

Open Geospatial Consortium, 2009. Geography Markup Language. http://www.opengeospatial.org/standards/gml

Panagiotidi, S., Katsiri, E., Darlington, J., 2005. On Advanced Scientific Understanding, Model Componentisation and Coupling in GENIE, *Proceedings of the UK e-Science All Hands Meeting 2005*, pp. 1163-1170, Nottingham, UK, Sep. 2005. ISBN 1-904425-53-4. http://www.allhands.org.uk/2005/proceedings/papers/559.pdf

Paredes, C. and Elorza, F.J., 1999. Fractal and multifractal analysis of fractured geological media: surface-subsurface correlation. *Computers & Geosciences*, **25**(9), 1081-1096.

Paton, D, Carr, M., Trudgill, B., Ortner, H., Medwedeff, D.A., 2007. Alpine-scale 3D geospatial modeling: Applying new techniques to old problems. *Geosphere*, **3**, 527-549.

Pearce, M.A., Jones, R.R., Smith, S.A.F., McCaffery, K.J.W., Clegg, P., 2006. Numerical analysis of fold curvature using data acquired by high-precision GPS. *Journal of Structural Geology*, **28**, 1640-1646.

Perugini, D., Poli, G., Prosperini, N., 2002. Morphometric analysis of magmatic enclaves: a tool for understanding magma vesiculation and ascent. *Lithos*, **61,** 225-235.

Physiome Project, 2009. http://www.physiome.org

Price, K., 2010. Annotated Computer Vision Bibliography. http://iris.usc.edu/Vision-Notes/bibliography/contents.html

PRINCE2, 2009. Projects in Controlled Environments. The Office of Government Commerce (London, UK). http://www.prince-officialsite.com/

Rafols, E.J., Ring, M.B., Sutton, R.S., Tanner, B., 2005. Using predictive representations to improve generalization in reinforcement learning. *Proceedings of the 19th international joint conference on Artificial Intelligence*, 835-840. Morgan Kaufmann, San Francisco, CA. http://www.ijcai.org/papers/1650.pdf

Ramsay, J.G., 1967. *Folding and fracturing of rocks.* McGraw-Hill, New York. 568pp.

Raper, J., 2000. *Multidimensional Geographic Information Science*. Taylor and Francis, London. 300pp.

Rashed, T., 2004. Quantifying the ecological patterns of urban densification through multiple endmember spectral mixture analysis, landscape metrics and fuzzy logic. *International Archives of the Photogrammetry, Remote Sensing and Spatial Information Sciences,* Vol XXXV-B7:503-508. http://parker.ou.edu/~rashed/Publications/TarekRashed-ISPRS2004.pdf

Raskin, R., 2006. Development of ontologies for earth systems science, *in* Sinha, A.K., ed., Geoinformatics: Data to knowledge. *Geological Society of America Special Paper* **397**, 195-199. doi: 10.1130/2006.2397(14)

Richtsmeier, J.T., Deleon, V.B. and Lele, S.R., 2002. The promise of geometric morphometrics. *Yearbook of physical anthropology,* **45**, 63-91. www.hopkinsmedicine.org/FAE/JTRVBDLS2002YPA.pdf

Royse, K.R., 2010. Combining numerical and cognitive 3D modelling approaches in order to determine the structure of the Chalk in the London Basin. *Computers & Geosciences,* **36**, 500-511.

SEED, 2011. CyberGeologist. https://www.planetseed.com/node/15211

STEP Consortium, 2007. Seeding the EuroPhysiome: A Roadmap to the Virtual Physiological Human. http://www.europhysiome.org/roadmap

SUNY, 2010. Morphometrics at SUNY Stony Brook. http://life.bio.sunysb.edu/morph/

Schneider, P.J. and Eberly, D.H., 2002. *Geometric tools for computer graphics.* Morgan Kaufmann Publishers, Amsterdam. ISBN-13: 9781558605947

ScienceDirect, 2005. ScienceDirect: digital library of the future. http://www.sciencedirect.com/

Scott, A., 2006. Peer review and the relevance of science. *University of Sussex, SPRU Electronic Working Paper Series*, No. 145. http://econpapers.repec.org/paper/srussewps/145.htm

Simpson, G.G., 1963. Historical science, in Albritton, C.C. (editor), 1963. *The fabric of geology*. Freeman, Cooper & Co, Stanford, pages 24- 48.

Smith, I.F. (editor), 2005. Digital Geoscience Spatial Model, Final Report. *British Geological Survey Occasional Publication* No. 9, 56pp. http://www.bgs.ac.uk/downloads/start.cfm?id=535

Sowa, J.F., 2000. *Knowledge representation: logical, philosophical, and computational foundations.* Brooks/Cole, Pacific Grove. 694pp.

Sowa, J.F., 2001. *Processes and causality*. http://www.bestweb.net/~sowa/ontology/causal.htm

Stewart, S.A., Hay, G.J., Rosin, P.L., and Wynn, T.J., 2004. Multiscale structure in sedimentary basins. *Basin Research*, **16**, 183-197. doi: 10.1111/j.1365-2117.2004.00228.x http://www.geog.umontreal.ca/gc/PDFs/New_PDFs/2004_stewart_Hay.pdf

Stewart, S.A., Podolski, R., 1998. Curvature analysis of gridded geological surfaces. *In* Coward, M.P., Dalbatan, T.S., Johnson, H., (eds) Structural geology in reservoir characterisation. *Geological Society, London, Special Publications*, **127**, 133-147.

Steiniger, S., Hay, G.J., 2009. Free and Open Source Geographic Information Tools for Landscape Ecology. *Ecological Informatics*, **4**, 183-195. http://www.geo.unizh.ch/publications/sstein/sstein_freegitools_ecoinf2009.pdf

Strang, W.G. and Fix, G.J., 1973. *An analysis of the finite element method.* Prentice-Hall, Englewood Cliffs. 306pp.

Strebelle, S., 2002. Conditional simulation of complex geological structures using multiple-point statistics. *Mathematical geology*, **34** (1), 1-22. doi: 10.1023/A:1014009426274 http://www.springerlink.com/content/8g2meagu5k0u07pk/

Strebelle, S., Levy, M., 2008. Using multiple-point statistics to build geologically realistic reservoir models: the MPS/FDM workflow. *Geological Society, London, Special Publication*, **309**, 67-74. doi: 10.1144/SP309.5

Sutton, R.S. and Barto, A.G., 1998. Reinforcement learning: an introduction. MIT Press, Cambridge, MA. http://webdocs.cs.ualberta.ca/~sutton/book/ebook/the-book.html and chapter 1 at http://webdocs.cs.ualberta.ca/~sutton/book/chapter1.pdf

Tearpock, D.J. and Bischke, R.E., 2003. *Applied Subsurface Geological Mapping*. 2nd ed. Prentice-Hall, Upper Saddle River, NJ. 822pp.

Tennyson, A., 1842. The Lady of Shalott. http://charon.sfsu.edu/tennyson/tennlady.html

Thompson D'A. W., 1942. *On growth and form* (revision of 1917 edition). Cambridge University Press. 1116pp.

Trifonov, G.F., 1984. Maps as stages of the cognition process in geology. *In*: Dudich, E. (editor), 1984. *Contributions to the history of geological mapping*. Proceedings of the Xth INHIGEO Symposium 1982, Budapest, Hungary. Akademiai Kiado, Budapest, 47-53.

Turcotte, D.L., 2002. Fractals in petrology. Lithos, **65**(3), 261-271.

Turk, G., O'Brien, J.F., 2002. *Modelling with implicit surfaces that interpolate*. ACM Transactions on Graphics **21**(4), 855-873. http://www.cc.gatech.edu/~turk/pubs.html

USGS, 2007, (Committee on Research Priorities for the USGS Center of Excellence for Geospatial Information Science, Mapping Science Committee, National Research Council). *A Research Agenda for Geographic Information Science at the United States Geological Survey.* National Academies Press, 156 pp. ISBN-10: 0-309-11154-4 http://books.nap.edu/catalog.php?record_id=12004

USGS, 2008. National Geospatial Program http://www.usgs.gov/ngpo/

US National Center for Supercomputing Applications, 2008. NCSA 2010: The future of NCSA. http://www.ncsa.uiuc.edu/AboutUs/NCSA_2010.pdf

US National Research Council, 1993. *Solid-Earth Sciences and Society*. National Academies Press. 368pp. ISBN-10: 0309047390 ISBN-13: 978-0309047395

Usenet Newsgroup comp.theory.self-org-sys, 2008. Self-Organizing Systems (SOS) FAQ: Frequently asked questions. http://www.calresco.org/sos/sosfaq.htm

Van Wagoner, J.C., Hoyal, D.C.J.D., Adair, N.L., Sun, T., Beaubouef, R.T., Deffenbaugh, M., Dunn, P.A., Huh, C., and Li, D, 2003. Energy dissipation and the fundamental shape of siliciclastic sedimentary bodies. *Search and Discovery* Article #40080. http://www.searchanddiscovery.com/

Voisard, A., 1998. Geologic Hypermaps are more than Clickable Maps! Proceedings of the International ACM GIS Symposium, ACM Press, New York, November 1998. http://citeseer.ist.psu.edu/150047.html

W3C, 2005. World Wide Web Consortium: http://www.w3.org/ OWL Web Ontology Language Overview: http://www.w3.org/TR/owl-features/ RDF Resource Description

Framework: http://www.w3.org/RDF/ Semantic Web: http://www.w3.org/2001/sw/ XML Extensible Markup Language: http://www.w3.org/XML/

Wawrzyniec, T.F., Jones, R.R., McCaffrey, K., Imber, J., Holliman, N., Holdsworth, R.E., 2007. Introduction: Unlocking 3D earth systems, harnessing new digital technologies to revolutionize multi-scale geological models. *Geosphere*, **3**, 406-407. doi:10.1130/GES00156.1

Watson, G.S., 1966. The statistics of orientation data. *Journal of Geology*, 74, 786-797.

Wealands, S.R., Grayson, R.B., Walker, J.P., 2005. Quantitative comparison of spatial fields for hydrological model assessment – some promising approaches. *Advances in Water Resources*, **28** (1), 15-32.

Wijns, C., Poulet, T., Baschetti, F., Dyt, C., Griffiths, C.M., 2004. Interactive inverse methodology applied to stratigraphic forward modelling. *In* Curtis, A. and Wood, R. (eds) 2004. Geological Prior Information: Informing Science and Engineering. *Geological Society, London, Special Publications*, **239**, 147-156. 186239-171-8/04/$15.00.

Wikipedia, 2005. Wikipedia, The Free Encyclopedia. http://www.wikipedia.org/

Wikipedia 'Augmented reality', 2010. http://en.wikipedia.org/wiki/Augmented_reality

Wikipedia 'Complex systems ', 2010. http://en.wikipedia.org/wiki/Complex_systems

Wikipedia 'Google Maps', 2010. http://en.wikipedia.org/wiki/Google_Maps#cite_note-2

Wikipedia 'Massively multiplayer online game', 2010. http://en.wikipedia.org/wiki/Massively_multiplayer_online_game

Wikipedia 'Mesh generation', 2011. http://en.wikipedia.org/wiki/Mesh_generation

Winchester, S., 2001. *The Map that changed the World*. Penguin Books, London.

Wu, J. and David, J.L., 2002. A spatially explicit hierarchical approach to modelling complex ecological systems: theory and applications. *Ecological Modelling*, **153** (1-2), 7-26.

Xu, X., Aiken, C.L.V., Bhattacharya, J.P., Corbeanu, R.M., Nielsen, K.C., McMechan, G.A., and Abdelsalam, M.G., 2000. *The Leading Edge*, **19**(2), 197-202.

Yang, C., Raskin, R., Goodchild, M., Gahegin, M. 2010. Geospatial Cyberinfrastructure: past, present and future. Computers, Environment and Urban Systems. **34**, 264-277

Zacharaki, E.I., Hogea, C.S, Biros, G., Davatzikos, C., 2008. A Comparative Study of Biomechanical Simulators in Deformable Registration of Brain Tumor Images. *IEEE Transactions on Biomedical Engineering*, **55**(3), 1233-1236. http://repository.upenn.edu/cgi/viewcontent.cgi?article=1146&context=meam_papers

Zhang, T., Switzer, P., Journel, A., 2005. Merging prior structural interpretation and local data: the Bayes updating of multiple-point statistics. *Proceedings of IAMG '05: GIS and spatial analysis,* **1**, 615-620.

Appendices

<<Table of contents 1
<<Introduction and overview 1
<<The emerging geoscience knowledge system 42
<<The future geological map 125
<<References 314
Appendices 327
Glossary 328
Some related initiatives 341
>>Index 359

Glossary

<<Table of contents 1
<<Introduction and overview 1
<<The emerging geoscience knowledge system 42
<<The future geological map 125
<<References 314
Appendices 327
>>Some related initiatives 341
>>Index 359

Definitions are given here of some terms as used in 'A Scenario for Systems geology: The emerging geoscience knowledge system and the future geological map'.

Abstraction: reducing the information content of a concept or an observable phenomenon, typically in order to retain only salient information, relevant for a particular purpose (see Scale, Granularity).

Agent: An autonomous entity, or a complex software entity that performs useful tasks, with a degree of autonomy, seeking to achieve defined goals in collaboration with its user.

Algorithm: A formal set of rules or instructions that can be followed to solve a problem or perform a specific task, such as the instructions of a computer program.

Analogy: The resemblance in some particulars between things otherwise unalike. In logic, reasoning from parallel cases, based on the assumption that if things have some similar attributes, other attributes may also be similar.

Attractor: In complex systems, a preferred position in state space, such that if the system starts from another state, it tends to evolve towards an attractor.

Augmented reality: A means of combining in real time images of the actual world with an overlay of computer-generated images registered in three dimensions.

Browser: A software application that assists the user in searching for, retrieving and presenting relevant information, typically from the internet or World Wide Web.

Business model: A summary of the strategies, resources, organisation, infrastructure and operational processes to be employed in achieving a defined set of objectives.

Cartoon: A preliminary design or sketch.

Classification: The systematic assignment of objects to categories based on their properties.

Cloud computing: Distributed computing supplying services, such as data and processes, to the desktop or mobile device from the 'cloud' of large, distributed data centres.

Codify: Create a representation or record of something in a form appropriate to the organised system of which it becomes a part.

Collaboratory: A networked system linking scientists for formal and informal communication across locations and organisations to share and discuss their investigations and collaborate in such tasks as system design or research projects.

Complex system: A complex, emergent system has many adjacent parts that may interact according to simple rules without central control. Feedback mechanisms may result in effect not being proportional to cause and the linear equations of physics may not apply.

Conceptual model: Descriptive model of a system clarifying the meaning of the terms and concepts, their interrelationships, and system boundaries, independently of the design and implementation of their computer representation.

Configuration: The spatial arrangement, pattern, form and shape of objects or their properties. Used by Simpson (1963) in contrasting 'The actual state of the universe or any part of it at a given time, its configuration, is not immanent and is constantly changing' with 'The unchanging properties of matter and energy and the likewise unchanging properties and principles arising therefrom are immanent in the material universe.'

Content: In this context, the representation of information, such as text, images, maps, algorithms or data that is held (contained) in an information source, such as a book, journal, web-site or database.

Continuous function: A function where a small change in input is matched by a small change in output. It is therefore possible to create a very small zone about a point within which the value does not significantly change. Continuity is a basic concept, discussed at length in mathematical texts.

Convolution: A mathematical operation on two functions occupying the same space, to produce a third function.

Correlation (statistical): Two random variables are positively correlated if high values of one are likely to be associated with high values of the other, and negatively correlated if high values of one are likely to be associated with low values of the other. The correlation coefficient is a widely used measure of correlation.

Correlation (stratigraphical): A means of piecing together information from separate observations or from separate outcrops to establish more widespread units with similarity in properties, age, or mode of formation, thereby aiming to classify the units, locally or globally.

Curvature: Intuitively, the extent to which a line or surface differs from a straight line or flat surface. Mathematically, curvature comes in many forms, measured in ways extending far beyond this scenario.

Cyber-: Derived from the Greek word meaning 'steersman' through cybernetics (the study of control mechanisms and feedback systems in animals and machines), used as a prefix (equivalent to e-) to indicate the electronic or computer based version of a conventional product or service.

Cyberenvironment: Aspects of the cyberinfrastructure assembled to meet requirements relevant to a particular field of enquiry, aiming to maintain global compatibility while providing access through interfaces that match users' working practices.

Cyberinfrastructure: An integrated assemblage of computing, information and communication facilities, deploying the combined capacity of multiple sites to provide a framework to underpin research and discovery, typically with broad access and end-to-end coordination. According to Wikipedia, the term cyberinfrastructure "was used by a United States National Science Foundation (NSF) blue-ribbon committee in 2003 in response to the question: how can NSF, as the nation's premier agency funding basic research, remove existing barriers to the rapid evolution of high performance computing, making it truly usable by all the nation's scientists, engineers, scholars, and citizens? The NSF use of the term focuses on the integrated assemblage of these information technologies with one another. Cyberinfrastructure is also called **e-Science;** in particular, the United Kingdom has a major e-Science initiative."

Cyber-strategy: A plan or scenario describing how an organisation or individual intends to respond to the current and future development of the infrastructure.

DBMS: See Database Management System.

DEM: A digital elevation model, typically a digital matrix of point values, representing the elevation of the ground surface.

Data: A collection of observations, measurements or other information about a set of variables, generally in the form of numbers, words, or images.

Database management system: a set of computer programs to create, maintain and use a database (an organised collection of digital data), typically providing facilities to select and retrieve items meeting criteria, such as values equal to, less than or more than values specified by the user for each of a set of variables. See also Relational database.

Dereferencing: Looking up a URI on the Web to obtain the referenced information.

Delaunay triangle: Components of a mesh of adjacent triangles (triangulated irregular network), joining known sampled values at their vertices in such a way that the circumcircle of each triangle contains no other points. It is used for modelling terrain and other surfaces, as the triangulation avoids narrow triangles and can be computed efficiently.

Dialectic: The theory and practice of weighing and reconciling juxtaposed or contradictory arguments for the purpose of arriving at the truth.

Dimension: The minimum number of coordinates needed to specify a point within a space or object.

Dimension (Euclidean): The three dimensions of familiar everyday space, where a point can be located by distances from a chosen origin on three axes at right angles (x, y, and z) and the distance between two points can be calculated by the root of the sum of squares of the differences of their x, y and z values (the theorem of Pythagoras); also the extension to any number of dimensions (axes) of 'Euclidean' space, with distance calculated in this way.

Dimension (Fractal): The fractal dimension (D) indicates statistically how completely a fractal function fills space on zooming down to finer granularity (scale).

Discontinuity: In geology, a break or abrupt change in the physical properties of a rock body. In mathematics, a point where there is a break or gap in a continuous function.

Documentation: The process of storing and retrieving information in a field of knowledge.

e-book: (also eBook, ebook) a book published in electronic form or a digital version of a paper book, readable by computer, possibly with interactive content including hyperlinks and multimedia.

e-Science: See cyberinfrastructure.

Earth system processes: The forces for change which operate now and shaped the past evolution of successive configurations of the solid Earth, as disentangled by geologists in their record of Earth history. Their outcome depends on the input, not on when and where they took place (they are invariant under specifiable time and space transformations).

Earth systems science: The unified study of the physical, chemical and biological components, processes and their interactions that determine states and changes in the planet. See Earth System Science Partnership (ESSP) at http://www.essp.org/.

Emergence: Complex patterns, properties and systems resulting from relatively simple interactions.

Entity: Something with a distinct, separate existence, usually in the real world, as distinct from an object that is regarded as an abstract representation of the real-world entity.

Experiment: Observations made in circumstances over which the scientist has control.

Feedback: The process whereby part of the output of a system is returned to an input control mechanism that regulates its further output. Positive feedback intensifies the process, negative feedback brings deviations back towards the average value.

Filtering: A process that selectively enhances or reduces specified components of the information stream.

Fitness function: A means of measuring the desirability or suitability of a particular state for a specific purpose.

Forward model: The model of a process affecting objects with known properties and composition and predicting the outcome.

Fractal: A geometrical form with a pattern that is repeated at ever-decreasing scales. If split into parts, each part resembles a reduced-size copy of the whole.

Framework: A logical structure and guidelines giving a broad overview for classifying and organizing complex information, within which detail can be added as required.

Function (mathematical): An abstract entity that associates an input to a corresponding output according to some rule.

GIS: See Geographic Information System.

GML: (Geography Markup Language) an XML encoding for the transport and storage of geographic information

gce: The geological cyberenvironment (see cyberenvironment)

Generalisation: In cartography, showing salient features, possibly in a simplified form, and removing unnecessary detail. In a narrative document, providing an abstract, summary or general outline of the essential content.

Generalisable prediction: The ability to extend a local predictive model to have wider relevance – emphasised in the reinforcement learning approach.

Generalised vertical section: A vertical diagram, typically in the margin of a map, representing graphically the sequence of rock units occurring within a particular area.

Geographical Information Science: Goodchild (1992) defined Geographic Information Science as "a multidisciplinary research enterprise that addresses the nature of geographic information and the application of geospatial technologies to basic scientific questions". GIScience relies on expertise from many allied fields and has intimate ties to geospatial technology and applications (USGS, 2007).

Geographical Information System (GIS): An integrated system for the capture, storage, management, retrieval, analysis, manipulation and display of geographically referenced spatial data and its attributes.

Geoinformatics: The application of information science and technology to geography and geoscience. Sinha et al. 2010: "an informatics framework for the discovery of new knowledge through integration and analysis of Earth science data and applications".

Geological cyberenvironment: The cyberinfrastructure for end-to-end support of geological investigation, for example, in the context of a solid Earth systems model.

Geological framework model: A model depicting and clarifying the principal relationships among the findings of geology, linking aspects of the content and organisation of the geoscience knowledge system.

Geological knowledge system: The geological knowledge system collects, organises, evaluates, assembles and supplies knowledge of the solid Earth. The developing network of support services based on computing, information and communications technology (the cyberinfrastructure) is an important external influence on future directions of the geological knowledge system, and leads to the concept of systems geology.

Geological surveying: A process, based on available existing knowledge, of selectively observing the properties and relationships of located geological objects, bringing appropriate representations of them into a broader knowledge structure, and testing the consequences.

Geological survey organisation: A state, national or federal institution employed to maintain and advance the knowledge of geosciences, traditionally centred on production of geological maps, reports, and archives of records and specimens.

Geology: Geology is defined in the AGI Glossary as: "The study of the planet Earth – the materials of which it is made, the processes that act on these materials, the products formed, and the history of the planet and its life forms since its origin. Geology considers the physical forces that act on the Earth, the chemistry of its constituent materials, and the biology of its past inhabitants as revealed by fossils...The knowledge thus obtained is placed in the service of man – to aid in the discovery of minerals and fuels of value in the Earth's crust, to identify geologically stable sites for major structures, and to provide foreknowledge of some of the dangers associated with the mobile force of a dynamic Earth." (Bates and Jackson, 1980).

Geostatistics: An approach to the statistical analysis and interpolation of spatial patterns, originally developed in mining geology.

GIS: See Geographical Information System.

Granularity: The level or degree of specific detail or resolution at which information is observed or presented.

Grid: "The grid integrates services across distributed, heterogeneous, dynamic 'virtual organizations' formed from the disparate resources within a single enterprise and/or from external resource sharing and service provider relationships in both ebusiness and e-science." (Foster et al., 2002). The Grid aims to provide seamless access to computational resources, data and the services to process it.

Hierarchy: An organised body of things (ranked in classes one below the other) branching downwards as an inverted tree structure.

Holistic: A view of a system that emphasises its properties and interrelationships acting as a whole, as opposed to the reductionist approach of studying its components in isolation as distinct entities.

Hypertext: As on the World Wide Web, information units organised as a network, through which the reader can navigate by following links embedded in it by the author. It typically represents a body of written, pictorial and other material interconnected in a complex way that cannot be represented on paper, implemented in html (HyperText Markup Language).

IT: See Information technology.

Idiopathic: [Of a disease] arising spontaneously from an unknown cause.

Immanent: Something naturally inherent and intrinsic within and throughout its domain.

Information and **knowledge**: As used here, information is a representation of knowledge, which is regarded as what is known about a topic (and possibly recorded), gained through learning, experience and familiarity.

Information technology (IT): The application of computers, communications and software to manage, process and disseminate information

Information trail: A sequence of items of information, typically connected by hyperlinks, tracing the development of ideas or interpretations, or a sequence of actions and operations followed in a workflow, for example during field survey.

Information type: The manner in which information is represented and processed, for example, as spatial images, narratives, data, algorithms, tacit and background knowledge.

Infrastructure: The basic facilities, services and installations needed for a system to function.

Interface: The shared boundary between systems or parts of a system, or the means of interaction across the boundary that makes joint operation possible.

Interoperability: Interoperability of information is the ability of concepts, terms or models from various sources to work together, by meeting standards that enable sharing and reuse of information.

Interpolation: The estimation of values, for example at a point or along a line or surface, in order to predict a value or complete a visualisation.

Invariant: An object with the property of invariance, that is, it does not change under a specific set of transformations or sequence of operations.

Inverse model: A model addressing the 'inverse problem' of looking at the outcome of past events, and trying to work back to understand the processes by which it came about.

Investigation model: A model of how an investigation (such as a geological survey of an area) is carried out, including the tools, techniques and procedures by which information is collected, assembled and communicated.

Knowledge: See Information.

Knowledge base: A dynamic repository for information and methods for accessing and processing it. It is generally machine-readable and online, and may include the means to access expert knowledge.

Kriging: A method of interpolating a surface with a least-squares algorithm, calculating surface values from nearby data points according to distance, weighted on the basis of a semi-variogram.

Linked data: A style of publishing and interlinking structured data by looking up URI's on the Web. As opposed to seeing a fixed set of data sources, 'Linked Data' aims to create an unbound global data space (Bizer et al., 2009**).** It links things in the world through their descriptions, by means of structured data on the Web.

Magnifying or dilating (geometry): Multiplying all distances by the same factor.

Mapping (Geological): Conventionally, geological mapping leads to a graphical depiction, usually on a flat surface, of spatial relationships and forms of geological features or properties in a selected area of the Earth's surface or subsurface. As defined in mathematics, mapping relates the elements of one set to those of another. A broader definition of geological mapping could be 'relating elements of geological observation or interpretation of the solid Earth to corresponding elements in appropriate models in the geoscience knowledge system'.

Mark-up: Symbols inserted in a document in a mark-up language such as SGML, HTML or XML, to tag the beginning and end of character sequences that can be interpreted by machine, and can be omitted for displaying to the user.

Mechanism: A group of objects or parts that interact together in a predetermined manner through a chain of causation to bring about a particular product or result.

Meme: According to Dawkins (The Selfish Gene, 1976): "a unit of cultural inheritance, hypothesized as analogous to the particulate gene and as naturally selected by virtue of its 'phenotypic' consequences on its own survival and replication in the cultural environment."

Examples are ideas and concepts passed from mind to mind by imitation or explanation, evolving through variation, selection and heredity.

Metadata: Metadata is a description of data (often used in a broad sense of representations of information or knowledge) and its context. It is structured to assist the user or computer to find, manage, control and understand the data.

Metamodel: A metamodel is a description of the organisation and function of a model, to assist the user or computer to find, manage, control and understand its contents.

Metric space: A set where a metric, or notion of distance between elements of the set, is defined. For example, in the familiar three-dimensional Euclidean metric space, the distance between two points, which are elements of the set, is defined by the length of a straight line joining them.

Microdocument: A short note or module of information, typically referenced by a URI and seen here as a system component, such as a minimum revisable unit.

Minimum revisable unit: A self-contained subset of information in a documentation system: a component of the system designed to be revised as necessary without endangering the integrity of the system as a whole

Model: A formalised representation giving a simplified view of aspects of the real (or of an imaginary) world relevant to the purposes in hand. [See also forward, inverse, conceptual, multiresolution models]

Module: A subdivision of a system that can be combined with others in various ways to perform different functions, can be independently replaced or upgraded, or can be plugged into the system to extend its functionality.

Morphing: A seamless transition that changes one image into another by, for example, applying geometrical transformations to adjust the images so that corresponding points coincide.

Multiresolution model: a model representing an object or process at various levels of detail, in order to display selected scales on request or to investigate interactions of phenomena at different scales.

Narrative: A message, story or discourse relating a sequence of real or imagined events, descriptive features, or both.

Objects: Representations of real-world or conceptual things or entities of interest in a particular context.

Object class: An abstraction giving a general description of the expected properties and behaviour of the objects belonging to that class. They are a means of categorising object instances within larger groupings.

Object instances: Representations of specific, identified, real-world or hypothetical objects.

Object-oriented: An approach to analysis, design, and classification, which can support many aspects of thinking about objects and their relationships including linking them with interweaving threads.

Octree: A three-dimensional equivalent of the quadtree, providing a compact representation of a property sampled at varying spatial density.

Ontology: A formal representation describing concepts, entities and relationships in a domain of knowledge, typically providing a more detailed and rigorous machine-readable specification than a thesaurus or taxonomy.

Orthogonal: Orthogonal variables refer to axes that are at right angles to one another, or are independent of one another.

Paradigm: The set of common beliefs and agreements shared between scientists about how problems should be understood and addressed (Kuhn, 1962).

Patch: In ecology, represents a discrete area of relatively homogeneous environmental conditions, with reference to a phenomenon of interest at a particular scale.

Perspective projection (geometry): Diminution of size (of objects in a 2D projection) with distance from the viewpoint, mimicking the effect of perspective

Pixel: A picture element, the smallest element of an image that can be separately accessed.

Platform: A structure designed to provide a stable base enabling a range of diverse entities and activities to collaborate for a shared purpose.

Prediction: Drawing conclusions from incomplete evidence. Predictions can result from reasoning about a hypothesis (a suggested explanation of a phenomenon), and are 'useful' (throw light on the likely truth of the hypothesis) if they can be tested by observation or experiment. Also, predictions of as yet unobserved phenomena can stem from a theory, in the sense of a comprehensive explanation supported by facts gathered over time.

Predictive: Providing a probabilistic estimate of properties, situations, behaviour, or events. See also Generalisable Prediction.

Predictive reinforcement learning: A means of characterising a learning problem in terms of an agent seeking to achieve a goal by interacting with an uncertain environment.

Prior information: What is already known before setting out to solve a problem.

Process: A particular course of action intended to achieve a result, or a series of natural occurrences that bring about change.

Project: An activity undertaken for a particular purpose within a particular business setting, generally with its own objectives, priorities, operational definitions, and sampling schemes.

Project plan: The project plan identifies what is to be achieved, by whom, how, where, and when, and may be broken down into component tasks, their interactions and duration.

Project profile: A brief outline of the contents of the intended project results, including their provenance, to enable others to establish the relevance of the project to their own work.

Projection (geometry): Reduction of the number of dimensions.

Provenance (of information): The source, origin or derivation of items of information, which might be formalised in terms of, for example, project, originator, date, place, collection method, archive or database identifier, authorisation.

Quadtree: A hierarchical data structure in which geographical space is subdivided (to varying levels) into nested square tiles.

Reasoning: The process of deducing inferences or conclusions from assumptions, observations, or other evidence.

Reconciliation: Kent (1978, pp. 202-203) points out that people have different views of reality, and that these change with time. But the views overlap and so can be *reconciled* with varying degrees of success to serve different purposes. "By reconciliation, I mean a state in which the parties involved have negligible differences in that portion of their world views which is relevant to the purpose at hand."

Reductionism: The 'reductionist' mode of explaining complicated phenomena reduces them to simple parts controlled by mechanical processes governed by the deterministic laws of physical science (in contrast to the systems approach).

Reinforcement learning: Approaching a problem by exploiting existing knowledge of the situation to guide the explorative process – learning more by problem-solving to add to that knowledge and thereby improving the process. In an artificial intelligence context, the goal is defined by a reward function, which might reflect the objectives of a geological investigation. The method is considered to be well suited to generalisable prediction.

Relational database: A collection of computer-readable data organised as tables ('relations'), generally 'normalised' to avoid duplication ('redundancy').

Resolution: In the geographical sense, regarded as the minimum distance between two points on a map or image that can be distinguished by eye or other sensor.

Resource: A fundamental entity of the Web architecture, such as a document, file or multimedia object, identified by a Uniform Resource Identifier (URI) or locator (URL).

Rotation (geometry): Turning an object about an axis through the origin.

Scale-space: In GIS, scale is the ratio of the distance between two points on a map or image to the corresponding distance on the ground (see also Resolution). In the study of scale-space, however, scale may refer to the level of detail detected by eye or instrument across a range of scale from the finest detail discriminated, to the entire image or field of view. Scale-space theory regards this range as a multi-resolution continuum, zooming in or out to reduce or increase the amount of detail using a filtering process. It studies the range of scales over which objects exist and geological processes operate.

Scattergram, scatter plot: A graphic presentation in which measurements are shown at points along axes representing the values of variables.

Scenario: A description of a plausible, though uncertain, outcome.

Script: A program or set of instructions that is interpreted by another program (as opposed to a compiled program that is interpreted by the computer processor).

sEsm: See Solid Earth system model.

Self-organization: the spontaneous emergence of global coherence out of local interactions (Heylighen, 2001)

Self-similarity: When part of a self-similar object is viewed at another scale (reduced or magnified), it retains its shape characteristics. The statistical shape properties of parts of the object are the same as those of the whole over a range of scale change. See **Fractal**.

Semantic Web: Berners-Lee et al. (2001) described the Semantic Web as an extension woven into the structure of the existing Web, in which information is given well-defined meaning, improving the ability of computers and people to work in cooperation.

Simulation: Imitation of aspects of internal processes of a system and their results; usually to visualise, statistically compare with, or predict real-world occurrences.

Solid Earth systems model (sEsm): An approach to structuring distributed knowledge of the science of geology to provide an integrated view in the context of sciences of the solid Earth as a whole. A model of the systems of the solid Earth, organised within a framework or metamodel that depicts and clarifies the principal relationships among the findings of geology, providing a multidimensional map to locate and connect ideas, concepts, workflows of investigation and threads of reasoning. The content of the model is distributed information referring to: the three-dimensional disposition and configuration of the present-day observable objects of the solid Earth (where things are and how they are arranged); their observed and interpreted properties, composition and relationships, at all scales; geological processes and the outcomes of their interactions with configurations of objects; events and historical changes throughout geological time.

Space, metric: A set of points, the distance between any two of which can be defined by a non-negative real number. For example, in the familiar three-dimensional Euclidean space, the distance between two points is defined as the square root of the sum of squares of differences in their x, y and z coordinates.

Spatial model: A model interpreting spatial data and relationships to clarify and understand spatial forms and processes and thereby predict real world attributes.

Spline: A class of mathematical functions (such as piecewise polynomials) used for interpolation or smoothing. They mimic the operation of the flexible ruler used by draftsmen to draw smooth curves.

Standards: Established norms or requirements generally set out in formal documents by a recognised authority.

State space: Phase or state space is a mathematical abstraction where each variable (measurable property) of a system is regarded as a separate dimension.

Strategy: A plan of action designed to achieve a specific outcome.

Stratigraphical unit: A stratum or body of strata recognised as a unit for description, mapping, or correlation.

Stratigraphy: The systematic definition and description of natural divisions of stratified rocks and their arrangement, including their classification, nomenclature, correlations, composition, mutual relationships, interpretation, and distribution.

Stretching (geometry): Applying different magnification along different axes.

Symbiosis: A close interdependence or association (in the literal sense, of animals or plants of different species) often of mutual benefit.

Syndrome: A set of properties, features, sequences or relationships that tend to occur together, and may characterise a particular interpretation. Detecting one may therefore lead the observer to look for the others.

Synergy: The enhanced result of interaction among parts of a system that mutually reinforce one another, so that the whole is greater than the sum of the parts.

System: A set of interacting parts that function as a whole. The systems approach involves study of linkages or interfaces between the component activities.

Systems geology: A view of geology re-based on the developing cyberinfrastructure and regarded as a system (a set of interacting parts that function as a whole) embedded in the wider knowledge system.

Tacit knowledge: Knowledge which is acquired through practice and is not or cannot be articulated explicitly.

Technology: The application of scientific and other technical knowledge to practical tasks.

Template: A file (or paper form) providing pre-prepared elements in repetitive documents and guiding their completion.

Thiessen polygon: (Also known as a Voronoi diagram) A form of decomposition of a metric space determined by distances to a discrete set of points. The polygon sides are perpendicular bisectors of the sides of Delaunay triangles in a triangulated irregular network (see Delaunay triangle).

Tiling: Division of space into regular or irregular polygons (tiles).

Transform: Applying a function that changes the position, direction or scales of the axes of a coordinate system.

Translation (geometry): Bodily movement of an object in a defined distance and direction.

Typed: Data are typed by placing them in a category or classification, typically indicating their possible values, how they are stored, and the range of valid operations that can be applied to them. Typing may apply to datasets and other objects.

URI (Uniform Resource Identifier): a name (string of characters) identifying an object (resource). For example, a URL might refer to its location on the internet.

Utility: A directly and immediately service with established functionality, performance and dependability.

Variable: A quantity that can assume any of a set of values

Variogram: In geostatistics, a graph relating the difference in elevation (or values of some other property) of two points on a surface to the distance between them.

Visualisation: Transforming quantitative data (including the results of interpolation) into sensory information – images that the eye and brain can interpret and visualise.

Wavelet: A local function, similar to a bounded segment of a Fourier series, representing a pattern. Wavelet analysis compares the pattern at various scales to each area of an image (by convolution), providing a mechanism to locate patterns and measure their contribution to the overall variation in the image.

Web GIS: Web mapping refers to the process of creating and delivering maps on the World Wide Web. Web GIS adds GIS procedures to analyse and process geographical enquiries and data.

Workflow: The representation of a process or procedure in terms of a sequence of operations to be carried out to complete a task.

World Wide Web (Web, WWW): Web pages and other hyperlinked internet resources retrievable by Hypertext Transfer Protocol (HTTP).

Worldview: The overall perspective and beliefs from which a person or group sees, understands and interprets reality, or a particular aspect of it.

Zero-crossing: A point where the sign of a function changes (as it crosses the axis or zero value of the function's graph), or (in the case of a set of scanned lines) a line joining such points.

Some related initiatives

<<Table of contents 1
<<Introduction and overview 1
<<The emerging geoscience knowledge system 42
<<The future geological map 125
<<References 314
<<Appendices 327
<<Glossary 328
Some related initiatives 341
Internal BGS 341
Other geological initiatives 343
Hydrocarbons geology 347
Earth Sciences 349
Other relevant fields 355
>>Index 359

This is a rather arbitrary and very incomplete selection from many initiatives related to the Earth Sciences. It consists largely of brief extracts from Web sites that describe their content, and aims to give a flavour of developments relevant to systems geology.

Internal BGS

<<Some related initiatives 341

IDA

"The (Intranet Data Access) application provides routine searching and data management functionality for in-house users of a wide range of BGS geoscience data (http://bgsintranet/resources/data/ida/idamain.htm). The IDA is designed to address the issue of the lack of a uniform user interface. It uses a single suite of web technologies to provide a common interface that is available on every machine in BGS whether desktop PC, notebook, UNIX workstation or Apple. Users no longer have to wander the site looking for the PC with the interface to a given database mounted on it and ask the operator to get them the data. The IDA has a common look and feel so that a user always knows where to look for the 'Search' button etc and they are always named the same."

GeoIndex

"is a web map-based index of information that BGS has collected or has obtained from other sources. The British Geological Survey's GeoIndex (http://www.bgs.ac.uk/geoindex/home.html) allows you to see the extent of many of the important data holdings of the BGS. Launched in 2000, GeoIndex helps users to discover which of these datasets may assist them in their business or scientific interests. It is implemented using ArcServer WebGIS technology."

GSI3D

"GSI3D (Geological surveying and investigation in three dimensions) is now available on general release as part of the not–for–profit GSI3D Research Consortium (http://www.gsi3d.org.uk/consortium.html). It is a methodology and associated software tool for 3D geological modelling which enables you to quickly and intuitively construct 3D solid models of the subsurface for a wide range of applications. The methodology and software has been developed jointly by the British Geological Survey (BGS) and INSIGHT GmbH and is being applied by the BGS, where it is the modelling tool of choice.

"GSI3D is designed for the geoscientist, rather than expert software users. The model is built by enabling the user to construct traditional cross sections by correlating boreholes and outcrop data to produce a network of interlocking sections, or geological fence diagram. Together with a digital elevation model, this geological interpretation is then used by the software engine to produce a 3D solid model of the subsurface — a single click operation. Geoscientists can draw their sections based on facts such as borehole logs or geophysical sections correlated by intuition — the shape 'looks right' to a geologist. This 'looks right' element pulls on the geoscientists' wealth of understanding of earth processes, examination of exposures and theoretical knowledge. GSI3D enables the efficient capture of tacit and implicit knowledge which until now has been difficult to tap into using existing modelling methodologies."

3D Geology in BGS

"BGS is developing 3D models at range of scales to meet the needs of our users (http://www.bgs.ac.uk/services/3Dgeology/home.html). A 1:1 million scale model of the whole of the UK has been completed and mainly serves as an educational tool used to give an overview of UK geology, including major faults, and can show other features such as the magnitude and depth of earthquakes. A range of regional models have also been developed in an ongoing programme of work. These models will extend to up to 5km depth and provide a well constrained structural framework for regional, strategic assessment of groundwater and energy resources, and for deep underground storage and waste repositories. Our detailed models concentrate on the near surface and are used for planning and development, archaeological investigations and site characterisation. We can develop 3D models of urban areas to meet your requirements."

NGDC

"The National Geoscience Data Centre (NGDC) (http://www.bgs.ac.uk/services/ngdc/home.html) holds a comprehensive collection of geological and environmental information on the surface and subsurface of Great Britain, and offshore, which is available to the public, industry and academia.

"The Data Centre manages earth science datasets, physical collections, records and other information gathered or generated by the BGS, or its precursors, in addition to data provided by external organisations.

"It is also the [UK] Natural Environment Research Council's designated data centre for the earth sciences and maintains and makes available the results of academic research from its grant holders and other researchers."

SIGMA

"The objectives of SIGMA (System for Integrated Geoscience Mapping) (http://www.bgs.ac.uk/research/sigma/home.html) are to design, develop and implement a structured and consistent methodology for the complete geological surveying process in BGS. Implementation has necessitated the development of digital systems for the acquisition and management of geoscientific data and the construction of spatial models, maps and GIS. The project is adopting a holistic, systems engineering approach to address all aspects of the implementation process, including development of equipment and software, re-engineering of management systems and work processes and staff training and support."

SIGMA Mobile

"BGS SIGMAmobile (http://www.bgs.ac.uk/research/sigma/download.html) is a Tablet PC system that is designed to allow field staff to collect digital data/information in the field. One of its main strengths is the ability to bring practically any digital dataset to the field including DigMap, historic OS maps, scanned field slips, NEXTMap etc. SIGMA Mobile uses a combination of heavily customised ArcGIS, MS Access and InfiNotes. While originally designed with the 'mapping geologist' in mind, other uses such as recording / describing landslides, near surface geology and minerals etc. can be added with support from the relevant Programmes."

Other geological initiatives

<<Some related initiatives 341

American Mineralogist Crystal Structure Database

(http://rruff.geo.arizona.edu/AMS/amcsd.php) "This site is an interface to a crystal structure database that includes every structure published in the American Mineralogist, The Canadian Mineralogist, European Journal of Mineralogy and Physics and Chemistry of Minerals, as well as selected datasets from other journals. The database is maintained under the care of the Mineralogical Society of America and the Mineralogical Association of Canada, and financed by the National Science Foundation."

CGI

"The Commission for the Management and Application of Geoscience Information is a Commission of the International Union of Geological Sciences. Our mission is to enable the global exchange of knowledge about geoscience information and systems. Our website (http://www.cgi-iugs.org/home.html) will be one of the most important ways that CGI will achieve its mission. We hope you will find the website worth visiting! If you have any feedback, comments, or suggestions you are welcome to send them to us. If you would like to know more about us or perhaps become a member of CGI, please see our participation pages. A major initiative of the CGI is the development of GeoSciML. More information about GeoSciMl, as well as downloads of GeoSciML materials, are available." [More on GeoSciML below.]

CHRONOS

(http://www.chronos.org/) "Geologic time is the intellectual theme that connects a wide variety of research endeavors in geoscience – missing is the corresponding cyberinfrastructure that allows the resources of all these endeavors to be pooled. CHRONOS's purpose is to transform Earth history research by seamlessly integrating geoscience databases and tools.

"CHRONOS is a team of geoscientists and information technology specialists creating a cyberinfrastructure that will deliver open access to a global federation of Earth history databases, tools, and services, thus providing: For academic, government, and industrial scientists – access to multiple, disparate databases on Earth history; data evaluation and conversion services; and powerful analytical tools. For autonomous databases, affiliated science initiatives, and data and tool contributors – a larger user community, greater visibility and acknowledgment, and access to tools and best practices, without the cost and burden of reproducing interoperability. For educators, students, and policy makers – a convenient source of Earth history data, visualization tools, expert opinion, and educational materials."

EarthChem

(http://www.earthchem.org/) "is a community-driven effort to facilitate the preservation, discovery, access and visualization of the widest and richest geochemical datasets. Search seamlessly across multiple databases. The EarthChem Portal offers a "one-stop-shop" for geochemistry data of the solid earth with access to complete data from multiple data systems. The portal features mapping and visualization tools. Access comprehensive igneous geochemistry and topical data collections. Earthchem builds and maintains topical data collections and provides access to topical datasets developed and maintained by partner projects. For example, the EarthChem Deep Lithosphere Dataset contains geochemical and petrological data from lower crust and upper mantle xenoliths."

GEON

Geosciences Network (GEON): "The project (http://www.geongrid.org/) is a collaboration among a dozen PI institutions and a number of other partner projects, institutions, and agencies to develop cyberinfrastructure in support of an environment for integrative geoscience research. GEON is funded by the NSF Information Technology Research (ITR) program.

"The key integrative science theme in GEON is a more quantitative understanding of the 4-D evolution of the North American lithosphere. The cyberinfrastructure in GEON is required to support an inherently distributed system—since the scientists, who are users as well as providers of resources (e.g., data, tools, and computing and visualization capabilities), are themselves distributed. Furthermore, GEON is required to tackle the extreme heterogeneity among data and tools, across a wide range of earth science sub-disciplines and disciplines.

"A number of integrative science themes provide the initial guiding applications for realizing this cyberinfrastructure. These include (1) gravity modeling of 3D geological features such as plutons, using semantic integration of (igneous) rock and gravity databases, and other geological and geophysical data, (2) study of active tectonics via integration of LiDAR data sets, data on distribution of faults and earthquakes, and geodynamics models, and (3) study of lithospheric structure and properties across diverse tectonic environments via the integration of geophysical, petrologic, geochronologic, and structural data and models.

"The GEON distributed system is based on a "service-oriented architecture (SOA)". Advanced information technologies have been developed to support "intelligent" search, semantic data integration, and visualization of multidisciplinary information spaces and 4D earth science data. The environment provides access to high performance computing platforms for data intensive analysis as well as for compute-intensive model execution. The GEON Portal provides users with a convenient, Web-based interface to access the various resources.

"The core GEON cyberinfrastructure is generic in nature and broadly applicable beyond the Geosciences to a variety of other science disciplines as well as other application domains. Indeed a number of geosciences and other projects are significantly leveraging a range of technologies that have been developed in GEON."

GEOROC

"GEOchemistry of Rocks of the Oceans and Continents (http://georoc.mpch-mainz.gwdg.de/georoc/) is a searchable collection of more than 50000 analyses from different tectonic settings."

GeoSciML

"The GeoSciML application (http://www.geosciml.org/) is a standards-based data format that provides a framework for application-neutral encoding of geoscience thematic data and related spatial data. GeoSciML is based on Geography Markup Language (GML – ISO DIS

19136) for representation of features and geometry, and the Open Geospatial Consortium (OGC) Observations and Measurements standard for observational data. Geoscience-specific aspects of the schema are based on a conceptual model for geoscience concepts and include geologic unit, geologic structure, and Earth material from the North America Data Model (NADMC1, 2004), and borehole information from the eXploration and Mining Markup Language (XMML). Development of controlled vocabulary resources for specifying content to realize semantic data interoperability is underway."

IGME 5000

"The 1:5 Million International Geological Map of Europe and Adjacent Areas. A major European GIS project: the 1:5 Million International Geological Map of Europe and Adjacent Areas (IGME 5000) (http://www.bgr.de/karten/igme5000/igme5000.htm) is being managed and implemented by the Federal Institute for Geosciences and Natural Resources (BGR) under the aegis of the CGMW (Commission of the Geological Map of the World). The project involves over 40 European and adjacent countries and the final area covered will reach from the Caspian Sea in the east, to the Mid-Ocean Ridge in the west, and from Svalbard to the southern shore of the Mediterranean Sea. The aims of the project are to develop a GIS underpinned by a geological database, and also a printed map providing up-to-date and consistent geological information.

"The GIS will hold significantly more information than the previous printed maps could ever provide. It will also offer versatility, e.g. to retrieve and present for the whole of Europe, information on age, petrography and structural and metamorphic features. More importantly the IGME 5000 GIS will provide the essential foundation for pan-European applied geo-environmental thematic mapping. While the main theme of the GIS is the pre-Quaternary geology of both the land and offshore areas of Europe, it is planned to include additional themes, such as Quaternary geology – a key factor influencing the natural landscape. In the course of the project also a CD-ROM will be produced with a subset of the GIS and the related database. The project is dependent on the numerous contributions of the many countries. An extensive multinational project like the IGME 5000 requires meticulous preparation and establishment of standards and protocols in order to provide the essential structure and guidelines for the data compilation e.g. common term dictionaries for the database. In addition a standard topographic base map was an essential prerequisite. So in many areas the IGME 5000 is establishing basic standards where none exist." See also Asch (2003[496]).

NAVDAT

"The North American Volcanic and Intrusive Rock Database (http://www.navdat.org/) is intended as a web-accessible repository for age, chemical and isotopic data from Mesozoic

---

[496] Asch, K., 2003. The 1:5 Million International Geological Map of Europe and Adjacent Areas: Development and Implementation of a GIS-enabled Concept. Geologisches Jahrbuch; SA 3, BGR, Hannover, 190 p. ISBN: 3-510-95903-5

and younger igneous rocks in western North America. This region has long been a natural laboratory for efforts to test the links between igneous activity, tectonics, and ore deposition. These efforts have relied on ad hoc databases that became obsolete or abandoned once the project was completed. NAVDAT represents an effort to provide a permanent and publicly available database for existing and new age and geochemical data from igneous rocks in western North America. The database allows a continent-wide look at complex space-time patterns of magmatism."

OneGeology

"OneGeology (http://www.onegeology.org/) is an international initiative of the geological surveys of the world and a flagship project of the 'International Year of Planet Earth'. Its aim is to create dynamic geological map data of the world available via the web. This will create a focus for accessing geological information for everyone. Thanks to the enthusiasm and support of participating nations the initiative has progressed rapidly and geological surveys and the many users of their data are excited about this ground-breaking project. "

PetDB

"Information System for Geochemical Data of Igneous and Metamorphic Rocks from the Ocean Floor (http://www.petdb.org/). PetDB archives and serves analytical data for whole rocks, glasses, minerals, melt inclusions, with emphasis on basalts and abyssal peridotites. PetDB contains major, trace-element, and isotope ratios for samples from mid-ocean ridge basalts, back-arc basins, young near-ridge seamounts, and old oceanic crust."

USGS National geologic map database

(http://ngmdb.usgs.gov/) "A geoscience resource for maps and related data about geology, hazards, earth resources, geophysics, geochemistry, geochronology, paleontology, and marine geology."

Hydrocarbons geology

<<Some related initiatives 341

In the 1990's, the Petrotechnical Open Software Corporation developed the object-oriented Epicentre Model, which included aspects of geology and geophysics, and could be implemented in relational database management software for computing efficiency. This ambitious attempt to achieve standards for interoperability in the context of upstream oil exploration data is now being continued by Energistics. They and the PPDM Association are largely concerned with modelling subsurface information for hydrocarbons exploration, including geological and geophysical data and the tools for data collection.

Energistics

Energistics – Standards Resource Centre (http://www.energistics.org/home) "As it develops, the Centre will become a one-stop reference source on information and process Standards and Best Practices, fostering greater communication among industry users of open standards...The scope of the Centre is the energy industry.

"The value proposition for our members, as well as for participants in all other industry collaborative groups, is to invest in collaborative efforts in a non-commercial or competitive environment that lead to the creation and ongoing support for industry Standards and Best Practices. These *products* become internalized and incorporated in energy company practices and procedures as well as supplier products and services thereby benefiting all participants with higher quality and lower cost information-related operations.

"The Centre can be thought of as a place to harvest the essential identification and descriptive information from these collaborations and to present that information in an integrated, easy to use manner. The Centre goes only so far as to help visitors identify and understand Standards and Best Practices at a high level. Links and pointers are provided to visitors to guide them to the original sources -- on the Energistics Web sites or in a location managed by the host organization."

The PPDM Association

The Professional Petroleum Data Management Association (PPDM Association) (http://www.ppdm.org/) "is a global, not for profit standards organization that works collaboratively with industry to create and publish data management standards for the resource industry. Through the PPDM Association, world-wide petroleum data experts gather together in a collaborative, round table approach to engineer business driven, pragmatic data management standards that will meet industry needs. PPDM Version 3.8 is an open, practical and usable standard that is supported by over 100 members. [It is] supported by members from a broad range of petroleum companies, government agencies, software application and data vendors, and service companies. Together, we are identifying new business opportunities through multidisciplinary information sharing. [Its mission is as] a global not-for-profit organization which collaborates with the petroleum industry to develop and promote information standards that enhance profitability."

Earth Sciences

<<Some related initiatives 341

DataGrid

Natural Environment Research Council (UK) 'DataGrid' (http://www.nerc.ac.uk/research/programmes/escience/results/ourdata.asp) : "New technologies are driving an explosion in the volumes of environmental data available. The volume, type, and range of data are increasing beyond the scope of any one institution to manage and make available. In the past, scientists struggled to find out what data existed and where they were held. If they could track down a dataset, they still needed to know enough about the format to be able to use it. The NERC DataGrid team has developed the infrastructure to allow scientists to find, understand, manipulate and visualise data from many institutions around the world. By using and extending international standards, the NERC DataGrid provides access to some major data holdings in the United States, Germany and Australia. Like the web, the NERC DataGrid has no owner or central control, data remains with data providers, and users can access all the data from a single location. Currently focused on atmospheric and marine science, the team expect to develop the DataGrid to cover all of NERC science."

EarthScope

"The EarthScope (http://www.earthscope.org/) scientific community is conducting multidisciplinary research across the Earth sciences utilizing the freely accessible data collected and maintained by EarthScope facilities. In-depth collaboration between scientists and educators bring the excitement of cutting-edge Earth science research into classrooms, museums and parks. EarthScope provides freely accessible data and data products from thousands of geophysical instruments that measure motions of the Earth's surface, record seismic waves, and recover rock samples from depths at which earthquakes originate. EarthScope is funded by the [US] National Science Foundation."

Earth System Curator

Earth System Curator (http://www.earthsystemcurator.org/) "The Earth System Curator team is prototyping a software environment for assembling, running, and archiving information about climate models. The idea is to make it easier for scientists to perform modeling experiments, and to coordinate with each other on efforts such as Model Intercomparison Projects (MIPs) (http://www-pcmdi.llnl.gov/projects/model_intercomparison.php) and Intergovernmental Panel on Climate Change (IPCC) (http://www.ipcc.ch) assessments."

Earth System Science Partnership

"The Earth System Science Partnership (http://www.essp.org/) is a partnership of four international global change research programmes (DIVERSITAS, IGBP, IHDP and WCRP) for the integrated study of the Earth System, the changes that are occurring to the system and the implications of these changes for global sustainability.

"The central activities of the ESSP are projects on issues of global sustainability, designed to address the global change aspects of four critical issues for human well-being: energy and carbon cycles, food systems, water resources and human health. Capacity building is also a central part of the ESSP activities.'

"The Earth System is the unified set of physical, chemical, biological and social components, processes and interactions that together determine the state and dynamics of Planet Earth, including its biota and its human occupants.

"Earth System Science is the study of the Earth System, with an emphasis on observing, understanding and predicting global environmental changes involving interactions between land, atmosphere, water, ice, biosphere, societies, technologies and economies."

eGY

The Electronic Geophysical Year, 2007-2008 (http://www.egy.org/index.php) (*e*GY) "provides an opportunity for the international geoscientific community to focus effort on a 21st Century e-Science approach to issues of data stewardship: open access to data, data preservation, data discovery, data rescue, capacity building, and outreach. The development of Virtual Observatories and Laboratories is a central feature of *e*GY.

"*e*GY is an internationally-recognized resolve by the science community to achieve a step increase in making past, present, and future geoscientific data readily, rapidly, conveniently, and openly available. *e*GY provides the international framework and a target for stimulating and coordinating activities to make this happen. *e*GY focuses on themes of electronic data location and access, permission and release of data, conversion of data into modern digital form, data preservation, capacity building, particularly in developing countries, and outreach. Promoting the development of a network of virtual observatories is a central feature of *e*GY."

EOSDIS

"The Earth Observing System Data and Information System (EOSDIS) (http://nasadaacs.eos.nasa.gov/) manages and distributes data products through the Distributed Active Archive Centers (DAACs) (http://nasadaacs.eos.nasa.gov/about.html). The centers process, archive, document, and distribute data from NASA's past and current research satellites and field programs. Each center serves one or more specific Earth science disciplines and provides data products, data information, services, and tools unique to its particular science."

ESMF

"The Earth System Modeling Framework (ESMF) (http://www.earthsystemmodeling.org/) collaboration is building high-performance, flexible software infrastructure to increase ease of use, performance portability, interoperability, and reuse in climate, numerical weather prediction, data assimilation, and other Earth science applications. The ESMF defines an architecture for composing complex, coupled modeling systems and includes data structures and utilities for developing individual models.

"The basic idea behind ESMF is that complicated applications should be broken up into smaller pieces, or components. A component is a unit of software composition that has a coherent function, and a standard calling interface and behavior. Components can be assembled to create multiple applications, and different implementations of a component may be available. In ESMF, a component may be a physical domain, or a function such as a coupler or I/O system."

ESSC

The UK Environmental Systems Science Centre (ESSC) (http://www.nerc-essc.ac.uk/index.php) is "concerned with gaining a better understanding of the environment by developing new ways of modelling earth system processes using spatial information, information that is often sensed remotely. We use techniques from the mathematical and physical sciences to solve problems which affect the natural world atmosphere, the oceans, the land surface and the solid earth."

FGDC

"The FGDC *[Federal Geographic Data Committee]* http://www.fgdc.gov/dataandservices coordinates the sharing of geographic data, maps, and online services through an online portal, geodata.gov, that searches metadata held within the NSDI Clearinghouse Network. The geodata.gov portal (http://geo.data.gov/geoportal/catalog/main/home.page) is operated in support of the Geospatial One-Stop Initiative to provide 'one-stop' access to all registered geographic information and related online access services within the United States. Geographic data, imagery, applications, documents, web sites and other resources have been catalogued for discovery in this portal. Registered map services allow casual users to build online maps using data from many sources. Registered data access and download services also exist for use by those interested in downloading and analyzing the data using GIS or viewer software."

GENIE

The Grid-enabled integrated Earth system model (GENIE) (http://www.genie.ac.uk/) is funded by the UK Natural Environment Research Council (NE/C515904) through the e-Science programme.

"The project will deliver both a flexible Grid-based architecture, which will provide substantial long-term benefits to the Earth system modelling community, and also new scientific understanding from versions of the ESM generated and applied in the project. Our scientific focus is on long-term and paleo-climate change, especially through the last glacial maximum (~21kyr BP) to the present interglacial, and the future long-term response of the Earth system to human activities. A realistic ESM for this purpose must include models of the atmosphere, ocean, sea-ice, marine sediments, land surface, vegetation and soil, ice sheets and the energy, biogeochemical and hydrological cycling within and between components." For details of the GENIE scalable modular platform, see Panagiotidi et al. 2005[497].

GML

"The Geography Markup Language (GML) http://www.opengeospatial.org/standards/gml is an XML grammar for expressing geographical features. GML serves as a modeling language for geographic systems as well as an open interchange format for geographic transactions on the internet. As with most XML based grammars, there are two parts to the grammar – the schema that describes the document and the instance document that contains the actual data. A GML document is described using a GML Schema. This allows users and developers to describe generic geographic data sets that contain points, lines and polygons. However, the developers of GML envision communities working to define community-specific application schemas [en.wikipedia.org/wiki/GML_Application_Schemas] that are specialized extensions of GML. Using application schemas, users can refer to roads, highways, and bridges instead of points, lines and polygons. If everyone in a community agrees to use the same schemas they can exchange data easily and be sure that a road is still a road when they view it. Clients and servers with interfaces that implement the OpenGIS® Web Feature Service Interface Standard [http://www.opengeospatial.org/standards/wfs] read and write GML data. GML is also an ISO standard (ISO 19136:2007) [www.iso.org/iso/iso_catalogue/catalogue_tc/catalogue_detail.htm?csnumber=32554]. See also the GML pages on OGC Network: http://www.ogcnetwork.net/gml ."

IGBP

"The International Geosphere-Biosphere Programme (IGBP) (http://www.igbp.net/) is a research programme that studies the phenomenon of Global Change... The vision of IGBP is to provide scientific knowledge to improve the sustainability of the living Earth. IGBP studies the interactions between biological, chemical and physical processes and interactions with human systems and collaborates with other programmes to develop and impart the understanding necessary to respond to global change.

[497] Panagiotidi, S., Katsiri, E., Darlington, J., 2005. On Advanced Scientific Understanding, Model Componentisation and Coupling in GENIE, *Proceedings of the UK e-Science All Hands Meeting 2005*, pp. 1163-1170, Nottingham, UK, Sep. 2005. ISBN 1-904425-53-4. http://www.allhands.org.uk/2005/proceedings/papers/559.pdf

"IGBP's research goals are to:

- Analyze the interactive physical, chemical and biological processes that define Earth System dynamics
- The changes that are occurring in these dynamics
- The role of human activities on these changes"

OpenMI

OpenMI (http://www.openmi.org/reloaded/) [OpenMI stands for Open Modeling Interface and aims to deliver a standardized way of linking of environmental related models.]

"The OpenMI can be described at two levels. At the users level, the OpenMI provides a standard interface, which allows models to exchange data with each other and other modelling tools on a time step by time step basis as they run. It thus facilitates the modelling of process interactions. The models may come from different suppliers, represent processes from different domains, be based on different concepts, have different spatial and temporal resolutions and have different spatial representations including no spatial representation. A useful analogy is to consider the OpenMI as the modelling equivalent of a USB cable.

"At the IT level, OpenMI standard is a software component interface definition for the computational core (the engine) of the computational models in the water domain. Model components that comply with this standard can, without any programming, be configured to exchange data during computation (at run-time). This means that combined systems can be created, based on OpenMI-compliant models from different providers, thus enabling the modeller to use those models that are best suited to a particular project. The standard supports two-way links where the involved models mutually depend on calculation results from each other. Linked models may run asynchronously with respect to timesteps, and data represented on different geometries (grids) can be exchanged seamlessly.

"The OpenMI standard is defined by a set of software interfaces that a compliant model or component must implement. These interfaces are available both in C# and Java."

PRISM

Program for Integrated Earth System Modelling (PRISM) (http://www.prism.enes.org/) "provides the Earth System Modelling community with a forum to promote shared software infrastructure tools. The ever increasing complexity of Earth System Models and computing facilities is a heavy technical burden on the research teams developing them. The goal of PRISM is to help share the development, maintenance and support of standards and state-of-the-art software tools to assemble, run, and analyse the results of Earth System Models based on component models (ocean, atmosphere, land surface, etc..) developed in the different climate research centres in Europe and elsewhere. PRISM is organised as a distributed network of experts who contribute to five 'PRISM Areas of Expertise' (PAE): Code coupling and I/O, Integration and modelling environments, Data processing, visualisation and management, Meta-data, and Computing issues. PRISM was initially funded as a project under the European Union's Framework Programme V (2001-2004) and its long term support is now ensured by multi-institute funding via the PRISM Support Initiative (PSI)."

SEEGrid

Solid Earth and Environment Grid (SEEGrid) (https://www.seegrid.csiro.au/wiki/bin/view/Main/WebHome) "Sustainable management of mineral, energy and environmental resources is a knowledge-based process that relies upon continual access to accurate geo-spatial data in its many forms, data processing and analysis tools, and integration platforms. Over the past decade, the shift to geographic information systems (GIS), 3D and temporal modelling, process simulation and visualisation have transformed the way that earth scientist work. In order to achieve the next advance required to sustainably manage our resources, we must be able to easily, quickly and reliably access the huge volumes of complex geoscientific data as well as suitable processing and analysis tools required to generate terrain specific knowledge and visualise it in a mix of 2D to 4D environments. Grid technologies provide part of the solution by facilitating access to the different and non-centralised resources. Grid technologies have the capacity to make access to geoscientific data repositories, processing packages and computer power as easy as the web has made access to information.

"Generic grid technologies are not sufficient to achieve this objective. It is necessary for open standards and interfaces to be established by communities to be able to interoperate effectively. The Solid Earth and Environment Grid community has been established to bring together people in the earth, environmental and computing sciences to address the issues of "transparent access" to data and knowledge about the earth, and the available and potential technologies offered by the grid that enhance our ability to explore for and manage our natural and mineral resources.

"A workshop sponsored by the CSIRO Glass Earth Initiative, Geoscience Australia, the Predictive Mineral Discovery Cooperative Research Centre (pmd*CRC), Australia's Academic and Research Network (aarnet), and the Australian Research Council (ARC) was held in July of 2003 to propose this initiative. The response was immediately enthusiastic and this web site now serves as the community 'meeting place' for the establishment of the open standards and interfaces for this community."

SWEET

Semantic Web for Earth and Environmental Terminology (http://sweet.jpl.nasa.gov/) "This project provides a common semantic framework for various Earth science initiatives. The semantic web is a transformation of the existing web that will enable software programs, applications, and agents to find meaning and understanding on web pages. SWEET developed these capabilities in the context of finding and using Earth science data and information.

"SWEET ontologies are being used in many projects:

DOLCE (Descriptive Ontology for Linguistic and Cognitive Engineering)
GEON (Geosciences Network)
LEAD (Linked Environments for Atmospheric Discovery)
ESML (Earth Science Markup Language)
ESIP (Earth Science Information Partner) Federation
GENESIS (Global Environmental & Earth Science Information System)
IRI (International Research Institute for Climate and Society)
MMI (Marine Metadata Initiative)
PEaCE (Pacific Econformatics and Computational Ecology)
SESDI (Semantically Enabled Science Data Integration)
VSTO (Virtual Solar-Terrestrial Observatory)"

Other relevant fields

<<Some related initiatives 341

DAML

"A DAML Ontology of Time (http://www.cs.rochester.edu/~ferguson/daml/): This collaborative project, led by Jerry Hobbs, aims to develop a representative ontology of time that expresses temporal concepts and properties common to any formalization of time. The ontology is formulated as a set of first-order predicate calculus axioms. These axioms can be used as-is, or can be specialized to describe other, more specific, temporal theories. In several places ontological choices must be made. These are clearly indicated in the text and result in optional axioms that might or might not be part of any specific temporal ontology. This page provides documents and resources from the DAML-Time effort, part of the DARPA Agent Markup Language (http://www.daml.org/) project." [Note also CHRONOS in Other geological initiatives]

DITA

Darwin Information Typing Architecture (http://en.wikipedia.org/wiki/Darwin_Information_Typing_Architecture): "The Darwin Information Typing Architecture (DITA) is an XML-based architecture for authoring, producing, and delivering information. Although its main applications have so far been in technical publications, DITA is also used for other types of documents such as policies, procedures, and training.

"The DITA architecture and a related DTD and XML Schema were originally developed by IBM. The architecture incorporates ideas in XML architecture, such as modular information architecture, various features for content reuse, and *specialization*, that had been developed over previous decades. DITA is now an OASIS standard.

"DITA content is written as modular *topics*, as opposed to long 'book-oriented' files. A DITA *map* contains links to topics, organized in the sequence (which may be hierarchical) in which they are intended to appear in finished documents. A DITA map defines the table of contents for deliverables. *Relationship tables* in DITA maps can also specify which topics link to each other.

"Modular topics can be easily reused in different deliverables. However, the strict topic-orientation of DITA makes it an awkward fit for content that contains lengthy narratives that do not lend themselves to being broken into small, standalone chunks. Experts stress the importance of content analysis in the early stages of implementing structured authoring."

Flybrain

Flybrain (http://flybrain.neurobio.arizona.edu/) An online atlas and database of the *Drosphila* nervous system.

"The 'Basic Atlas' provides the user with a hypertext tour guide to the basic structural elements of the *Drosophila* nervous system. It links schematic representations, serial sections through the entire brain, and Golgi impregnations of individual cells. When appropriate, these are also linked to enhancer-trap images and to other gene expression data. We hope that the Basic Atlas will provide the novice with a usable overview of how the different parts of the nervous system are constructed and connected. It is intended that the Atlas links will provide a tour through the brain and its main structures, including JAVA applets and VRML manipulatable reconstructions. Teaching tools that can be used with the database are currently being developed by the University of Freiburg component of the database consortium."

INSPIRE

European Commission INSPIRE Directive (http://inspire.jrc.ec.europa.eu/index.cfm)

"In Europe a major recent development has been the entering in force of the INSPIRE Directive in May 2007, establishing an infrastructure for spatial information in Europe to support Community environmental policies, and policies or activities which may have an impact on the environment. INSPIRE is based on the infrastructures for spatial information established and operated by the 27 Member States of the European Union. The Directive addresses 34 spatial data themes needed for environmental applications, with key components specified through technical implementing rules. This makes INSPIRE a unique example of a legislative 'regional' approach.

"What is the INSPIRE Directive?
"The INSPIRE directive came into force on 15 May 2007 and will be implemented in various stages, with full implementation required by 2019.The INSPIRE directive aims to create a European Union (EU) spatial data infrastructure. This will enable the sharing of environmental spatial information among public sector organisations and better facilitate public access to spatial information across Europe. A European Spatial Data Infrastructure

will assist in policy-making across boundaries. Therefore the spatial information considered under the directive is extensive and includes a great variety of topical and technical themes. INSPIRE is based on a number of common principles:
Data should be collected only once and kept where it can be maintained most effectively.
It should be possible to combine seamless spatial information from different sources across Europe and share it with many users and applications.
It should be possible for information collected at one level/scale to be shared with all levels/scales; detailed for thorough investigations, general for strategic purposes.
Geographic information needed for good governance at all levels should be readily and transparently available.
Easy to find what geographic information is available, how it can be used to meet a particular need, and under which conditions it can be acquired and used."

ISO standards

International Standards Organisation (http://www.iso.org/iso/iso_catalogue.htm) "ISO has developed over 18 500 International Standards on a variety of subjects and some 1100 new ISO standards are published every year. The full range of technical fields can be seen from the listing International Standards. Users can browse that listing to find bibliographic information on each standard and, in many cases, a brief abstract. The online ISO Standards listing integrates both the ISO Catalogue of published standards and the ISO Technical programme of standards under development."

STEP

STEP Consortium. Seeding the EuroPhysiome: A Roadmap to the Virtual Physiological Human. 5 July 2007. http://www.europhysiome.org/roadmap

"The Virtual Physiological Human (VPH) is a methodological and technological framework that, once established, will enable collaborative investigation of the human body as a single complex system. The VPH provides the European research infrastructure that will make it possible for biomedical researchers to complement their conventional reductionist approach with what we call an *integrative* approach, where biological processes are described from a systems point of view. The present document, compiled through a consensus process that has involved more than 300 stakeholders from research, industry and clinical practice aims to provide a research roadmap that will become a blueprint for the realisation of the VPH. The document discusses in detail all aspects of this endeavour."

Systems biology

Systems biology: a Grand Challenge for Europe, European Science Foundation, 06.09.2007. (http://www.esf.org/publications/medical-sciences.html) 54 p.

"Biological and biomedical research is undergoing revolutionary developments that are likely to have a lasting impact on society. These developments involve scientific disciplines including physics, chemistry, mathematics, engineering and computer science among others. They enable us to know and measure the properties and interplay of the molecules that constitute life. In principle we are now capable of unraveling complete sets of chemical reactions, interactions and dynamic structures through which molecules, cells and organs carry out specific functions of living organisms, including humans. Integrating the explosively growing amounts of data available on these components and generating understanding on how their maze of interactions in time and space govern life is termed Systems Biology.

"Systems Biology evolved by recognising that biological systems are far too complex to be solved by classic biological approaches. Systems Biology tightly integrates expertise from physicists, mathematicians, engineers with biological knowledge. It gives a central role to predictive mathematical models that integrate all relevant data on the topic of investigation and exploits such models to decide which experiments are most effective. In this way, an effective and goal-oriented iterative cycle of model-driven experimentation and experiment-driven modelling is initiated. As Systems Biology progresses, multifactorial diseases, such as diabetes, arthritis, heart failure and cancer, may be understood in terms of failure of molecular components to cooperate properly. Consequently, complex diseases may be approached and treated in a much more rational and effective way. It should be Europe's ambition to be at the forefront of pinpointing the systemic causes of diseases, aiming at the rational design of targeted therapies and drugs."

UML

Unified Modeling Language-(http://www.uml.org/): "The Unified Modeling Language™ - UML - is OMG's most-used specification, and the way the world models not only application structure, behavior, and architecture, but also business process and data structure.

"UML, along with the Meta Object Facility (MOF™), also provides a key foundation for OMG's Model-Driven Architecture®, which unifies every step of development and integration from business modeling, through architectural and application modeling, to development, deployment, maintenance, and evolution.

"OMG is a not-for-profit computer industry specifications consortium; our members define and maintain the UML specification which we publish in the series of documents linked on this page for your free download. Software providers of every kind build tools that conform to these specifications. To model in UML, you'll have to obtain a compliant modeling tool from one of these providers and learn how to use it." [Links are provided from this page].

Index

<<Table of contents 1

abstraction...131, 143, 183, 196, 200, 222, 261
agent...80
allometry...271
ambiguities..148, 180, 189, 191, 197, 222, 232, 234, 250, 252, 259
American Mineralogist Crystal Structure Database...343
applications...193
apps...*See* applications
attractors...161
augmented reality...92, 301
background knowledge...*See* prior knowledge
basis functions...244
Bayesian probabilities...304
bicubic polynomials...237
blending...270
blobs...253
boundaries...147, 149, 155, 160, 218, 269, 279, 303
  edge-detector...266
browsers...197, 306, 311
canonical coordinates...263
cartoons...141
CGI...344
chaos theory...162
CHRONOS...344
classification...140, 157, 182, 234
classification space...111
collaboration...54, 128
collaboratories...9
complex systems...102
computing 'cloud'...109
concept development...19, 44, 45
concept modelling...21
configuration...244, 269, 274
configurations..57, 92, 111, 131, 157, 160, 161, 175, 229, 267, 289, 308
contingency planning...164
continuity...218, 237
contouring...*See* interpolation
convolution...258
costs...9
Critical Path Analysis...142
curvature...237
cyberenvironment...28, 88
  geological...85
cyberinfrastructure...12
DAML...355
data
  unevenly spaced...239
database...200
databases, relational...59
DataGrid...349
Delauney triangles...210, 237, 239
design requirements
  gce...90
  sEsm...83
developable surface...287
diagnosis...302
dialectic...130, 197, 209
differential geometry...272, 287
digital cartography...203
digital elevation model (DEM) *See* models: elevation
directed graphs...142
direction cosines...230
discontinuities...218, 264, 269
dissipative structures...163
DITA...121, 355
DIVERSITAS...350
diversity, subsystems...109
documentation...6, 75, 94, 96, 108, 113, 119, 305
Earth System Curator...349
Earth system processes...26
Earth systems...65
EarthChem...344
EarthScope...349
e-book...7
eGY...350
electronic notebook...300
encapsulation...186
Energistics...348
EOSDIS...***350***
Epicentre Model...347
ESMF...351
ESSC...351
ESSP...350

evaluation.......97, 101, 133, 169, 190, 310
exit strategy..199
extent (observations) 253
facets 239, 241
federated structure 109
feedback 131, 160, 300
fence diagrams 299
FGDC..351
filters...........................256, 260, 295, 312
Gaussian 92, 256, 261
Gaussian derivatives.........................265
Laplacian...265
spatial 271
finite element methods (FEM) 236, 277
finite elements....................................272
fitness for purpose 13
fitness function...................................161
Flybrain 356
fractals 217, 233, 253
fragments 297, 313
framework..... 15, 48, 78, 83, 96, 104, **106**, 110, 121, 138, 168, 221, 294, 299, 309, 310
stratigraphic....................................**139**
function
local 216
orthogonal.......................................216
overdetermined...............................215
underdetermined 215
gce .... *See* cyberenvironment, geological
generalisation..............152, 181, 255, 295
map.......................................131, 220
generalised vertical section (GVS) 294, 298
GENIE..351
GeoIndex 169, 291, 306, 341
geoinformatics....................................101
geological investigation.......................313
e-support..85
stages of 88, 103, 137, 192
geological mapping38, 101, 104, **125**, 130, 148, 173, 196, 252, 277, 288
geological processes.......27, 100, 264, 277
geological surveying 172, 189, 264
geological surveys......74, 78, 95, 102, 108, 109, 118, 143, 151, 153, 168, 199, 201, 283, 305, 310
geological systems.........................75, 227
GEON 91, 345
GeoRef 311
GEOROC..345
GeoSciML..............................96, 344, 345
geostatistics 213, 241, 244, 271
gestalt 49, 60
GIS.. 177
GPS.................................... 201, 272, 300
grain 253
granularity.. 111
Grid 51
GSI3D 101, 206, 342
holistic........................... 60, 108, 149, 313
holistic view 66, 74, 195, 226
homologous 269
hyperlinks.................... 189, 190, 193, 260
hypertext 302
IGBP 350, 352
IGME 5000 346
IHDP 350
immanent processes..................... **57**, 132
immanent properties......................... 233
inconsistencies..............*See* reconciliation
indexes............................... 171, 291, 310
information
community..................................... 169
legacy .10, 168, 195, 232, 250, 295, 305, 310, 311
salient 260
trails20, 46, 62, 124, 193, 226, 299, 302, 309
types22, **62**, 91, 131, 172, 189, 202, 222
infrastructure.................................... 8, 17
inheritance... 309
interfaces 49, **133**
user 171
interoperability 40, 78
interpolation ..92, 100, 146, 149, 152, 162, 205, 210, 212, 221, 231, 235, 243, 244, 259, 280, 284
discrete smooth 236
Fourier series 214
polynomials.................................... 213
quintic polynomial 239
shape.. 270
splining... 215
interpretations............................. 58, 132
invariance 56, 92, 163, **232**, 241, 245, 267, 270
invariant attributes 23
investigation 16
knowledge
unexpressed........................... 135, 227
knowledge base 134

knowledge representation .... *See* representations
knowledge systems 12, 115, 311, 313
geological.................102, 284, 296, 300
knowledge utility.....................................25
kriging 213, 271
Lady of Shalott......................................133
landmarks 269
landscape metrics...................................254
lidar...241
Linked Data.................................25, 55, 96
management 16
map generalisation.................................248
map projections......................................235
map scale..249
mapping processes.........................60, 284
maps, thematic..............................59, 200
mark-up **175**, 193, 309
mark-up, spatial.....................................176
matrix algebra 230
memes 58
memory
episodic...136
spatial 136
metadata 115, 136, **139**, 176, 197, 209, 234, 251, 260, 261, 277, 299, 304, 311
metamodel 109, 110
metric spaces...278
microdocuments..........137, 192, 287, 311
migration of information..........................96
model...81
geological business.......32, 93, 152, 158
geological framework.................35, 105
geological infrastructure 37, 117
geological investigation.......34, 98, 141, 143, 226, 292
inverse 298
models 38
conceptual 129, 307
deformable 274, 280
dialectic 129, 291
dynamic 40, 163, 229, 277, 290, 292, 307
elevation 207
faceted...297, 303
field survey 299, 300
forward 154
fractal...162
geological..132
geological surveying ..83, 146, 195, 196, 208
grid 208
imagery 274
imperfect 157
inverse 154, 280
multi-faceted 297
multi-scale 255, 259, 277, 303
process.....................111, 170, 252, 263
sharing 47, 108
spatial.40, 100, **128**, 131, 138, 140, 151, 170, 176, 204, 270, 278, 287, **293**, 306
morphing 270
morphometrics 40, **269**, 287
mosaic..253
moving average 212
multiple hypotheses 155, 189, 297
multi-resolution ... *See* models: multi-scale
NAVDAT 347
navigating information 301
networks, social 54
NGDC...342
normal science.................................13, 107
NSDI 351
object behaviour.................232, 277, 312
object boundaries......................................92
object classes 181, 289
object instances......................................181
object oriented 39
object reconstruction 274
object relationships 183
object store...137
object-oriented 225, 303
object-oriented approach24, 48, **178**, 191, 195
objects 24, 48, 111, 175, 179
octrees 177, 219
OneGeology 347
online games..54
ontologies 24, 48, 51, 83, 91, 96, 108, 111, 309
OpenMI 353
palinspastic *See* models: dynamic, process
paradigm.............12, 13, 49, 106, 116, 132
parametric representation 244
patch dynamics 254
patches.........................236, 253, 266, 279
PetDB 347
Petri Nets 142
Physiome Project 190
piecewise blending 237
pilot studies 82, 129, 139

policy 81, 225
POSC 347
PPDM Association 348
predictions 79, 133, 162, 163, 261
predictive machine 76, 97
predictive reinforcement learning 95
predictive systems 67
primal sketch 301
PRINCE2 93
print on demand 124
prior knowledge 79, 156, 158, 266, 279
PRISM 353
procedures 307
Procrustes methods 269
project design 114
project management 93, 96
projects 95, **167**
properties
  spatial 266
provenance 84, 92, 100
quadtree 219
quadtrees 177
quality assessment *See* evaluation
quaternions 230
quintic polynomials 237
RDF 50, 55
reasoning58, 113, 135, 140, 147, 189, 197, 226, 229, 231, 234, 267, 279, 286, 299, 300, 308
reconciliation 47, 84, 136, **187**, 188, 297
reductionist approach 65
reference frame 231, 236
reinforcement learning 79, 104
relationships 111
  spatial 174, 184
remote sensing 92, 100, 253
representations 75, 83, 86, 100, 134, **148**, 182, 280, 294, 301
resolution 249, 256
revisable units 7
reward function 81
risks 10, 25, 67
rules 197
scale **256**
scale-space 92, 100, 153, 241, 248, 255, 269, 279
scenario 4, 129, 199
scope 49
scripts 193
search criteria 137
search engines 90, 305
SEEGrid 354
self-affine 233, 271
self-organisation 160
self-similar 234
self-similarity 162, 218, 219
Semantic Grid 51, 236
semantic links 83, 176
Semantic Web 50, 91, 184
semi-variance 242
service-oriented knowledge utility 52
sEsm *See* solid Earth systems model
shape 162, 244, **267**, 279
SIGMA 343
SIGMA Mobile 343
simplification 157
simulation 77, 154, 243, 273
slope scattergram 242
slope variogram 242
slopes 237, 241
social networking 67
SOKU *See* service-oriented knowledge utility
solid Earth systems model 26, 69, 71, 76, 108, 299, 305
solid-Earth Sciences 27
standards 96
state-space 160, 161
stereogram 242
strategy 8, 9, 12, 38, 93
stratigraphy 107, 141, 184, 260, 277
sub-models 115, 142, 219, 273, 281, 298, 300, 303, 311
subsurface geology 74, 96, 147, 155, 171, 206, 225, 272, 285, 347
subsystems 157, 162, 192, 199
surface smoothing 213
surfaces
  implicit 270
surveying
  electronic 201, 300
SWEET 51, 354
symbiosis 21, 47, 77
symmetry 271
symptoms 302
syndromes 302
systems 49, **160**, 172, *See* geological systems
  complex 159, 250, 309
  emergent 180, 218, 236, 272
  geological 252
  object-oriented 199

open 163
systems approach 12, 65, 114, 121, 283
Systems Biology 27, 358
systems design 83, 101, 142, 259, 266
systems geology 3, 12, 100, 102
tacit 102, 115, 302
tagging .... *See* mark-up
templates 276, 280, 300
thematic maps 146
Thiessen polygons 210, 239
threads of reasoning *See* information:trails
tomography 219, 255
topological configuration 269
topology 233
transformations
affine 233
geometrical . 40, 229, **230**, 268, 274, 295
rigid-body 234, 244
spatial 244
topological 233
trend surface 216
uncertainty 92, 222, 277
unexpressed knowledge 23, 63, 232
user interface 305
USGS National geologic map database 347
value function 81
variograms 213, 241, 245
version control 83, 123
Virtual Physiological Human 27, 357
vision 256
human 301
visualisation . 170, 172, 205, 217, 221, 229, 234, 246, 250, 255, 263, 286, 291, 295, 306
voxels 219
VRML 176
WATERS 88
wavelets 220
WCRP 350
Web 2.0 54
Web of Data 55, 87, 312
web of thoughts .... *See* information:trails
websites 121
weighting function 213
workflows 193
workflows, scientific 46, 53, 114
workflows, surveying 61, 92, 113, 122
World Wide Web 168
worldview 131
XML 50, 175
zero-crossings 92, 264, 269, 279

<<Table of contents 1

Made in the USA
Monee, IL
15 November 2025

34943312R00205

ECONOMY AND CLASS IN SUDAN